Affinity Labelling and Cloning of Steroid and Thyroid Hormone Receptors

Edited by H. Gronemeyer

VCH

Ellis Horwood Series in Biomedicine

Series Editor: Dr. Alan Wiseman, Department of Biochemistry, University of Surrey

Published Titles

Cholecystokinin (CCK) in the Nervous System: Current Developments in Neuropeptide Research J. de Belleroche and G. J. Dockray (Eds.)

Sterolbiosynthesis Inhibitors: Pharmaceutical and Agrochemical Aspects D. Berg and M. Plempel

Chemiluminescence: Principles and Applications in Biology and Medicine A. K. Campbell

Steroid Hormone Receptors: Their Intracellular Localisation C. R. Clark (Ed.)

Brain 5-HT_{1A} Receptors: Behavioural and Neurochemical Pharmacology C. T. Dourish, S. Ahlenius and P. H. Hutson

Cellular and Molecular Basis of Cholinergic Function M. J. Dowdall and J. N. Hawthorne (Eds.)

Peptidergic Mechanisms in the Cerebral Circulation L. Edvinsson and J. McCulloch (Eds.)

Neuropharmacology and Pesticide Action M. G. Ford, G. G. Lunt, R. C. Reay and P. N. R. Usherwood (Eds.)

Combating Resistance to Xenobiotics: Biological and Chemical Approaches M. G. Ford, D. W. Holloman, B. P. S. Khambay and R. M. Sawicki (Eds.)

Fibrinogen, Fibrin Stabilisation and Fibrinolysis: Clinical, Biochemical and Laboratory Aspects J. L. Francis (Ed.)

Drug Delivery to the Respiratory Tract D. Ganderton and T. Jones (Eds.)

Molecular Genetics and Immunoanalysis in Blood Coagulation J. C. Giddings

Development of Drugs and Modern Medicines: A Conference to Honour Professor Arnold H. Beckett J. W. Gorrod, G. G. Gibson and M. Mitchard (Eds.)

Biological Oxidation of Nitrogen in Organic Molecules: Chemistry, Toxicology and Pharmacology J. W. Gorrod and L. A. Damani (Eds.)

Affinity Labelling and Cloning of Steroid and Thyroid Hormone Receptors H. Gronemeyer (Ed.)

Drug Delivery Systems: Fundamentals and Techniques P. Johnson and J. G. Lloyd-Jones (Eds.)

Monitoring Neurotransmitter Release During Behaviour M. H. Joseph, M. Fillenz, I. A. Macdonald and C. A. Marsden (Eds.)

Interaction of Steroid Hormone Receptors with DNA M. Sluyser (Ed.)

Growth Factors and Oncogenes in Breast Cancer M. Sluyser (Ed.)

The Molecular Biology of Receptors: Techniques and Applications of Receptor Research A. D. Strosberg (Ed.)

The Blood-Brain Barrier in Health and Disease A. J. Suckling, M. G. Rumsby and M. W. Bradbury (Eds.)

Neuropeptides and Their Peptidases A. J. Turner (Ed.)

Forthcoming Titles

The Blood-Brain Brain Barrier: Drug and Therapeutic Aspects
J. P. Bates and N. G. Burnet

Sulphur Drugs and Related Organic Compounds: Chemistry, Biochemistry and Toxicology L. A. Damani (Ed.)

Handbook of Quality Control in Pharmaceutical Products S. Denyer and R. Baird

Neurochemical Pathology of the Human Brain C. J. Fowler and J. A. Hardy

Clinical Pharmacy: Pathology and Therapeutics R. J. Greene and N. D. Harris

Electrochemical Detection: Techniques in the Fundamental and Applied Biosciences
G. A. Junter (Ed.)

Blood Substitutes: Physiology and Medical Applications K. C. Lowe (Ed.)

Handbook of Cardiac Glycosides: Chemistry, Pharmacology and Toxicology of Nonrenal Digitalis Substances N. Rietbrock and B. G. Woodcock

Leukaemia Cytochemistry C. S. Scott (Ed.)

Brain Imaging: Techniques and Applications N. F. Sharif and M. E. Lewis (Eds.)

Critical Factors in Haemostasis: Evaluation and Development C. E. Williams, P. E. Short, A. J. George and M. P. B. Entwistle

Distribution

Great Britain and Ireland: VCH Publishers (UK) Ltd., 8 Wellington Court, Wellington Street, Cambridge CB1 1HW (Great Britain)

USA and Canada: VCH Publishers, Suite 909, 220 East 23rd Street, New York, NY 10010-4606 (USA)

Switzerland: VCH Verlags-AG, P. O. Box, CH-4020 Basel (Switzerland)

All other countries: VCH Verlagsgesellschaft, P. O. Box 1260/1280, D-6940 Weinheim (Federal Republic of Germany)

ISBN 3-527-26557-0 (VCH Verlagsgesellschaft) ISBN 0-89573-579-2 (VCH Publishers)

Affinity Labelling and Cloning of Steroid and Thyroid Hormone Receptors

Edited by H. Gronemeyer

VCH

ELLIS HORWOOD
international publishers in science and technology

Dr. H. Gronemeyer
Faculté de Médicine,
Institut de Chimie Biologique
F-67085 Strasbourg Cedex
France

Series Adviser in Neurosciences:
Dr. C. Clark
Warner-Lambert Company
2800 Plymouth Road
Ann Arbor
Michigan 48105
USA

Cover photo: Visualization of chromatin binding sites of the ecdysone receptor(s) in *Drosophila melanogaster* polytene nuclei. The steroid hormone 2O-OH ecdysone was covalently linked to its chromosomal binding sites – via its receptor(s) – by *in situ* photocrosslinking of salivary glands. After fixation, the glands were squashed and photographed (top). Chromatin binding sites were detected by immunofluorescence microscopy (bottom) using anti-ecdysone antibodies and fluorescing second antibody.

Deutsche Bibliothek Cataloguing-in-Publication Data
Affinity labelling and cloning of steroid and thyroid hormone receptors / ed. by H. Gronemeyer. – Cambridge ; New York, NY ; Basel (Switzerland) ; Weinheim : VCH ; Chichester (England) : Horwood 1988
(Ellis Horwood series in biomedicine)
ISBN 3-527-26557-0 (VCH, Weinheim) Pp.
ISBN 0-89573-579-2 (VCH, Cambridge ...) Pp.
NE: Gronemeyer, Hinrich [Hrsg.]

ISSN 0930-3367

British Library Cataloguing in Publication Data
Affinity labelling and cloning of steroid and thyroid hormone receptors.
1. Man. Hormone receptors. Affinity labelling. Cloning
I. Gronemeyer, H.
612'.405

Library of Congress data available

Published jointly in 1988 by
Ellis Horwood Ltd., Chichester, England
and VCH Verlagsgesellschaft mbH, Weinheim, Federal Republic of Germany

Printed and Bound in Great Britain by
Hartnolls Limited, Bodmin, Cornwall.

Table of contents

Editor's introduction

This book describes the exploration of a pathway — sometimes stony and tedious — that leads from the identification and characterization of some members of a family of transcriptional regulatory proteins to the initiation of their functional analysis. Along the way some old questions have been answered but, more importantly, it has become possible to ask new types of questions which were hidden beyond the scientific horizon a few years ago. The genealogical analysis of what is now recognized as a superfamily of nuclear receptor genes began with the applications of new techniques to a specific class of transcription factors that become active by binding steroidal ligands. This binding is reversible and represented a major problem for various preparative and analytical procedures. In consequence, initially, several molecular weights were assigned to each receptor and it was not clear whether proteolysis, aggregation or complex formation was responsible for the different forms. Therefore the development of affinity labeling techniques, providing a means of converting reversible into covalent ligand binding, was a most important contribution to the study of steroid receptor proteins. Techniques nourish knowledge and therefore the first section of this book (Chapters 1 to 6) deals with a detailed experimental description of affinity labeling techniques and techniques used in conjunction with crosslinking. The results that have been obtained in applying these methods to study steroid and thyroid hormone receptors are described in the subsequent section (Chapters 7 to 13).

The idea of treating both steroid and thyroid hormone receptors in such a volume was not easily accepted when this book was conceived because they were considered structurally (ligand) and functionally (localization, activation, translocation) very different according to classical views. It turned out, however, that both of these receptors are children of the same family and that some classical concepts have to be modified. In fact, now it would be considered a mistake to exclude the thyroid receptor from such a treatise.

A specific type of analysis using crosslinking techniques, which is not considered in this book, is illustrated in Fig. 1. This shows that it is possible to localize steroid receptor binding sites on polytene chromosomes (for review and references see

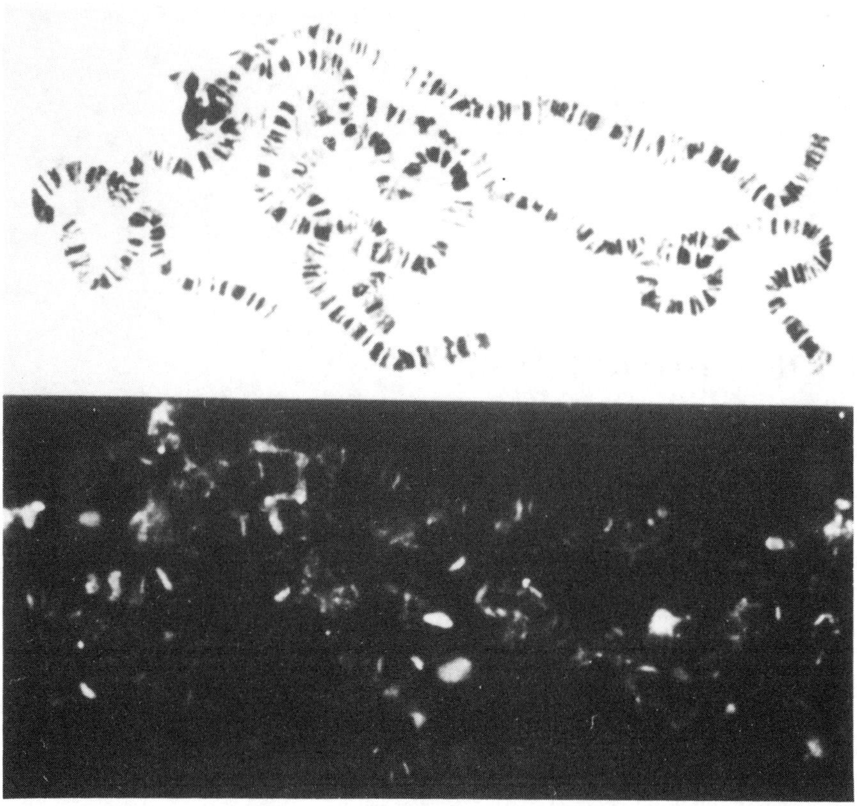

Fig. 1 — Visualization of chromatin binding sites of the ecdysone receptor(s) in *Drosophila melanogaster* polytene nuclei. The steroid hormone 2O-OH ecdysone was covalently linked to its chromosomal binding sites — via its receptor(s) — by *in situ* photocrosslinking of salivary glands. After fixation, the glands were squashed and photographed (top). Chromatin binding sites were detected by immunofluorescence microscopy (bottom) using anti-ecdysone antibodies and fluorescing second antibody.

Gronemeyer and Govindan (1986), *Mol. Cell. Endocrinol.* **46**, 1–19). With the long-awaited cloning of the ecdysonereceptor(s) this system will be ideal to analyze the temporal programming of hormone-inducing transcriptional activation.

A major milestone in our understanding of transcriptional regulation by nuclear receptors is the cloning of steroid hormone receptors. This is described in the third section which starts with a review of the general strategies that have been chosen to clone the individual receptors (Chapter 14). Various approaches have been used but the reader will realize that it is the story of the chicken progesterone receptor which threads its way through this book from the initial crosslinking to the final functional analysis. In keeping with the nomenclature of Chapter 14 the cloning strategies used were A1-3, 10–12, C1, D1 (unsuccessful) and A1-4, C1 (3S a, b; Ca), 4^1, 5^1, 11, 4^3, 5^2, 6, 7, 8^1, 9 (successful). The structural analysis and functional dissection of the cloned steroid receptors have revealed two important perspectives. First, specific segments in the primary structure correspond to functional blocks and second, common structural principles reveal an until recently unknown genealogy. Apparently steroid receptors are a subfamily, like thyroid receptors, of a superfamily of nuclear receptor genes. Further subfamilies, such as those of the retinoic acid receptor and hap, may be soon defined. All these proteins have three basic functional characteristics; DNA binding, ligand binding and the potential to activate transcription by an as-yet unknown mechanism. Together they are members of a recently defined class of transcription factors, the so-called 'finger proteins', which use a common structural feature for their DNA binding domain. Exemplified by the well-known GAL4, finger proteins are already present in yeast as the simplest eukaryote organism.

Two contributions in the third section of this book suggest directions for future walks. These are: the study of the transcriptional regulation of these transcription factors (Chapter 16) and the identification of the protooncogene c-erbA as the thyroid hormone receptor (Chapter 17).

My thanks go to all the authors, to my colleagues in the LGME and to Sue Horwood (Ellis Horwood, Ltd) for facilitating this book. I hope that the reader will enjoy the walk.

Authors' affiliations for correspondence

Pierre Chambon	L.G.M.E. du C.N.R.S./U. 184 de l'I.N.S.E.R.M.
Hinrich Gronemeyer	Institut de Chimie Biologique
Jean-Marc Jeltsch	Faculté de Médecine
	11 rue Humann
	67085 STRASBOURG Cédex, FRANCE
	Telephone: 88.37.12.55
Ulrich Gehring	Institut für Biologische Chemie
	Universität Heidelberg
	Im Neuenheimer Feld 501
	6900 HEIDELBERG, FRG
	Telephone: 06221–562514
Manjapra V. Govindan	MRC Group in Molecular Endocrinology
	Laval University Medical Center
	QUEBEC G1V 4G2, CANADA
	Telephone: 419-654-2296
Kathryn B. Horwitz	Departments of Medicine and Pathology
	Box B151
	University of Colorado Health Sciences Centre
	4200 East Ninth Avenue
	DENVER, CO 80262, USA
	Telephone: 303-394-8443
Benita S. Katzenellenbogen	Department of Physiology and Biophysics
	524 Burrill Hall
	University of Illinois at Urbana-Champaign
	407 South Goodwin Avenue
	URBANA, IL 61801, USA
	Telephone: 217-333-9769 or 333-7565
Herbert H. Samuels	Division of Molecular Endocrinology
	Department of Medicine
	New York University Medical Center

550 First Avenue
NEW YORK, NY 10016, USA
Telephone: 212-340-6279

S. Stoney Simons Jr.
Steroid Hormones Section
Bldg 8, Rm B2A-07
National Institute of Diabetes
Digestine and Kidney Diseases
National Institutes of Health
BETHESDA, MD 20892, USA
Telephone: 301-496-6796

Donald J. Tindall
Baylor College of Medicine
Texas Medical Center
Department of Cell Biology
HOUSTON, TX 77030, USA
Telephone: 713-799-6200

Björn Vennström
Differentiation Programme
EMBL
Postfach 10 22 09
6900 HEIDELBERG, FRG
Telephone: 06221–387415

Part I
Techniques used in the affinity labeling studies of steroid and thyroid hormone receptors

1

Estrogen receptor

B. S. Katzenellenbogen[a] and J. A. Katzenellenbogen[b]
[a]Department of Physiology and Biophysics and [b]Department of Chemistry,
University of Illinois, Urbana, USA

I. AFFINITY LABELING AGENTS FOR THE ESTROGEN RECEPTOR

A. Synthesis of aziridines

Complete experimental details for the synthesis of all the described aziridine-based affinity labeling agents have been published (Katzenellenbogen *et al.*, 1983; Simpson *et al.*, 1987; Salituro *et al.*, 1987; Zablocki *et al.*, 1987). All involve a total synthesis that culminates in the preparation of a methanesulfonate ester that is displaced with ethyleneimine in a final step.

Cautionary note. Ethyleneimine is a volatile, reactive, carcinogenic and actively toxic substance. All manipulations involved in its preparation, purification, and use should be performed in a well-ventilated fume hood, providing protection against explosion. A full-face gas mask with a fresh ammonia (silica gel) canister is recommended in the event of inadequate ventilation. The ammonia-like odor threshold is reported as 2 ppm, but a threshold limit of 0.5 ppm in air for continuous exposure has been set. Dry ethyleneimine may be stored in the freezer, either neat (over potassium hydroxide pellets) or a solution in an inert solvent (e.g. toluene) for several months without decomposition (Dermer *et al.*, 1969).

B. Radiochemical synthesis

Complete details for the radiochemical synthesis of the aziridine-based affinity labeling agents have been published (Katzenellenbogen *et al.*, 1983; Simpson *et al.*, 1987; Salituro *et al.*, 1987; Elliston *et al.*, 1987). For the preparation of the tritium-labeled agents Tam-Az, Naf-Az, and KN-Az, appropriate iodine-substituted derivatives of the methanesulfonate precursors were subjected to reductive tritiolysis with carrier-free tritium gas over a palladium on aluminium catalyst in dimethylformamide solvent on a 0.05–0.1 mmol scale. This treatment was performed by a commercial laboratory. After removal of exchangeable tritium label, a portion of the labeled material was returned for purification and aziridine preparation.

Since the tritium-labeled methanesulfonate is relatively stable to various manipulations, it was most convenient to purify this material rigorously at this stage. This was done generally by HPLC under either reversed phase (C-18 silica gel: water–acetonitrile) or normal phase conditions (silica gel: hexane–dichloromethane–isopropanol: 65–33–2) (Simpson *et al.*, 1987). At this point, material with radiochemical purity in excess of 95% was obtained.

The aziridines could then be prepared from this purified methanesulfonate by ethyleneimine displacement through exposure to excess ethyleneimine and triethylamine in acetonitrile at room temperature overnight. Yields in the displacement are moderate (30–40%) and the product is contaminated by polymerization products from ethyleneimine. Rapid purification by HPLC under normal phase conditions (silica gel: hexane–isopropanol–diethylamine: 65–34.5–0.5), however, generally suffices to produce material with high chemical purity and radiochemical purity of greater than 90–95%.

The iodine-labeled aziridine I-Tam-Az was radiolabeled directly (Salituro *et al.*,

1987): a vinyl tin precursor of the aziridine was treated with sodium [^{125}I]iodide and chloramine T and, after product isolation, purified by reversed phase HPLC to give the final aziridine directly.

C. Specific activity determination

Because the final aziridine products are relatively unstable to purification, specific activity determination of the tritium-labeled materials is carried out more conveniently at the stage of the tritium-labeled methanesulfonate precursor. Because it is this material that is subjected to careful purification by HPLC, its specific activity can usually be determined during preparative chromatographic purification runs.

The column and separation system is calibrated by injection of known quantities of unlabeled methanesulfonate, and the radiolabeled material is injected and collected. The quantity of eluted radioactivity is determined on an aliquot of the combined peak fractions, and the quantity of eluted material is estimated from the area on the UV trace from a standard curve constructed from the calibration runs.

D. Storage and manipulation

Although the aziridines are relatively labile compounds, they can be stored indefinitely under appropriate conditions. Ethanol (anhydrous) is a suitable solvent, and samples of tritium-labeled Tam-Az have been stored at −20°C for more than three years with less than 10% decrease in specific activity. Because the radiolysis products are still radiolabeled, the decline in radiochemical purity of multiply labeled aziridine radiochemicals (e.g. Naf-Az and KN-Az) will be faster. The iodine-labeled I-Tam-Az has been stored for up to two lifetimes (4 months) in ethanol with only minor decline in radiochemical purity.

Perhaps the most insidious problem in using the aziridine-based affinity labels concerns the instability of tracer-level quantities of aziridines in aqueous solutions and upon surfaces (Katzenellenbogen *et al.*, 1983). The simple act of drying down a trace level portion of aziridine as an ethanolic solution under a stream of nitrogen in a glass or plastic vessel, followed by redissolution, can lower radiochemical purity drastically, indicating major chemical change; in aqueous solution, the aziridines have half-lives of only hours (*ca* 20 h) at 25–37°C and are unstable even in the cold. Aqueous solutions that contain proteins, however, are more stable.

These stability problems can be avoided simply enough, however: stock solutions of the radiolabeled materials are kept in ethanol and are prepared at such concentrations that the aziridine can be added directly as an ethanol solution to the preparation to be radiolabeled — cytosol or cells in culture. In the case where aqueous solutions are needed (intralumenal injection of aziridines into the uterus, for example), samples of the aziridines in buffer containing bovine serum albumin can be prepared fresh and used immediately.

E. Chromatographic analysis and radiochemical purity determination

Radiochemical purity can be determined on an analytical scale either by HPLC using the same systems used for the preparative purifications or by thin layer chromatography. By HPLC, a sample of unlabeled aziridine as a standard is added to the aliquot of radiolabeled material to be analyzed and fractions are collected throughout the chromatogram and counted. The activity summed over the peak fractions of the eluting standard expressed as a percentage of the total activity in the chromatogram gives the radiochemical purity. Radiochemical purity analysis by TLC proceeds in a similar fashion, with the chromatogram being cut into 2–5 mm slices which are counted. Suitable chromatographic solvent systems are normal phase on silica gel (triethylamine–ethanol–ether: 10–20–70) and reversed phase on C-18 silica gel (water–acetonitrile: 30–70) (Simpson *et al.*, 1987).

II. COVALENT LABELING OF ESTROGEN RECEPTORS

A. Labeling of cytosol receptors

This protocol can be used in labeling cytosol receptors from a variety of tissue and cells. In tissue cytosols with high proteolytic activity, it may be necessary to include inhibitors such as molybdate (20 or 40 mM), leupeptin (1 mg/ml), phenylmethylsulfonylfluoride (PMSF, 1 mM), and soybean trypsin inhibitor (5 mg/ml) in order to obtain intact, undegraded receptor. The appropriate mixture and concentration of protease inhibitors will need to be determined for specific tissues. For studies with cytosol from immature (day 20–23) rat uterus or from MCF-7 cells, where proteolysis is low when preparations are maintained at 0–4°C, inclusion of 20 mM molybdate or 1 mg/ml leupeptin show >90% intact Mr 66000 receptor, but receptor degradation may be marked if preparations are incubated at higher temperatures (Monsma *et al.*, 1984).

Several buffer systems (based on Tris or phosphate) are appropriate for preparation of cytosol and labeling of receptors, and the choice of buffer may depend on the subsequent plans for the covalently labeled receptor. We have found TEG+Mo buffer, which contains 10 mM Tris–HCl, 1.5 mM EDTA, 10% (v/v) glycerol, and 40 mM sodium molybdate, pH 7.4 at 4°C, to be a good buffer for preparation of rat uterine or MCF-7 cell cytosol receptors.

B. Typical protocol for labeling cytosol estrogen receptor

The procedure given here describes the labeling of rat uterine estrogen receptors. It can be modified slightly by the investigator, as needed for other tissues.

Uteri (*ca* 30 mg wet weight per uterus) are excised from 19–23 day old Sprague-Dawley Holtzman rats (Holtsman Co., Madison, WI) and are homogenized at 5 uteri/ml or TEG+Mo buffer (*ca* 5 mg soluble protein per ml). Homogenates are centrifuged at 180000*g* for 30 min at 0–4°C, and the supernatant is carefully collected and then filtered through a Millipore 0.22 μm filter to remove lipids. Dimethylformamide (DMF) is then added to give a final concentration of 3%.

This uterine cytosol is then incubated for 2 h at 0°C with 20 nM [^{3}H]Tam-Az or for 3 h at 15°C with 20 or 40 nM [^{3}H]KN-Az in the absence and presence of a 100-fold excess of radioinert estradiol. (Cytosol labeling with [^{3}H]KN-Az can also be carried out at 0°C for 3 h, but labeling is less efficient at the lower temperature; see Elliston *et al.*, 1987). Ligands are prepared in ethanol solutions such that the final concentration of ethanol does not exceed 1%. Labeling is terminated and excess ligand removed by incubation with charcoal–dextran slurry (1 part slurry to 9 parts cytosol) for 10 min at 4°C, and charcoal is removed by centrifugation at 12000*g* (5 min), or samples are directly spotted onto filter paper disks and treated with organic solvents to measure covalent attachment (see below).

Cytosol receptor can also be partially purified (*ca* 4-fold) either before or after covalent labeling by precipitating the receptor with 30% ammonium sulfate. Ammonium sulfate precipitation also serves to separate the receptor from the bulk of the cytosol proteases, and has proven particularly beneficial with mammary tumor samples.

Preparation of cytosol receptors from cells follows a very similar protocol. In order to ensure that receptors in the cells are largely unoccupied, cells (such as MCF-7 breast cancer cells) which are grown in Eagle's Minimal Essential Medium (MEM) with 5% calf serum are transferred to Eagle's MEM with 5% charcoal dextran-treated calf serum (CDCS) for 7–10 days before harvesting.

MCF-7 cell cytosol is prepared by harvesting near-confluent T-150 flasks and washing the cells twice with TEG buffer containing 20 mM sodium molybdate. The cells are then homogenized in the same buffer and the homogenate is centrifuged for 30 min at 180 000*g*. The supernatant (cytosol) is then made 3% in DMF and labeled with 20 nM [^{3}H]Tam-Az (or with 10 nM [^{3}H]estradiol to determine receptor content of the cytosol) for 2 h at 0–4°C. Determination of non-specific binding is accomplished by including a 200-fold excess of unlabeled estradiol in parallel incubations. After the incubations, excess ligand is removed by incubation with charcoal–dextran slurry (1 part slurry to 9 parts cytosol) for 10 min at 0–4°C, and the charcoal is then removed by centrifugation at 12 600*g* (5 min). The radioactivity associated with receptor after charcoal–dextran treatment at 0°C measures total attachment which may be covalent or non-covalent. Measurement of covalent attachment is determined by an ethanol disk assay (see below).

C. Labeling of intact cells, subcellular fractionation and preparation of nuclear receptor extracts

Near-confluent T-150 flasks of cells in tissue culture media containing 5% CDCS are labeled with 10 nM [^{3}H]E_2 or 20 nM [^{3}H]Tam-Az for 1 h at 37°C. If labeling is with [^{3}H]KN-Az, 40 nM [^{3}H]KN-Az is utilized, and labeling is for 3 h at 37°C. For non-specific binding determinations, cells are preincubated for 30 min at 37°C with 200-fold excess of unlabeled E_2, although coincubation of [^{3}H]Tam-Az or [^{3}H]KN-Az with excess unlabeled E_2 gave the same results.

Cells are harvested (using Hank's balanced salt solution (HBSS) containing 1 mM EDTA), washed with TEG buffer and then homogenized with 1 ml TEG buffer. The homogenate is centrifuged (800*g*, 10 min) and the supernatant collected. The crude nuclear pellet is washed twice with TEG buffer (pH 7.4) and the supernatants combined. The supernatant fraction is centrifuged at 180 000*g* (30 min), and the resulting supernatant (cytosol) and high speed pellet are assayed for radioactivity.

For salt extraction of the nuclear pellet, the washed nuclear pellet from two T-150 flasks of cells is resuspended in 0.5 ml of Tris–extraction buffer (10 mM Tris–HCl, 1.5 mM EDTA, 10% glycerol, 0.6 M NaCl, pH 8.5 at 4°C). The suspension is incubated for 1 h at 0–4°C with resuspension every 15 min, and then centrifuged at 180 000*g* for 30 min to give the nuclear salt extract and nuclear pellet fractions.

Cells incubated with [^{3}H]Tam-Az in the presence and absence of a 200-fold excess of unlabeled E_2 can also be directly lysed with SDS sample buffer (1% SDS, 0.13 M mercaptoethanol, 20% glycerol), thereby avoiding subcellular fractionation.

Cells are incubated for 10 min or 30 min at room temperature with SDS sample buffer, and these whole cell extracts are then boiled for 2 min at 100°C and analyzed by SDS–polyacrylamide gel electrophoresis.

The estrogen receptors in intact MCF-7 cells can also be effectively labeled with [^{3}H]Tam-Az not only when cells are attached to the flasks but also in cells in suspension. Since less volume and, hence, less radiolabeled Tan-Az is needed for suspension labeling, this protocol is used frequently. For labeling of cells in suspension, MCF-7 cells are released from the flask with HBSS containing 1 mM EDTA, decanted into a 50 ml conical tube, and pelleted at 1000 rpm for 8 min. The intact cells are resuspended in MEM+5% CDCS growth media and labeled in suspension with 10 nM [^{3}H]estradiol or 20 nM [^{3}H]Tam-Az with and without a 200-fold excess of unlabeled estradiol for 1 h at 37°C. The cells are then washed with TEG buffer and homogenized in 10 mM Tris, 1.5 mM EDTA, 10% glycerol buffer, pH 7.4 (TEG buffer), with a Dounce homogenizer using a B-pestle. The homogenate is centrifuged at 800*g* for 10 min, washed once in TEG buffer, and the resulting nuclear pellet is extracted for 1 h at 0–4°C with TEG buffer (pH 8.5) containing 0.6 M NaCl, with resuspension every 15 min. The suspension is then centrifuged at 180 000*g* for 30 min to give the nuclear salt-extracted receptor preparation.

D. Assay of covalent attachment

Attachment is assayed by an ethanol disk assay (Katzenellenbogen *et al.*, 1977b). For the ethanol disk assay, samples of 50 μl or less are spotted onto 2.2 cm Whatman 3 filter paper disks. These are collected and washed in boiling 95% ethanol twice for 15 min, rinsed twice at room temperature in 1:1 ether:95% ethanol, rinsed twice in pure ether and dried; all wash and rinse solutions are used at 10 ml/disk. The protein, together with any covalently attached compound, is precipitated by the ethanol and is retained within the fibres of the paper, while the non-covalently attached material is extracted from the disk. The disks are counted in mini-vials in 5 ml aqueous counting solution (ACS) scintillation cocktail (Amersham).

E. Scintillation counting

All samples in which [^{3}H]Tam-Az or [^{3}H]KN-Az are protein bound are counted in ACS scintillation cocktail (Amersham). This cocktail proved superior in solubilizing proteins which otherwise remained in an aqueous phase and were counted only at low efficiency. [^{3}H]estradiol-labeled samples are counted in triton–xylene scintillation cocktail (Katzenellenbogen *et al.*, 1973a). When counting was carried out in our Nuclear-Chicago Isocap 3000 refrigerated scintillation counter, an efficiency of 33–45% was obtained.

III. ANALYSIS OF COVALENTLY LABELED ESTROGEN RECEPTORS

A. Analysis of receptor size by SDS–polyacrylamide gel electrophoresis

Samples to be analyzed, after mixing with an equal volume of 2-fold concentrated SDS–sample buffer, are incubated at 100°C for 2 min. The 2-fold concentrated sample buffer for electrophoresis contains 0.125 M Tris–HCl (pH 6.8), 4% SDS, 20% glycerol, 10% 2-mercaptoethanol and bromophenol blue. Gels either are prepared according to Laemmli (1970) and contain 10% total acrylamides (T) with

2.7% bis-acrylamide (C) in the resolving gel, or are SDS-containing 10–20% polyacrylamide gradient slab gels (Sheen and Katzenellenbogen, 1987; Katzenellenbogen *et al.*, 1987). The acrylamide gradient (10–20%) separating gel contains 0.373 M Tris (pH 8.8) and *N,N'*-dimethylene-bis-acrylamide (0.675–1.8%), and the stacking gel contains 0.4% acrylamide, 1% agarose, 0.1% SDS and 0.125 Tris, pH 6.8.

Samples are electrophoresed (30 mA/gel) at 10°C in a 16 cm × 18 cm × 1.5 mm Protean gel apparatus (BioRad Laboratories). It takes 3–4 h for the tracking dye to reach within 1 cm of the bottom of the gel. At the conclusion of the run, processing is as follows. Gels to be used for autoradiography of tritium-labeled receptor are fixed and stained in Destain 1 (50% v/v MeOH–10% v/v acetic acid) plus 0.125% w/v Coomassie Blue R250 overnight, then destained in Destain 1, followed by Destain 2 (5% v/v MeOH, 7% v/v acetic acid). After photographing, the stained gel is treated for 30 min with Enlighten Rapid Autoradiography Enhancer and then dried at 60°C under vacuum with a BioRad Gel Slab Drier. The dried gels are exposed to Kodak X-Omat AR film at −80°C for 7–30 days (Katzenellenbogen *et al.*, 1987). In the case of receptor labeled with [^{125}I]I-Tam-Az, stained gels are dried but Enlightening is omitted. The dried gels need only be exposed to Kodak X-Omat AR film at −80°C for 1–2 days, which is sufficient to give a dark signal (Salituro *et al.*, 1987).

In cases where the receptor content of samples is low, or where one wishes to determine the receptor mobility using a tritiated affinity labeling ligand in 1 day, gels may be sliced and counted. Lanes can be cut out and then sliced into 2 mm pieces with a razor blade slicer. Slices are incubated overnight in 5 ml Fluorosol scintillation cocktail (National Diagnostics, Manville, NJ) and then counted for radioactivity. The mobility of the estrogen receptor is compared with that of the molecular weight standards from BioRad covering the range from 92 500 to 14 400 (Monsma *et al.*, 1984; Katzenellenbogen *et al.*, 1987).

B. Analysis of receptor charge by isoelectric focusing in 8 M urea and by 2-dimensional gel analysis

Isoelectric focusing is performed in 6 mm × 85 mm polyacrylamide tube gels containing 5% total acrylamides, 15% crosslinker, 2% ampholytes (pH 3.5–10) and 8 M urea (Monsma *et al.*, 1984; Katzenellenbogen *et al.*, 1987). The upper buffer contains 0.4% ethanolamine and the lower reservoir contains 0.5 M acetic acid. Samples are brought to 8 M in urea and either loaded on the top of the polymerized gel or mixed with the gel components prior to polymerization. Focusing is carried out at 1 mA/gel with a voltage limit of 400 V, for 18 h at 4°C. After focusing, the gels are either frozen and sliced for determination of pH and radioactivity profiles, or placed on top of the stacking gel of an SDS–polyacrylamide slab gel of the same composition as described above. The isoelectric focusing gel is sealed in place with 1% agarose in sample buffer and electrophoresed until the dye front reaches 1 cm from the bottom of the gel. Gels are processed for fluorography as described in the section above.

C. Analysis of receptor sedimentation properties in sucrose gradients

The sedimentation properties of covalently labeled estrogen receptors can be compared with those of reversibly ([^{3}H]estradiol) labeled receptors on linear 5–20% sucrose gradients. Gradients (3.6 ml) are prepared in polyallomer tubes in either

TEA buffer (10 mM Tris, 1.5 mM EDTA. 0.02% azide, pH 7.4 at 4°C) or TEA buffer containing 0.4 M KCl. Nuclear or cytosol receptors (0.2–0.3 ml) can be directly applied to the gradients, which are then centrifuged at 246 000*g* (50 000 rpm, SW 60 Ti Beckman rotor) for 16 h at 4°C for salt-containing gradients and for 13 h at 4°C for low salt gradients (prepared in TEA buffer). Sedimentation coefficients of the receptor are determined relative to the internal markers [^{14}C]ovalbumin (3.5 S), [^{14}C]γ-globulin (6.6 S), and [^{14}C]β-amylase (9.25 S). Gradient fractions containing ligand covalently associated with receptor must be counted in a scintillation cocktail, such as ACS scintillation fluid, that allows adequate association of aqueous and organic phases in the scintillation cocktail (see cautionary notes regarding scintillation counting; Section II.E and Chapter 7, Section III.F).

IV. DETAILS FOR SPECIFIC STUDIES ON THE ESTROGEN RECEPTOR

A. Limited proteolytic digestion of covalently labeled estrogen receptors to analyze estrogen receptor structure

Cytosol and nuclear receptors affinity labeled with [^{3}H]Tam-Az or [^{3}H]KN-Az as described above can be used for most studies without subsequent purification. If partial purification of nuclear or cytosol receptors is desired, this can be achieved by precipitation with $(NH_4)_2SO_4$ (30% saturation, 20 min, 0–4°C). The precipitate is collected by centrifugation at 48 000*g* for 20 min and the pellet dissolved in an appropriate volume of TEG buffer. Protein concentration of the preparations is determined from the 280 nm/260 nm absorbance, and protease treatments are then conducted at 22°C with varying protease concentrations and times (Katzenellenbogen *et al.*, 1987).

At specified time points, aliquots of samples digested with trypsin or chymotrypsin are removed and added to tubes containing soybean trypsin inhibitor (final concentration 50 μg/ml) on ice. Samples digested with V8 protease are added to tubes containing PMSF (final concentration 1 mM) on ice. Some aliquots of samples are mixed with an equal volume of 2-fold concentrated SDS-sample buffer and incubated at 100°C for 2 min. All tubes are quick-frozen on dry ice and stored at −80°C for subsequent analysis by SDS-PAGE.

B. Chemical crosslinking of covalently labeled nuclear estrogen receptors to study the monomeric–dimeric nature of the nuclear estrogen receptor

The nuclear salt-extracted estrogen receptor complexes of MCF-7 (Miller *et al.*, 1985) and presumably other target cells can be crosslinked with the reversible, mercaptan-cleavable crosslinker, 2-iminothiolane, to yield a stable dimeric receptor form. Following the exposure of intact cells to [^{3}H]Tam-Az and labeling of the nuclear receptors, receptor crosslinking is effected by incubating 500 μl of nuclear salt extract with 15 μl of 10 M triethanolamine buffer pH 8.5, 2 μl of 500 mM magnesium acetate and 86 μl of freshly made 2-iminothiolane (1 mg/100 μl) in 20 mM triethanolamine buffer, for 60 min at 0–4°C. The unreacted 2-iminothiolane is removed from the nuclear salt extract by dialysis for 1.5 h against 20 mM triethanolamine, 1 mM EGTA with 0.6 M NaCl, pH 8.5. The sample is then incubated with 3 μl of 30% H_2O_2 for 20 min at 0–4°C. Control samples are treated exactly the same as crosslinked samples except that 2-iminothiolane is not added.

The extent of receptor crosslinking can be evaluated by centrifuging receptor on 3 M urea-containing sucrose gradients which dissociates non-crosslinked receptors to the 4 S form. The molecular weight of crosslinked estrogen receptors can be analyzed on SDS–polyacrylamide gels without and with reversal of the 2-iminothiolane crosslinking. Reversal of the 2-iminothiolane crosslinking is achieved by adding 0.13 M β-mercaptoethanol to the sample prior to heating and application to the SDS gel. Since the crosslinking is mercaptan sensitive, the SDS gels used in these studies and the SDS sample buffer are made up without β-mercaptoethanol.

V. PULSE-CHASE EXPERIMENTS TO DETERMINE RATES OF ESTROGEN RECEPTOR TURNOVER IN TARGET CELLS

A. Pulse-chase experiments in MCF-7 breast cancer and rat uterine cells *in vitro*

MCF-7 cells or primary cultures of rat uterine cells (Monsma *et al.*, 1984; Nardulli and Katzenellenbogen, 1986) are labeled with 20 nM [^{3}H]Tam-Az for 1 h at 37°C with and without 100-fold excess E_2 for 30 min pretreatment and throughout the 1 h labeling period. At the end of the incubation, labeling medium is decanted and the cells are rinsed three times with 15 ml media over a period of 30 min. 20 ml of media, which is 1×10^{-6} M in unlabeled E_2, is then added and the cells are incubated for various periods of time. Zero hour controls were neither washed nor transfered to chase media after labeling. The cells are then harvested and cytosol and nuclear receptor salt extracts are prepared as described above. The high speed pellets from both cytosol and nuclear extract fractions are resuspended in 0.5 ml buffer using small Teflon–glass homogenizers. The fractions are assayed for total and covalently bound radioactivity. Covalently bound radioactivity is assayed by the ethanol disk assay. The cytosol and nuclear extracts are also assayed for bound radioactivity by the charcoal–dextran method. All fractions are frozen at −80°C for subsequent SDS–gel electrophoretic analysis of receptor size and amount.

B. Pulse-chase experiments in rat ureri *in vivo*

26-day-old rats are anesthetized with 10 mg ketamine and 0.1 mg acepromazine per 100 g of body weight. A small abdominal incision is made and the uterine horns are exposed. [^{3}H]Tam-Az is used to radiolabel estrogen receptors in uteri (Nardulli and Katzenellenbogen, 1986). The [^{3}H]Tam-Az is prepared in saline containing 0.2 μg BSA per μl of saline (BSA–saline) in order to ensure that the [^{3}H]Tam-Az remains in solution. For determination of non-specific binding, 5 μl of 10^{-4} M radioinert E_2 in BSA–saline is injected into each uterine horn at 20 min prior to instillation of [^{3}H]Tam-Az. For determination of total [^{3}H]Tam-Az binding, 5 μl of the BSA–saline vehicle is injected into each uterine horn at 20 min prior to [^{3}H]Tam-Az. [^{3}H]Tam-Az (5 μl of 10^{-6} M [^{3}H]Tam-Az in BSA–saline) is then injected into each uterine horn to label receptors. These injections are performed using 10 μl Hamilton syringes. At 1 h after [^{3}H]Tam-Az injection, 12.5 μl of 10^{-4} M E_2 in BSA–saline is injected into each horn using a 25 μl Hamilton syringe (unlabelled E_2 chase). The muscle layers and skin of the rats are sutured separately with 2–0 nylon filaments.

After designated times of unlabeled E_2 chase, the animals are killed and their uteri excised. Two uteri are combined in a ground glass homogenizer with 500 μl of 50 mM Tris–HCl, 7.5 mM EDTA, 10 mM thioglycerol, 10% glycerol, pH 7.4

containing 5 mg/ml soybean trypsin inhibitor, 1 mg/ml leupeptin, and 1 mM PMSF as protease inhibitors. After chilled samples are homogenized, the homogenate is brought to 1 ml and then centrifuged at 800*g* for 10 min. The pellet is resuspended in 360 μl of buffer containing protease inhibitors and 0.6 M NaCl. The sample is incubated at 4°C for 1 h with vortexing every 15 min. This salt-extracted nuclear pellet is then centrifuged at 180 000*g* for 30 min. A portion of the supernatant is assayed for covalently bound radioactivity using the ethanol disk assay. Sample buffer is added to another portion of the salt extract and the sample is boiled at 100°C and then frozen at −80°C until analyzed on SDS-polyacrylamide gels.

REFERENCES

Dermer, O. D. and Ham, G. E. (eds.) (1969) *Ethyleneimine and Other Aziridines*, Academic Press, New York.

Elliston, J. F., Zablocki, J. A., Katzenellenbogen, B. S. and Katzenellenbogen, J. A. (1987) Ketononestrol aziridine, an agonistic estrogen receptor affinity label: study of its bioactivity and estrogen receptor covalent labeling. *Endocrinology*, in press.

Katzenellenbogen, J. A., Myers, H. N. and Johnson, H. J. (1973a) Reagents of photoaffinity labeling of estrogen binding proteins. Synthesis of some diazo and azide derivatives of estradiol, estrone, and hexestrol. *J. Org. Chem.* **38**, 3525–3533.

Katzenellenbogen, J. A., Johnson, H. J. and Myers, H. M. (1973b) Reagents for photoaffinity labeling of estrogen binding proteins. The binding affinity of some azide and diazo derivatives of estradiol, estrone, and hexestrol for the estrogen binding protein of rat uterus. *Biochemistry* **12** 4085–4092.

Katzenellenbogen, J. A., Myers, H. M., Johnson, H. J., Kempton, R. J. and Carlson, K. E. (1977a) Estrogen photoaffinity labels I: Chemical and radiochemical synthesis of hexestrol diazoketone and azide derivatives: Photochemical studies in solution. *Biochemistry* **16**, 1964–1970.

Katzenellenbogen, J. A., Carlson, K. E., Johnson, H. J. and Myers, H. N. (1977b) Estrogen photoaffinity labels II: Reversible binding and covalent attachment of photosensitive hexestrol derivatives to the uterine estrogen receptor. *Biochemistry* **16**, 1970–1976.

Katzenellenbogen, J. A., Carlson, K. E., Heiman, D. F., Robertson, D. W. and Katzenellenbogen, B. S. (1983) Efficient and highly selective covalent labeling of the estrogen receptors with ^{3}H-tamoxifen aziridine. *J. Biol. Chem.* **258**, 3487–3495.

Katzenellenbogen, B. S., Elliston, J. F., Monsma, F. J., Springer, P. A., Ziegler, Y. S. and Greene, G. L. (1987) Structural analysis of covalently labeled estrogen receptors by limited proteolysis and monoclonal antibody reactivity. *Biochemistry* **26**, 2364–2373.

Laemmli, K. (1970) Cleavage of structural proteins during the assembly of the head of bacteriophage T4. *Nature* **227**, 680—684.

Miller, M. A., Mullick, A., Greene, G. L. and Katzenellenbogen, B. S. (1985) Characterization of the subunit nature of nuclear estrogen receptors by chemical crosslinking and dense amino acid labeling. *Endocrinology* **117**, 515–522.

Monsma, F. J., Katzenellenbogen, B. S., Miller, M. A., Ziegler, Y. S. and Katzenellenbogen, J. A. (1984) Characterization of the estrogen receptors and its dynamics in MCF-7 human breast cancer cells using a covalently-attaching antiestrogen. *Endocrinology* **115**, 143–153.

Nardulli, A. M. and Katzenellenbogen, B. S. (1986) Dynamics of estrogen receptor turnover in uterine cells *in vitro* and in uteri *in vivo*. *Endocrinology* **119**, 2038–2046.

Salituro, F. G., Elliston, J. F., Carlson, K. E., Katzenellenbogen, B. S. and Katzenellenbogen, J. A. (1987) [^{125}I]Iododesethyl tamoxifen aziridine: synthesis and covalent labeling of the estgrogen receptor with an iodine-labeled affinity label. *Steroids*, submitted.

Sheen, Y. Y. and Katzenellenbogen, B. S. (1987) Antiestrogen stimulation of the production of a 37,000 molecular weight secreted protein and estrogen stimulation of the production of a 32,000 molecular weight secreted protein in MCF-7 human breast cancer cells. *Endocrinology* **120**, 1140–1151.

Simpson, D. M., Elliston, J. F. and Katzenellenbogen, J. A. (1987) Desmethylnafoxidine aziridine: An electrophilic affinity label for the estrogen receptor with high efficiency and selectivity. *J. Steroid Biochem.* **28**, 233–245.

Zablocki, J. A., Katzenellenbogen, J. A., Carlson, K. E., Norman, M. J. and Katzenellenbogen, B. S. (1987) Estrogenic affinity labels: Synthesis, irreversible receptor binding and bioactivity of aziridine-substituted hexestrol derivatives. *J. Med. Chem.*, in press.

2

Glucocorticoid receptor

S. S. Simons, Jr.
Steroid Hormones Section, National Institutes of Health, Bethesda, USA

I. INTRODUCTION

Three general methods of affinity labeling exist: photoaffinity labeling, electrophilic affinity labeling, and chemoaffinity labeling (for review, see Simons and Thompson, 1982). The first two methods have been successful in yielding affinity labeled glucocorticoid receptors. However, most studies of the physical and biological properties of covalent receptor–steroid complexes have been conducted with electrophilic affinity labeled receptors. This chapter covers the most experimental techniques that have been employed to label glucocorticoid receptors covalently and to study the properties of the resultant covalent complexes.

II. PREPARATION OF AFFINITY LABELING STEROIDS

A. Electrophilic affinity labels

1. Dexamethasone 21-mesylate (Dex-Mes; M. Pons and S. S. Simons, Jr., unpublished results)

Dexamethasone (1.044 g, 2.66 mmol) was dissolved in 23 ml of tetrahydrofuran (THF; Aldrich, 99.9%), cooled to 0°C under argon, and treated with 0.457 ml of methanesulfonyl chloride (Aldrich, 5.90 mmol). A solution of triethylamine (0.875 ml, 6.28 mmol) in 5.3 ml of THF at 0°C was rapidly added (10 s) to the dexamethasone–methanesulfonyl chloride solution. A white precipitate appeared

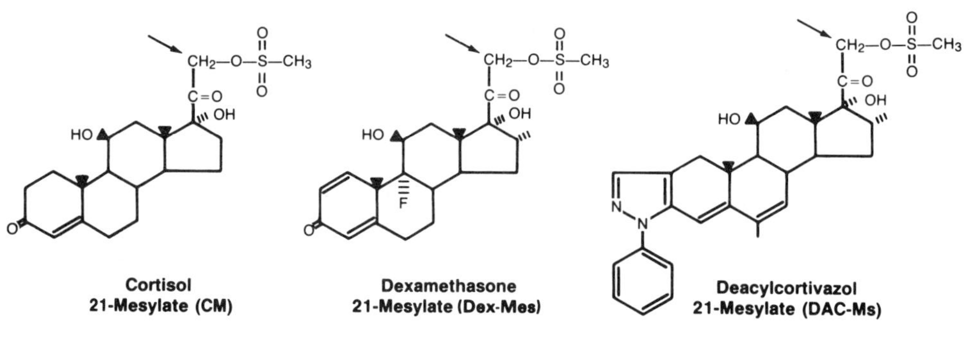

Fig. 1 — Structures of electrophilic affinity labeling steroids. Arrows indicate the reactive position of each molecule.

after 20 s. The reaction is over after 1 min at 0°C, as determined by thin-layer chromatography analysis (35:10 chloroform:acetone or 8% methanol in chloroform on silica gel). Water (20 ml at room temperature) was added and the THF was removed by rotary evaporation. An additional 10 ml of water was added and the solution cooled to 0°C for ⩾ 1 h before collecting the solid by filtration, which was dried overnight ($CaCl_2$, vacuum desiccator) to give 1.22 g (98% yield) of a white solid (decomposition, 228–229°C). Two recrystallizations from ethyl acetate afforded the TLC pure and analytically pure, solvated steroid Dex-Mes·EtOAc (decomposition, 231–232°C). The analysis calculated for $C_{23}H_{31}SFO_7{\cdot}C_4H_8O_2$ (molecular weight, 558.65) was as follows: C, 58.05; H, 7.04: F, 3.40. The analysis found was C, 57.91; H, 7.15; F, 3.37. UV spectroscopy (95% EtOH) of Dex-Mes·THF (formed by recrystallization in THF) gave $\lambda_{max} = 239$ nm, $\varepsilon = 1.67 \times 10^4$.

A slightly less convenient, but almost as high yield (91%), preparation of Dex-Mes in anhydrous pyridine is also available (Simons *et al.*, 1980a).

2. Cortisol 21-mesylate (CM; Simons et al., 1980a)

Methanesulfonyl chloride (1.2 ml, 15.5 mmol) in 20 ml of THF was added dropwise to a stirred solution of 5.0 g of cortisol (13.8 mmol) and 2.1 ml triethylamine (15.4 mmol) in 180 ml THF under nitrogen, which was maintained at 0°C for 15 min and then warmed to room temperature. 3 h later, an additional 1 ml of triethylamine (7.2 mmol) and 0.54 ml methanesulfonyl chloride (7.0 mmol) were added. After stirring overnight, the mixture was filtered, concentrated to 25 ml and added to 300 ml H_2O. The precipitated solid (5.85 g; 96% yield; decomposition, 158–159°C) was dissolved in acetone, filtered, precipitated with petroleum ether and recrystallized from ethyl acetate (yield, 65%; decomposition, 167–168°C). A second crop from ethyl acetate gave 1.03 g (decomposition, 164–165°C). Additional recrystallizations from ethyl acetate gave the analytical sample (decomposition, 179–181°C). UV spectroscopy (95% EtOH) gave $\lambda_{max}=242$ nm, $\varepsilon=1.68\times10^4$. The analysis calculated for $C_{22}H_{32}SO_7$ (molecular weight, 440.56) was as follows: C, 59.98; H, 7.32; S, 7.28. The analysis found was C, 60.06, H, 7.46; S, 7.01.

3. *Deacylcortivazol 21-mesylate (DAC-Mes; Lamontagne et al., 1984)*

To 159.7 mg deacylcortivazol (0.327 mmol) in 2.44 ml anhydrous pyridine (dried over 4 Å molecular sieves) at 0°C under argon was added 50.6 μl methanesulfonyl chloride (0.654 mmol; 2.0 eq) with stirring. After 3 h at 0°C, another 10 μl methanesulfonyl chloride (0.129 mmol; 0.4 eq; total, 2.4 eq) was added. After a total reaction time of 5 h at 0°C, the solution was added to 50 ml ice water. The resulting precipitate was stirred at 0°C for 1 h, filtered, washed three times with 16 ml 0°C water, and dried under high vacuum to give 175.7 mg of white powder (yield, 94.8%). This crude mesylate was dissolved in $CHCl_3$ with a few drops of MeOH and applied to two 2000 μm silica gel plates (Analtech, Newark, DE), which were developed with 3% MeOH in $CHCl_3$. Repetition of this chromatographic procedure, using ethyl acetate to extract the mesylate, afforded a 64% yield of the analytically pure product as a solvate containing 1 mol eq ethyl acetate (decomposition, >190°C): IR spectra (Nujol) exhibited peaks at 3550, 3330, 1731, 1366 and 1178 cm^{-1}; UV spectroscopy (absolute EtOH) gave λ_{max} = 315 nm (ε = 2.10×10^4). Low temperature EI mass spectrum gave large peaks at *m/e*=88 (EtOAc) and 73 ($EtO{-}C{=}O^{\cdot+}$), which rapidly decreased with time. In the CI mode with ammonia, mass spectral peaks were observed at *m/e* 567 (MH^+, 69%) and 471 (100%). The analysis calculated for $C_{31}H_{38}N_2O_6S \cdot C_4H_8O_2$ (molecular weight, 654.81) was as follows: C, 64.19; H, 7.08; N, 4.28. The analysis found was C, 64.22; H, 6.73; N, 4.48.

Deacylcortivazol mesylate exhibited the expected reactivity of a β-hydroxy-α-keto mesylate (Simons *et al.*, 1980a). Deacylcortivazol mesylate is stable in hot MeOH–H_2O but is consumed by thiols in basic acetone solutions in less than 10 min at 0°C and gives positive reaction with *p*-nitrobenzylpyridine (*p*NBP) (Vogel *et al.*, 1970). Furthermore, the shift of the C-20 carbonyl in the IR spectrum of deacylcortivazol mesylate ($C_{20}=0$ at 1731 cm^{-1}) *vs.* deacylcortivazol ($C_{20}=0$ at 1710 cm^{-1}) is identical to that seen for CM and Dex-Mes (Simons *et al.*, 1980a).

B. Photoaffinity labels

1. *Dexamethasone, triamcinolone acetonide, and R5020*

All are commercially available.

2. *21-diazodexamethasone (Manz and Govindan, 1980)*

Dexamethasone (0.2 g) was dissolved in 300 ml methanol. Cupric acetate (25 mg in 100 ml methanol) was added and air was bubbled through the well-stirred solution.

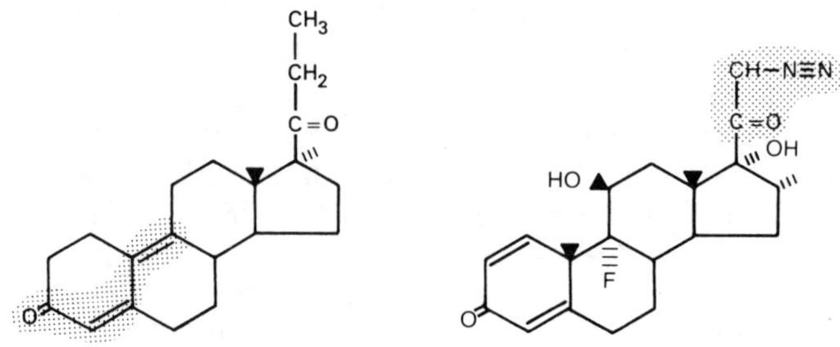

Fig. 2 — Structures of some photoaffinity labeling steroids. Stippled areas cover the photoactivatible portion of each steroid.

The reaction was stopped after 30 min by adding 400 ml H_2O. The solvents were evaporated under reduced pressure and the remaining slightly yellow oil was dissolved in ethyl acetate. The organic phase was washed twice with sodium bicarbonate (1%) and water. The solvent was evaporated and the remaining oil dissolved in acetone. The crude C-21 aldehyde hydrate was isolated as a white powder by the addition of water. The melting point was 125–128°C (literature, 127–130°C (Marx and Smith, 1962)); mass spectrum (high resolution) m/e = 390.1834; expected molecular ion, 390.1842 for $C_{22}H_{27}FO_5$.

The C-21 aldehyde hydrate (400 mg) was treated on a steam bath for 2 h with 75 mg hydroxylamine hydrochloride in 5 ml of pyridine and 5 ml of ethanol to give 280 mg of the C-21 oxime, which was recrystallized from methanol (decomposition 220°C; mass spectrum (high resolution) m/e = 405.1957; expected molecular ion, 405.1951 for $C_{22}H_{28}NFO_5$).

The C-21 oxime (50 mg, 0.12 mmol) was dissolved in 1.5 ml 1 N NaOH, diluted with 3 ml distilled water, and cooled in an ice bath. Some pieces of ice and a 5-fold excess of freshly prepared chloroamine (prepared by combining 3.6 parts of a 13% solution of sodium hypochlorite with 1 part of concentrated ammonium hydroxide) were added and the mixture was kept in the dark for 14 h at 0–4°C. The crude product was collected and washed with ice-cold distilled water. Preparative thin-layer chromatography on silica gel plates with 9:1 (v/v) chloroform:methanol yielded 32 mg (59%) of 21-diazodexamethasone (decomposition, 174°C); UV spectroscopy (methanol) gave λ_{max} = 246 nm ($\varepsilon = 2.6 \times 10^4$); mass spectrum ($m/e$) exhibited no parent ion, 374 ($M-N_2$), 354, 279, 122 (base).

III. AFFINITY LABELING OF CELL-FREE RECEPTORS BY DEX-MES

A. Preparation of HTC cell receptors (Reichman *et al.*, 1984)

Growth of HTC cells in spinner and monolayer cultures has been described (Thompson, 1979). The warm cells were harvested by centrifugation at 0°C (1600 g for 10 min) and washed three times in phosphate-buffered saline (total of 500 ml for

4 l of cells at 5×10^5/ml). The cell plates were quick frozen in dry ice for 45 min and then stored at −20°C. Long-term storage of the cells at −20°C reduces the yield of receptor in the cytosol preparations (Simons, unpublished data) so that the pellets were usually used within 1 week. After the pellet was thawed, an equal volume of pH 9.5 $TAPS_0$ buffer (25 mM TAPS, 1 mM EDTA, and 10% (v/v) gylcerol adjusted to pH 9.5 at 0°C) was added, and the solution was refluxed 20 times in a pipet at 0°C. At this point, >95% of the cells were ruptured, as determined by staining with Azur A. The lysate was centrifuged at 0°C for 15 min at 17 000 g followed by 90 min at 200 000 g. The 200 000 g supernatant, or cytosol, was used immediately or quick frozen in liquid nitrogen. Cytosol frozen in this manner retained its original receptor binding capacity for at least 24 months when the frozen cytosol was rapidly thawed and used immediately (S. S. Simons, Jr., unpublished results).

B. Preparation of rat liver receptors (Eisen *et al.*, 1981)

Rats were killed by decapitation and the livers were perfused *in situ* with HEG buffer (25 mM HEPES, 1.5 mM EDTA, 10% (v/v) glycerol, pH 7.6 at room temperature) containing 0.1 M NaCl. All subsequent operations were performed at 4°C. The livers were removed and homogenized in 2 volumes of HEG (0.1 M NaCl). The homogenates were centrifuged at 100 000 g for 1 h. The supernatant (cytosol) fractions were quickly frozen in 10 ml aliquots and stored in liquid nitrogen.

The frozen liver cytosol (5 ml) was slowly thawed, adjusted to contain 20 mM sodium molybdate, and treated with 3.4 ml of saturated $(NH_4)_2SO_4$ (pH 7.5) over 20 min to give a final concentration of 40%. The protein precipitate was collected by centrifugation (10 000 g for 20 min) and resuspended in the desired buffer containing 20 mM sodium molybdate. This procedure is necessary to remove components (probably low molecular weight thiols such as glutathione) which otherwise prevent the covalent labeling of liver receptors by Dex-Mes. Sephadex G-100 chromatography can be used instead of $(NH_4)_2SO_4$ precipitation to remove these interfering components. The Na_2MoO_4 is required to preserve steroid binding activity.

C. Covalent labeling of HTC cell receptors (Simons and Miller, 1984)

Cytosol is treated for 2.5 h at 0°C in 0.67 volumes of pH 9.5 $TAPS_0$ buffer (25 mMTAPS, 1 mM EDTA, 10% (v/v) glycerol, pH adjusted at 0°C: note that no added thiols should be used since they selectively and rapidly react with Dex-Mes (Simons *et al.*, 1980a; Simons, 1987) with [^{3}H]Dex-Mes in the absence or presence of 80-fold excess of [^{1}H]dexamethasone (final pH 8.8 at 0°C). In those experiments, employing 2×10^{-7} M [^{3}H]Dex-Mes, the amount of covalently labeled 98K receptor equalled 98% ± 11% (SD, $n=8$) of the total availability cell-free receptors, as measured by the specific charcoal-resistant binding of 5×10^{-8} M [^{3}H]dexamethasone ± 600-fold excess [^{1}H]dexamethasone in a parallel assay solution. The yield of covalent receptors is decreased by incubating for less time, with less Dex-Mes, with more dilute receptor solutions, or at lower pH. Higher pH values and temperatures can be counterproductive owing to receptor instability. The optimal pH for cytosols of other cells will vary, depending on receptor stability (Simons *et al.*, 1983).

It has been stated that using frozen livers or cells to prepare steroid receptors increases the proteolytic degradation of the receptors (Sherman *et al.*, 1983). We do see considerable proteolysis of receptors from frozen cells that are also mechanically homogenized in the presence of 2 mM Ca^{2+} (Fig. 3A). However, receptors prepared by freeze–thaw lysis in buffer without Ca^{2+} (see Section III.A) exhibited no proteolytic fragments (Fig. 3B).

D. Covalent labeling of rat liver receptors (Eisen *et al.*, 1981)

$(NH_4)_2SO_4$ precipitated liver cytosol (5 ml) was resuspended in 5 ml of TAPS buffer (25 mM TAPS, 50 mM NaCl, 20 mM Na_2MoO_4, 10% (v/v) glycerol, pH=8.0 at 0°C) containing $(1\text{–}2)\times10^{-7}$ M [^{3}H]Dex-Mes ± 100-fold excess of [^{1}H]dexamethasone for 2 h at 0°C. The yield of covalently labeled receptors is ~80% of the total available specific [^{3}H]dexamethasone binding to cell-free receptors. Higher pH buffers can be used to increase the efficiency of covalent labeling but the steroid binding activity of liver receptors decreases at higher pH so that the total yield of affinity labeled receptors also decreases (Simons *et al.*, 1983).

E. Exchange binding

1. To detect preformed covalent complexes (Simons and Thompson, 1981)

Crude HTC cell receptors were preincubated with 6.0×10^{-8} M [^{1}H]cortisol or 7.1×10^{-8} M [^{1}H]Dex-Mes with or without 6 μM [^{1}H]cortisol, for 3.2 h at 0°C (final protein concentration, 4.7 mg/ml). After addition of activated charcoal to remove free steroid, and to inactivate steroid-free receptors (Simons *et al.*, 1980b), followed by centrifugation to pellet the activated charcoal, the preincubated cytosols were adjusted to 1.9×10^{-8} M [^{3}H]dexamethasone with or without 11 μM [^{1}H]dexamethasone. After subsequent incubation at 0°C, activated charcoal was again added to remove free steroid and the amount of specifically bound [^{3}H]dexamethasone (= total−non-specific binding) that was formed by exchange binding in cytosols preincubated with cortisol, Dex-Mes, or Dex-Mes plus 85-fold excess cortisol was determined. Maximal exchange binding of Dex-Mes pretreated receptors (~20°% of cortisol pretreated receptors) was obtained after incubation for about 2 h.

2. Covalent labeling of previously bound receptors (Housley et al., 1985)

L-cell cytosol (1.0 ml) containing 10 mM sodium molybdate and 2 mM dithiothreitol was stirred on ice for 2.5 h with 1/2 volume of deoxycorticosterone–agarose (Grandics *et al.*, 1984) that was equilibrated at 4°C in 10 mM Hepes, 10 mM molybdate and 2 mM dithiothreitol, pH 7.35. The agarose-bound components of cytosol were collected by centrifugation and washed six times with 10 volumes of cold Buffer A (10 mM Hepes, 10 mM molybdate, 10% (w/v) glycerol, 0.5 mM EDTA, pH 7.6 at 4°C). Receptor was eluted from the resin by suspension in an equal volume (0.5 ml) of Buffer A containing 2.5 μM [^{3}H]Dex-Mes and stirring on ice for 4–6 h. The suspension was centrifuged in a syringe containing a porous polyethylene disk to recover the eluate containing the affinity labeled receptors.

IV. PHOTOAFFINITY LABELING OF CELL-FREE RECEPTORS

While photoaffinity labeling of glucocorticoid receptors does not require prior protein purification, most of the reports have employed partially purified receptor (Westphal *et al.*, 1981; Payvar *et al.*, 1983; Gronemeyer *et al.*, 1983; Gehring and

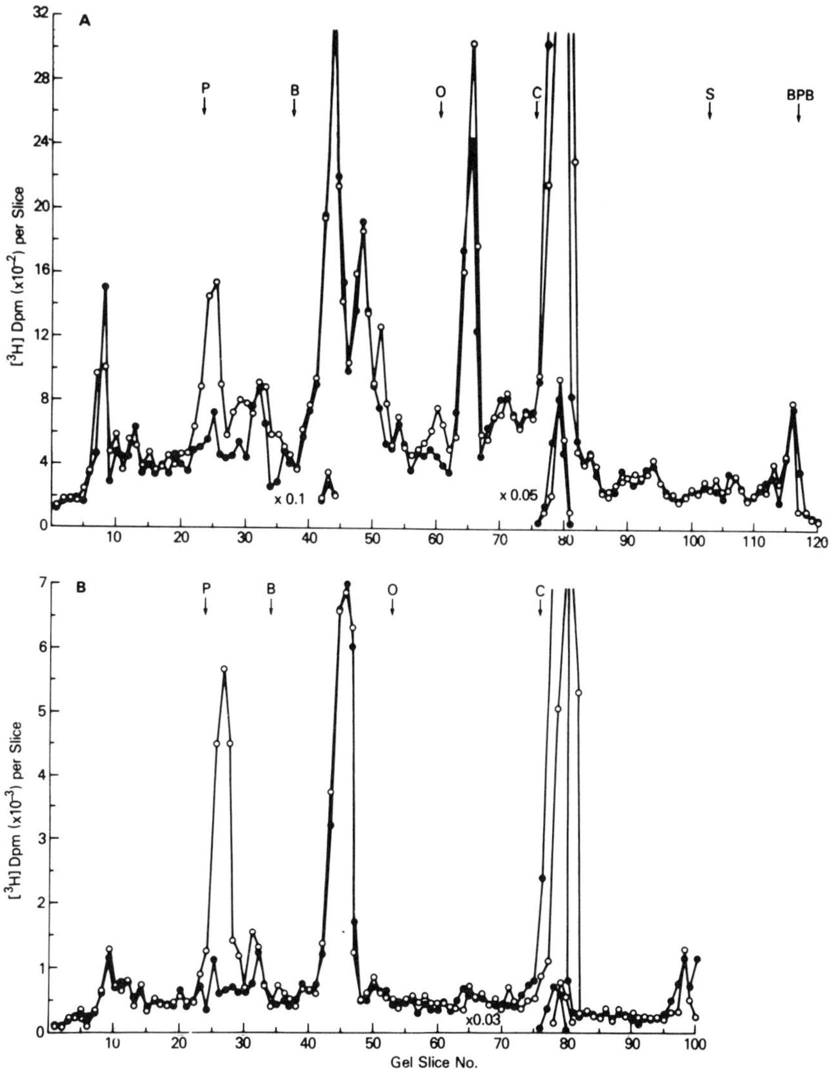

Fig. 3 — [^{3}H]Dex-Mes labeling of HTC cell cytosol receptors prepared by different methods. (A) Cytosol was prepared by mechanical homogenization of frozen cells in buffer A (20 mM Tricine, 2 mM $CaCl_2$, 1 mM $MgCl_2$, pH 8.0 at 0°C) as previously described (Simons *et al.*, 1979). (B) Cytosol was prepared by freeze–thaw lysis in pH 9.5 $TAPS_0$ buffer as described in Section III.A. For both cytosol preparations, covalent labeling of the receptors (i.e. 2×10^{-8} M [^{3}H]Dex-Mes without (○) and with (●) 1.6×10^{-6} M [^{1}H]Dexamethasone at 0°C, pH ~8.8, for 2–3 h) and analysis of the solutions by SDS–polyacrylamide gel electrophoresis were conducted under similar conditions (see Sections III.C, VII.A and VII.B for methods]. P, phosphorylase b, 97 400; B, BSA, 66 300; O, ovalbumin, 45 000; C, carbonic anhydrase, 30 600; S, soybean trypsin inhibitor, 21 500; BPB, bromphenol blue. (From Reichman *et al.*, 1984.)

Holtz, 1983; Hoschutzky and Pongs, 1985). Nordeen *et al.* (1981) have photoaffinity labeled crude receptors as follows. Cytosols (from HTC cells or S49 cells) were incubated on ice for 2–3 h with R5020 and then irradiated directly or immediately after removing unbound steroid by adding 0.2 volumes of 100 mg/ml of dextran-coated charcoal, blending on a Vortex mixer for a few seconds, and then centrifuging for 20–30 s in a microfuge (12 000 g). 1 ml samples, containing 2–10 mg of protein in borosilicate glass tubes (12 mm×75 mm) maintained at 4°C, were irradiated in a Rayonet model RPR-100 photochemical reactor (Southern New England Ultraviolet Co.) equipped with 350 nm 'black light' lamps with an intensity of approximately 9 mW/cm^2 at the center of the chamber. After irradiation, samples were immediately desalted on columns of Sephadex G-25 and quickly frozen in ethanol–dry ice. The activated charcoal treatment reduces non-specific labeling by 80-95%, but specific labeling is decreased by about 40%, apparently as a result of the rapid dissociation of R5020 from glucocorticoid receptors. The covalent labeling efficiency was 2–5%, which is typical for photoaffinity labeling in general.

V. ELECTROPHILIC AFFINITY LABELING OF CELL-FREE, NON-RECEPTOR MOLECULES BY DEX-MES AND DAC-MES

In addition to the glucocorticoid receptor, other macromolecules with lower affinity and higher capacity for binding Dex-Mes can be covalently labeled. DAC-Mes does react with thiols (see Section II.A(3)) but the labeling of proteins in whole cells at 37°C (and of cell-free proteins at pH 8.8, 0°C) is negligibly small, perhaps because of increased hydrophobic binding resulting from the phenylpyrazole substituents on the A-ring (Simons and Miller, 1986). Under special conditions, DAC-Mes does label bovine serum albumin (see Section V.B).

A. Labeling of rat serum albumin (RSA: Eisen *et al.*, 1981; Simons *et al.*, 1983)

The labeling of RSA in $(NH_4)_2SO_4$ precipitated rat liver cytosol by [^{3}H]Dex-Mes is conducted exactly as for the glucocorticoid receptor (see Section III.D). The only difference is that maximal labeling was obtained at a lower pH (~7.4) than with the glucocorticoid receptor.

B. Labeling of bovine serum albumin (BSA; Simons and Miller, 1986; Simons, 1987)

TAPS$_0$ buffer (25 mM TAPS, 1 mM EDTA, 10% (v/v) glycerol, adjusted to pH 9.5 at 0°C; 945 μl) and 40 μl of an aqueous solution of BSA were incubated at 22°C with 15 μl of [^{3}H]Dex-Mes in EtOH to give a final solution (pH 8.8 at 22°C) of 600 μg/ml BSA (9.05×10^{-6} M) and 3.0×10^{-7} M [^{3}H]Dex-Mes. At various times of incubation, 50 μl aliquots were adjusted to 0.55 M in β-mercaptoethanol and further incubated at 22°C for 30 min in order to consume unreacted Dex-Mes (Simons *et al.*, 1983). After 120 h of incubation, 20–40% of the added [^{3}H]Dex-Mes usually became covalently attached to BSA; 60% of the total labeling occurred during the first 7 h.

[^{3}H]DAC-Mes covalently labels BSA in a pH- and time-dependent reaction that is more extensive at 20°C than at 0°C. Analysis on SDS–polyacrylamide gels of experiments conducted at 20°C showed that 1.6–5% of the added [^{3}H]Des-Mes formed a covalent BSA adduct at pH 7.0 after 48 h compared with 4–5% with [^{3}H]Dex-Mes. However, BSA labeling by [^{3}H]DAC-Mes decreased by a factor of 2 when the pH of the reaction solution was raised to 8.9 while [^{3}H]Dex-Mes labeling of BSA increased to 23–33%.

C. Labeling of Y_b-glutathione-*S*-transferase (Homma and Listowsky, 1985)

Cytosol from livers of male Sprague-Dawley rats (175–200 g) (105 000 *g* supernatants) was prepared in 10 mM Tris–HCl+1.0 mM EDTA, pH 8.0 (Tris–EDTA) buffer. Cytosol preparations (3 ml) were adjusted to a protein concentration of 38 mg/ml with Tris–EDTA buffer and incubated with 10 μl of [^{3}H]Dex-Mes at a final concentration of 4×10^{-9} M. Some cytosol preparations were diluted 1:100 in the Tris–EDTA buffer and labeled with [^{3}H]Dex-Mes in the presence or absence of 1 μM dexamethasone, which revealed dexamethasone competition of [^{3}H]Dex-Mes binding–labeling. Reactions were allowed to proceed for 12 h at 4°C.

VI. ELECTROPHILIC AFFINITY LABELING OF PROTEIN BY CORTISOL AND DEXAMETHASONE

It is generally accepted that cortisol and dexamethasone bind to steroid receptors, and proteins in general, in a completely reversible manner to give only non-covalent complexes. However, under certain non-photochemical conditions, covalent steroid–protein adducts have been observed.

A. Labeling of BSA via Heyns rearrangement (Bucala *et al.*, 1982; Bucala *et al.*, 1986)

130 pmol of [^{3}H]cortisol was placed into tubes and evaporated under nitrogen. The steroid was dissolved in 60 μl of ethanol, followed by the addition of 0.54 ml of 50 mM potassium phosphate buffer (pH 7.4) and 0.4 ml of a solution of human serum albumin (100 mg/ml) dissolved in 50 mM potassium phosphate buffer, pH 7.4, containing 10% (v/v) ethanol. The final incubation mixture contained 0.13 μM steroid and 40 mg/ml of albumin. At time zero, the mixtures were transferred from 0°C to a 37°C water bath. At various times, 0.1 ml aliquots were removed, placed inside dialysis tubing (M_r cutoff, 12 000–14 000) and dialyzed for 24 h against 2 l of 50 mM potassium phosphate (pH 7.4)–10% ethanol. The remaining unreacted steroid was removed by chromatography over a column (1 cm×29 cm) of Sephadex G-100, which was equilibrated with 50 nM potassium phosphate (pH 7.4)–10% ethanol. The flow rate was 3 ml/h, and the protein eluate was monitored by absorbance at 280 nm. Fractions (0.5 ml) were collected and the moment of radioactivity was determined by scintillation counting. After 48 h of incubation at 37°C, about 30% of the added cortisol was bound to albumin.

Similar reactions of [^{3}H]dexamethasone at 0.4 μM with BSA under slightly different conditions (BSA concentration, 0.6 mg/ml; total EtOH = 1.5% (v/v); temperature, 21°C; time 0–5 days) were analyzed by SDS–polyacrylamide gel electrophoresis. Here much less covalent attachment was observed in a reaction that was pH dependent (~0.14% labeling after 5 days at pH 7.0; ~0.57% labeling after 5 days at pH 8.8) and independent of the free thiol content of the BSA (S. S. Simons, Jr., unpublished results).

VII. SDS–POLYACRYLAMIDE GEL ANALYSIS OF AFFINITY LABELED PROTEINS

The non-specific binding, and covalent labeling, of crude receptor preparations with [^{3}H]Dex-Mes increases dramatically with higher concentrations of Dex-Mes. Also, some of the non-specifically labeled proteins are more extensively labeled in the presence of competing [^{1}H]dexamethasone (Fig. 3 and Simons *et al.*, 1983). To date, SDS–polyacrylamide gel electrophoresis has proved to be the only method that can reliably be used to quantitate covalently labeled receptors. However, even this method does not work with very high levels of [^{3}H]Dex-Mes (e.g. $\geq 10^{-6}$ M in whole HTC cell labeling experiments) because the non-specific labeling in the region of the 98K receptor is so extensive (S. S. Simons, Jr., *et al.*, unpublished results).

A. Preparation of samples (Simons *et al.*, 1983)

For samples containing large amounts of [^{3}H]Dex-Mes labeled receptors (i.e. $\gtrsim$5000 dpm/20 μl), the sample is diluted with an equal volume of 2×SDS sample buffer (0.6 M Tris (pH 8.85), 2% SDS, 0.2 M dithiothreitol, 20% (v/v) glycerol, and bromphenol blue). Dilute solutions of labeled receptor are concentrated by precipitation in 10% trichloroacetic acid (TCA) at 0°C for at least 2 h (samples can be left overnight at 0°C at this stage). If the total protein concentration after addition of TCA is <100 μg/ml, soybean trypsin inhibitor (at a final concentration of 100 μg/ml) is added as carrier protein. The precipitate is collected by centrifugation (2000g for 10 min), resuspended in 1 ml of 10% TCA for at least 1 h, and recentrifuged as before, all at 0°C. The supernatant is removed by aspiration to give pellets which can be stored at 0 to −20°C for a few days. To prepare the acid-precipitated material for polyacrylamide gels, the pellet is dissolved in a minimum amount of 1 N NaOH (usually 10–15 μl) and then diluted with up to 100 μl of 2×SDS sample buffer.

Acetone has sometimes been used to wash acid-precipitated pellets (Wrange *et al.*, 1979). However, we have found that TCA precipitated, covalently labeled receptors are quite soluble in acetone, even at 0°C. Thus, for directly applied samples, the recovery of specific [^{3}H]Dex-Mes labeled receptor at M_r=98 000 on denaturing polyacrylamide gels was 5– to 6-fold higher than for acid-precipitated–acetone-washed samples of the same incubation solution (e.g. 6000 dpm *vs.* 900 dpm).

Cytosol solutions can be stored at −20°C, or −78°C, for several months before gel analysis. However, once 2×SDS sample buffer is added to the samples, the abundant proteins and labeled receptor undergo considerable degradation over time, even when stored at −20°C (S.S. Simons, Jr., unpublished results).

B. SDS–polyacrylamide gel electrophoresis of [^{3}H]Dex-Mes labeled receptors (Reichman *et al.*, 1984; Simons, 1987)

Samples were diluted with an equal volume of 2×SDS sample buffer, heated at 100°C for 5 min, and applied to constant percentage acrylamide gels (10.5–11% or 15% overlaid with a 3.6% stacking gel, both with a 1:38.9 ratio of bis-acrylamide to acrylamide), which were run in a water-cooled (15°C) Protean or Protean II slab gel apparatus (Bio-Rad). The gels were run at constant current (20 mA/gel while in the stacking gel, then 30 mA/gel for the 10.5–11% gels and 25 mA/gel for the 15% gels). Gels were fixed and stained in 50% methanol–7.5% acetic acid containing 0.02% Coomassie Blue (R-250), destained in 10% methanol–7.5% acetic acid, incubated for 1 h in Enhance (New England Nuclear) and 30–60 min in 10% Carbowax PEG 8000 (formerly PEG 6000; Fisher; recommended by New England Nuclear) with constant shaking at room temperature, dried on a Bio-Rad model 443 slab gel drier at

60°C with a sheet of dialysis membrane backing (Bio-Rad) directly over the gel to prevent cracking, and fluorographed for 7–12 days with Kodak X-OMAT XAR-5 film. The positions of the non-radioactive molecular weight markers were visualized on the fluorographs by overlaying the dried gel with a phosphorescent marker (UltEmit, New England Nuclear). For the slice and counting of gels, gels were sliced (Bio-Rad gel slicer), digested in 0.5 ml of 30% hydrogen peroxide at 60°C for about 4 h, and counted in 5 ml of Hydrofluor (National Diagnostics).

C. Non-equilibrium pH gradient electrophoresis (NEPHGE) of [^{3}H]Dex-Mes labeled-receptors (Smith and Harmon, 1985, 1987)

Samples to be subjected to NEPHGE were diluted with an equal volume of 2D sample solution (9.5 M urea, 2.0% ampholytes (1.6% pH 5–8, 0.4% pH 3–10), 5% β-mercaptoethanol, 2.0% NP–40). NEPHGE was performed, with minor modifications, by the method of O'Farrell *et al.* (1977). Polyacrylamide gels (3% with a 1:17.7 ratio of bis-acrylamide to acrylamide) containing 9.2 M urea, 2.0% NP–40, and a mixture of 1.6% pH 5–7 and 0.4% pH 3–10 ampholytes were cast in glass tubes (125 mm×3 mm) to a height of 11.5 cm. Electrophoresis of samples was performed on non-pre-electrophoresed gels from anode to cathode at 400 V for 2–12 h followed by a final 30 min of electrophoresis at 800 V. Isoelectric equilibrium was achieved after 6 h.

The pH gradient formed by ampholyte migration during NEPHGE was measured immediately after electrophoresis. Gels were sliced into 6 mm sections and incubated in distilled H_2O for 30 min. The pH of the water was then determined.

Immediately after electrophoresis, NEPHGE gels were sliced into 6 mm sections and incubated for 18 h with 0.5 ml of Protosol at 37°C. Acetic acid (17 μl) was added to neutralize the solution, and scintillation fluid was then added for quantitation of radioactivity.

D. Isoelectric focusing of [^{3}H]Dex-Mes labeled receptors on agarose gels (Danze *et al.*, 1987)

Agarose gel slabs (0.5 mm×245 mm×110 mm) were prepared 1 or 2 days before use according to the LKB instruction manual 1818A. The gel composition was 0.8% (w/v) isogel agarose EF, 0.5% (v/v) glycerol and 0.6% (v/v) ampholines (pH range 3.5–9.5). The slab was placed in an LKB multiphor 2117 apparatus connected to an LKB 2103 power supply. Samples (0.02 ml) were applied 1.5 cm from the cathode on a paper sample applicator. The gel was cooled at 4°C throughout the experiment and the running conditions were as follows: current and power unlimited, voltage going from 500 to 1500 V. After a 40 min run, the pH gradient was estimated along a control track which had been loaded with a sample of the low p*I* calibration kit (p*I*= 2.5–6.5; Pharmacia) and then stained with Coomassie brilliant blue G250. Tracks containing the labeled receptor samples were either cut into 1 mm slices for radioactivity counting or used for further chromatographic characterization. In some experiments a rather large amount of complex was applied nearly right across the agarose gel. After focusing, the position of the labeled receptor was checked on a control track and the bulk of the agarose corresponding to this position was cut out. The focused receptor could be eluted from the agarose for subsequent biochemical characterization (e.g. binding to DNA–cellulose columns).

VIII. COVALENT LABELING OF RECEPTORS IN WHOLE CELLS BY DEX-MEX AND LOCALIZATION OF THE COVALENT COMPLEXES

A. Labeling at 37°C (Simons and Miller, 1986)

A 1 ml suspension of HTC (or Fu5-5) cells at $\sim 5 \times 10^6$ (or 10×10^9) cells/ml in Swim's S77 medium (pH 7.75 at room temperature) containing 5% fetal calf serum, 5% newborn calf serum, 0.1% $NaHCO_3$ and 1.4% EtOH plus [^{3}H]Dex-Mes ± a 100-fold excess of [^{1}H]dexamethansone (or [^{3}H]dexamethasone ± a 600-fold excess of [^{1}H]dexamethasone) was incubated with occasional agitation at 37°C for 30 min. The binding reaction was terminated by the addition of 3 ml of phosphate-buffered saline (PBS; pH 7.8 at room temperature) and the cells were washed with PBS at room temperature. The drained cell pellets were ruptured by freezing for 10 s in liquid N_2. The frozen, lysed cells were brought to 0°C and then treated with 0.5 ml of pH 8.8 $TAPS_0$ buffer (25 mM TAPS, 1 mM EDTA, 10% (v/v) glycerol, pH adjusted at 0°C), resuspended by vortexing and centrifuged (500*g* at 0°C for 10 min) to give cytosol (supernatant) and nuclear (pellet) fractions. The nuclear fraction was resuspended in 0.5 ml of pH 8.8 $TAPS_0$ buffer and centrifuged to give a supernatant fraction containing some [^{3}H]dexamethasone-bound receptor (~10% of those in the above cytosol fraction) which sometimes was combined with the cytosol fraction for further analysis (see below). The washed nuclear pellet was extracted for 30 min at 0°C with 0.5 ml of pH 8.8 $TAPS_0$ buffer containing 0.3 M NaSCN and centrifuged (2100*g* at 0°C for 10 min) to give the nuclear extract fraction. The salt-extracted pellet was washed at 0°C with 0.5 ml of TCM buffer (20 mM Tricine, 2 mM $CaCl_2$, 3 mM $MgCl_2$, pH 7.7 at room temperature) and then incubated at 10°C for 60 min with 0.5 ml of TCM buffer containing 100 μg/ml of DNase I. The pellets of the DNase I treated nuclei were extracted twice with 0.1% SDS (100 μl of 0.1% SDS for 30 min followed by 200 μl of 0.1% SDS for 30–45 min, all at 0°C). For cells treated with [^{3}H]dexamethasone or [^{3}H]Dex-Mes, the combined SDS extracts contained about 34% or 84%, while the residual pellets contained $\leqslant$1% or $\leqslant$5% respectively of the nuclear non-NaSCN extractable dpm.

The covalently labeled 98K receptors (plus 82K fragment (Reichman *et al.*, 1984) when present) in the above cytosol, nuclear NaSCN extract and nuclear SDS extract fractions were concentrated by precipitation in 10% TCA, chromatographed on SDS–polyacrylamide (10.8%) gels, and quantitated by slice-and-counting as described in Section VII.B. These quantities of covalently labeled receptors were used to compute the ratio of cytoplasmic:nuclear complexes and to compare with the amount of specific [^{3}H]dexamethasone-bound receptors in parallel samples as determined by direct scintillation counting.

B. Labeling at 0°C (Foster *et al.*, 1983; Miller and Simons, in preparation)

Fu5-5 cells (10^7 cells in 1 ml) were labeled as described for 37°C except that the incubation (for 1 h) and washing of the cells was conducted at 0°C. The cell pellet was frozen in liquid nitrogen and then thawed in 0.5 ml of pH 8.8 $TAPS_0$ buffer containing 20 mM Na_2MoO_4. After centrifugation (600*g* at 0°C for 10 min), the supernatant (or cytosol) was collected. The covalent, unactivated complexes were precipitated by the rapid addition of saturated ammonium sulfate (pH 8.0 at 0°C) to a final concentration of 40% (Reichman *et al.*, 1984). In some preparations, a 20%

ammonium sulfate precut was used to decrease the amount of non-receptor proteins present. After 30 min of incubation at 0°C, the precipitation was collected by centrifugation at 2320*g* for 10 min at 0°C. The precipitate was resuspended in a minimal volume of pH 8.8 $TAPS_0$ buffer containing 20 mM sodium molybdate to preserve the unactivated status of the complexes (Reichman *et al.*, 1984) and stored at −20°C. Recovery of covalent receptor–Dex-Mes complexes after ammonium sulfate precipitation was generally greater than 95%.

It has been reported that the efficiency of 'whole cell' covalent labeling of HeLa S_3 cell receptors at 0°C by [^{3}H]Dex-Mes is increased from 10% to greater than 90% if the pH of the homogenization buffer is raised from 8.0 to 9.0 (Cidlowski and Richnon, 1984). Assuming that the higher pH does not destroy the steroid binding of the HeLa S_3 receptors (*cf.* Section III.C, and Simons *et al.*, 1983), this indicates that most of the covalent labeling of the whole cell receptor occurs after cell rupture. In contrast, Miller and Simons (in preparation) find no difference (±7%) in the absolute amount of Fu5-5 cell receptors that are labeled in whole cells at 0°C when the pH of the lysis buffer (see above) is varied from 6.8 to 9.5.

IX. PARTIAL PURIFICATION OF UNACTIVATED AFFINITY LABELED RECEPTORS

A. $(NH_4)_2SO_4$ precipitation of [^{3}H]Dex-Mes labeled receptors (Reichman *et al.*, 1984)

HTC cell cytosol receptors in TAPS buffer (final pH, ~8.8 at 0°C; see Section III.C) containing 20 mM Na_2MoO_4 were labeled with [^{3}H]Dex-Mes ± 80-fold excess of [^{1}H]dexamethesone for 2.5 h at 0°C. The solutions were then adjusted to contain 50 mM β-mercaptoethanol (to consume unreacted Dex-Mes (Simons *et al.*, 1983)) and 80-fold excess of [^{1}H]dexamethesone in both the competed and the uncompeted samples (to prevent possible binding of [^{3}H]Dex-Mes reaction products). After a further 45 min incubation at 0°C, 0.65 volumes of a saturated $(NH_4)_2SO_4$ solution (in water; pH adjusted to ~8.6 at 0°C with concentrated NH_4OH) was rapidly added (final $(NH_4)_2SO_4$ concentration, 39.4%). After incubation at 0°C for 30 min, the precipitate was centrifuged (0°C for 1 min in a microfuge), resuspended in pH 8.2 $TAPS_0$ buffer (25 mM TAPS, 1 mM EDTA, 10% (v/v) glycerol, pH adjusted at 0°C) containing 20 mM Na_2MoO_4, quick frozen at −78°C, and stored at −20°C or −78°C. By the criterion of binding to DNA–cellulose columns in pH 8.2 $TAPS_0$ buffer containing 20 mM Na_2MoO_4, $(NH_4)_2SO_4$ precipitated, unactivated HTC cell complexes contain <2% activated complexes.

The recovery of precipitated, [^{3}H]Dex-Mes labeled receptor is very high (>90%; S. S. Simons, Jr., *et al.*, unpublished results). Chromatography of the labeled receptors on Sephadex G-100 or G-150 prior to $(NH_4)_2SO_4$ precipitation reduces the final recovery of labeled receptors to ~65% with no appreciable further purification (S. S. Simons, Jr., preliminary results).

B. Immunoadsorption of [^{3}H]Dex-Mes labeled receptors (Eisen *et al.*, 1981)

Polyclonal rabbit anti-rat glucocorticoid receptor antibody (IgG) (Eisen, 1980) was purified from sera by chromatography on protein A–Sepharose. The purified IgG was then immobilized by adding 200 μg of IgG in 2.5 ml of HEG buffer (25 mM HEPES, 1.5 mM EDTA, 10% (v/v) glycerol, pH adjusted to 7.6 at room temperature) to a pellet (0.25 ml) of protein A–Sepharose. The mixture was rotated end-over-end for 1 h and then centrifuged at 200*g*. The pellet was washed twice with HEG buffer containing 0.1 M NaCl. Solutions of [^{3}H]Dex-Mes labeled receptors were mixed with the pellet and rotated end-over-end for 2 h. The mixture was centrifuged and washed as described above, all at 0°C. The pellet was treated with 2.5 ml of 0.1 M acetic acid to release radioactivity and protein (including IgG) from the protein A–Sepharose. The labeled receptors were precipitated with 10% TCA for gel analysis.

C. Affinity chromatography (Kovacic-Milivojevic and Vedeckis, 1986)

AtT-20 pituitary tumor cell line cytosol (3–6 ml) prepared in TMTg buffer (20 mM Tris–HCl, 10 mM sodium molybdate, 10 mM 1-thioglycerol, pH 7.0 at 22°C) was immediately mixed with 3–4 ml of Sterogel A affinity resin (Grandics *et al.*, 1984) and incubated for 2 h at 0°C with gentle stirring. After incubation, the resin was packed into a column and washed extensively with 125 ml of TMG buffer (20 mM Tris, 10 mM sodium molybdate, 10% glycerol, pH 8.0 at 22°C), and the receptor was eluted from the resin by suspension in a half-volume of TMG buffer containing 0.5 μM [^{3}H]Dex-Mes. After a 4 h incubation with gentle stirring, 1-thioglycerol was added to a final concentration of 10 mM. The affinity resin eluate was then chromatographed in TMTg buffer on a column (2.6 cm×60 cm) of Agarose A–1.5 M, 200–400 mesh (Bio-Rad), at a flow rate of 27 ml/h. 3 ml fractions were collected, and aliquots (0.3 ml) were taken for counting. The receptor peak fractions were pooled and applied to a 2 ml DEAE–cellulose column equilibrated in TMTg buffer. The column was washed with 10 volumes of TMTg buffer, and the labeled receptor was eluted with a salt gradient of 0–0.5 M KCl in TMTg buffer at flow rate 50 ml/h. The peak fractions containing the receptor-associated radioactivity were used immediately.

X. ACTIVATION AND DNA BINDING OF AFFINITY LABELED RECEPTORS AND PURIFICATION OF ACTIVATED COVALENT COMPLEXES

A. Activation of [^{3}H]Dex-Mes labeled receptors (Simons *et al.*, 1983; Simons and Miller, 1984)

HTC cell cytosol was treated (see Section III.C) for 2.5 h at 0°C in 0.67 volumes of pH 9.5 $TAPS_0$ buffer, either with $(1.5–2.0)\times10^{-7}$ M [^{3}H]Dex-Mes in the absence or presence of 80-fold excess [^{1}H]dexamethasone or with 5×10^{-8} M [^{3}H]dexamethasone in the absence or presence of 600-fold excess [^{1}H]dexamethasone. This mixture was then diluted with an equal volume of pH 8.8 $TAPS^0$ buffer. β-mercaptoethanol was added to a final concentration of 50 mM to prevent further covalent labeling by [^{3}H]Dex-Mes; an 80-fold excess of [^{1}H]dexamethasone was added to cytosols labeled just with [^{3}H]Dex-Mes in order to block the non-covalent binding of the reaction product of free [^{3}H]Dex-Mes and β-mercaptoethanol. The diluted cytosol was incubated at 20°C for 30 min to activate receptor–steroid complexes.

Slow activation, even at 0°C, occurs when the above labeled receptors are further diluted 5:8 in pH 8.8 $TAPS_0$ buffer instead of being heated at 20°C. The $T_{1/2}$ for this activation at 0°C is 1.32 ± 0.37 h (SD, $n = 3$). This activation is slower at lower pH values (Simons and Miller, 1984).

Rat liver receptors that were $(NH_4)_2SO_4$ precipitated and then labeled by [^{3}H]Dex-Mes in the presence of 20 mM Na_2MoO_4 can be activated simply by passing the labeled complexes through a Sephacryl S-200 column (Simons *et al.*, 1983).

B. Binding of [^{3}H]Dex-Mes labeled receptors to DNA–cellulose columns and purification of activated complexes (Simons *et al.*, S. S. Simons, Jr., and H. J. Eisen, unpublished results)

HTC cell cytosol (4.56 ml) was labeled with [^{3}H]Dex-Mes and activated by dilution (total volume, 15.29 ml) at 20°C as described above. These receptors were loaded (flow rate, ≤1 ml/min) onto a DNA–cellulose column (7.7 ml of DNA–cellulose (Pharmacia, ~1.5 mg calf thymus DNA/ml of cellulose) on top of an ~1 mm bed of Sephadex G-25 in a PD-10 column (Pharmacia)) which had been pre-washed with about 10 column volumes of pH 8.2 $TAPS_0$ buffer containing 50 mM NaCl. The column was then washed with 10 volumes of pH 8.2 $TAPS_0$ buffer containing 50 mM NaCl and the labeled receptors were eluted with pH 8.2 $TAPS_0$ buffer containing 500 mM NaCl (flow rate, ~1.5 ml/min). The peak fractions (0.7 ml fractions containing 2000–5000 dpm of 49 Ci/mmol [^{3}H]-steroid per microliter of eluant) were quick frozen and stored at −78°C until needed. The recovery of covalent, activated complexes (i.e. efficiency of activation) in this preparation was 53.8%.

Note (1). Addition of any NaCl to the activated complexes before DNA–cellulose chromatography can result in a decrease in the amount of complexes retained by the DNA–cellulose column.

Note (2). Presumably as a result of the relatively rapid activation of HTC cell receptor–steroid complexes in pH 8.8 $TAPS_0$ buffer at 0°C (see above), about half of those receptors that can be activated in a pH 8.8 $TAPS_0$ buffer preparation of complexes are found to bind to DNA–cellulose columns without formally being activated. Thus a two-step, 'double DNA–cellulose chromatography' purification of [^{3}H]Dex-Mes labeled receptor (Eisen and Glinsmann, 1978) does not work efficiently in these $TAPS_0$ buffers.

C. Purification of Dex-Mes labeled receptors by 'double DNA–cellulose chromatography' (Govindan and Gronemeyer, 1984)

Because of the limited amount of available [^{3}H]Dex-Mes, cytosol of six rat livers was prepared in 20 mM Tris–HCl, pH 8.0, 50 mM KCl, 1 mM EDTA, and 10% glycerol and incubated with 100 mM [^{3}H]Dex-Mes (9 Ci/mmol) for 1 h at 0°C. In parallel, the cytosol of 10 rat livers was prepared and incubated with 100 nM [^{1}H]dexamethasone. Both cytosol fractions were combined and the pH was adjusted to 7.0 with 1 N HCl. The labeled cytosol (250 ml) was filtered through 400 ml of phosphocellulose in a 10 cm diameter column previously equilibrated in buffer A (1 mM EDTA, 20 mM sodium phosphate, 50 mM NaCl, 10% glycerol, and 2 mM β-mercaptoethanol at pH 7.0; flow rate, 25 ml/min). The column was washed with buffer A and the combined flow-through (400 ml) was adjusted to pH 7.8 with 1 N NaOH. For regeneration, the phosphocellulose was washed immediately with 1 l of 1 M NaCl in buffer A and re-equilibrated in buffer A.

The phosphocellulose flow-through was applied onto a DNA–cellulose column (100 ml in a 5 cm diameter column) with a flow rate of 4 ml/min. The DNA–cellulose I flow-through containing unactivated complexes was activated (30 min at 20°C) and chromatographed on a DNA–cellulose II column (40 ml in a 5 cm diameter column). The flow-through of this column was reactivated (15 min at 20°C) and rechromatographed on a third DNA–cellulose column (30 ml in a 5 cm diameter column). All three DNA–cellulose columns, which contained DNA-bound activated complexes, were washed with 3 volumes of buffer C (1 mM EDTA, 20 mM Tris-HCl, 90 mM NaCl, 10% glycerol, and 2 mM β-mercaptoethanol at pH 7.8) and 3 volumes of 0.5X TBE buffer (0.09 M Tris–base, 0.09 M H_3BO_3, and 2.5 mM EDTA) followed by elution with 0.5 M NaCl in buffer C. The DNA–cellulose eluates were pooled and immediately precipitated in 50% ammonium sulfate overnight in ice and collected by centrifuging for 30 min at 25 000 rpm in SW 27 rotors (Beckman). After careful drying of the walls of the tubes, the precipitated receptors were stored as wet pellets at −80°C.

The ammonium sulfate precipitated [^{3}H]Dex-Mes labeled receptor preparation from the DNA–cellulose column eluate was dissolved in 1 ml of SDS sample buffer containing bromphenol blue as marker, by heating for 2–5 min at 80°C. Preparative gel electrophoresis was performed in a Bethesda Research Laboratories preparative gel electrophoresis system. The separation was performed on a 4 ml 7.5% SDS–polyacrylamide gel with a 2 ml 4% stacking gel. Electrophoresis was carried out at 3 mA and 70 V, and fractions of 1 ml at 5 min intervals were collected. Fraction number 1 contained the bromphenol blue. Radioactivity was determined in 10 μl of each fraction and aliquots of 2.5–10 μl were applied on a 7.5% SDS–polyacrylamide analytical gel. The radioactive fractions containing the glucorticoid receptor band were pooled and precipitated with cold 20% TCA, and the purity of this pool was analyzed again by SDS–polyacrylamide gel electrophoresis and fluorography.

D. Binding to DNA–cellulose pellets (Simons and Miller, 1984)

The background binding of [^{3}H]dexamethasone bound receptors to DNA–cellulose is often determined from the binding of [^{3}H]complexes to cellulose (Le Fevre *et al.*, 1979). Because so much more protein sticks to DNA–cellulose than to cellulose (Fig. 4), the binding to DNA–cellulose of receptors labeled with [^{3}H]+excess [^{1}H] steroid was used as a measure of non-receptor binding to DNA. This is particularly important in the case of labeled receptor solutions where proteins other than the receptor are covalently labeled in various amounts (Simons and Thompson, 1981; Eisen *et al.*, 1981; Reichman *et al.*, 1984).

DNA–cellulose pellets that were incubated with crude cytosol labeled with [^{3}H]Dex-Mes ± 100-fold excess [^{1}H]dexamethasone (Fig. 5A) can be extracted with pH 8.2 $TAPS_0$ buffer containing 500 mM NaCl (2×125 μl) to yield ⩾86% of the radioactivity bound to DNA–cellulose. Greater than 99% of the extracted, specifically bound [^{3}H]Dex-Mes was acid precipitable (S. S. Somins, Jr., unpublished results). More than 80% of the extracted specific binding is present in the 98K receptor (Fig. 5B; Simons *et al.*, 1983; Reichman *et al.*, 1984).

The actual pellet binding assay is conducted as follows at 0°C. Aliquots of

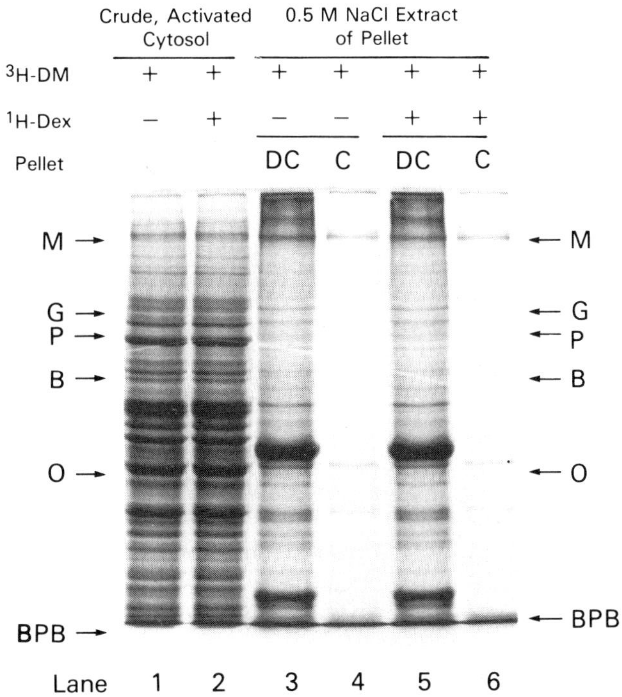

Fig. 4 — Coomassie Blue stained SDS–polyacrylamide slab gel of [^{3}H]Dex-Mes±[^{1}H]dexamethasone labeled cytosol bound to DNA–cellulose or to cellulose. Crude HTC cell receptors were labeled with 1.6×10^{-7} M [^{3}H]Dex-Mes±1.6×10^{-5} M [^{1}H]dexamethasone and then activated as described in Sections III.C and X.A. Aliquots of the activated solutions were diluted 1:2 with 2× SDS sample buffer for subsequent analysis on gels (Section VII.B). Four washed pellets (20 μl each) of DNA–cellulose, or of cellulose, in 300 μl of pH 8.8 $TAPS_0$ were each incubated with 500 μl of the above activated solutions for 2.5 h at 0°C and then washed with pH 8.8 $TAPS_0$ as described in Section X.D. Each group of four pellets was combined and extracted (2×0.5 ml) with pH 8.8 $TAPS_0$ containing 0.5 M NaCl. The protein in the combined two salt extracts was precipitated with TCA, dissolved in NaOH, and diluted with 2× SDS sample buffer. SDS–polyacrylamide slab gels (9%) were then run of the activated cytosols (lanes 1 and 2), of the TCA precipitated 0.5 M NaCl extracts of [^{3}H]Dex-Mes labeled cytosol that bound to DNA–cellulose (DC) and cellulose (C) (lanes 3 and 4), and of the TCA precipitated 0.5 M NaCl extracts of [^{3}H]dexamethasone labeled cytosol that bound to DNA-cellulose (DC) and cellulose (C) (lanes 5 and 6). The positions of the molecular weight markers are indicated by the arrows: M, myosin, 200 000; G, β-galactosidase, 116 250; P, phosphorylase b, 97 400; B, BSA, 66 300; O, ovalbumin, 45 000; BPB, bromophenol blue. (From Simons and Miller, 1984.)

DNA–cellulose slurry were added to 4.5 ml conical polystyrene centrifuge tubes (Sarstedt) and washed with 1.5–2.0 ml of pH 8.8 $TAPS_0$ buffer. After centrifugation (1250*g* for 2 min at 0°C), the supernatant was removed by aspiration. The final pellet volume was one-third that of the initial aliquot. $TAPS_0$ buffer, pH 8.8 (150 μl), and labeled cytosol containing activated complexes (250 μl) were added to the washed DNA–cellulose pellets. Tubes were vortexed to resuspend DNA–cellulose and then incubated on a Bellco roller drum apparatus at 13 rpm at 0–4°C. Cytosol labeled with

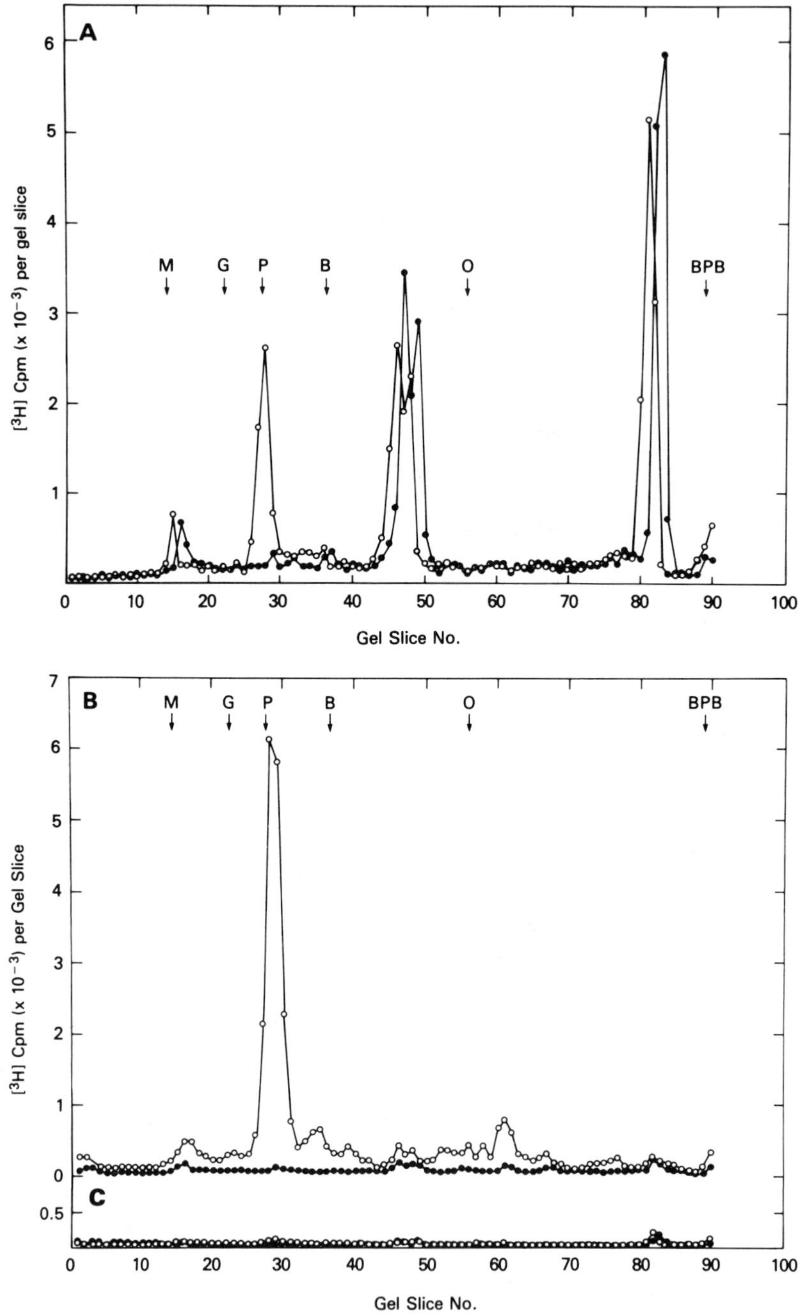

Fig. 5 — Binding of [^{3}H]Dex-Mes labeled cytosol to DNA–cellulose and to cellulose. The lanes of the gel in Fig. 4 were sliced and counted (○, cytosol labeled with [^{3}H]Dex-Mes; ●, cytosol labeled with [^{3}H]Dex-Mes+100-fold excess [^{1}H]dexamethasone) as follows: (A) activated cytosol (lanes 1 and 2 of Fig. 4); (B) 0.5 M NaCl extracts of DNA–cellulose pellets (lanes 3 and 5 of Fig. 4); (C) 0.5 M NaCl extracts of cellulose pellets (lanes 4 and 6 of Fig. 4). The arrows indicate the positions of the molecular weight markers used in Fig. 4. (From Simons and Miller, 1984.)

[^{3}H]dexamethasone was incubated with DNA–cellulose for 0.5–1.25 h; cytosol labeled with [^{3}H]Dex-Mes was incubated for 1–1.75 h. In both cases, >80% of the maximal binding is seen after 15 min and is stable for 1–4 h. Cytosol labeled with either steroid in the presence of excess [^{1}H]Dex-Mes was incubated for 1.5–3 h. Binding was terminated at equilibrium by centrifugation (1250*g* for 2 min at 0°C) and the supernatant was removed by aspiration. A single wash of the DNA–cellulose pellets dramatically reduced the non-specific binding but had little effect on the observed specific binding; a second wash caused little further change and was omitted. Thus, the pellets were routinely washed with 1.0 ml of $TAPS_0$ buffer, pH 8.8, centrifuged as described above, aspirated to complete dryness with a flame-tapered Pasteur pipet, and incubated for at least 1 h at room temperature with 0.2 ml of pH 8.8 $TAPS_0$ buffer (to dissociate bound steroid and/or receptors) before being transferred to a scintillation vial with two 0.1 ml H_2O washes and counted in 10 ml of Hydrofluor. Duplicates generally varied by less than 5%. Specific binding of [^{3}H]steroid–receptor complexes to DNA–cellulose was calculated as the difference between binding of cytosol labeled in the absense (total binding) and presence (non-specific binding) of excess [^{1}H]dexamethasone. Non-specific binding was usually less than 7% or 25% of the total binding in [^{3}H]dexamethasone, or [^{3}H]Dex-Mes labeled cytosols respectively.

E. DNA competition binding assay (Miller *et al.*, 1984)

This assay is a modification of the above DNA pellet binding assay and can be used with non-covalent [^{3}H]dexamethasone, and covalent [^{3}H]Dex-Mes, receptor–steroid complexes. Activated [^{3}H]labeled receptor–steroid complexes (250 μl) were incubated with 1 unit of DNA–cellulose (in 100 μl of pH 8.8 $TAPS_0$ buffer; 1 unit of DNA–cellulose=33.3 μl of DNA–cellulose pellet at a concentration of 1.2 mg DNA/ml of cellulose) and 50 μl of TE buffer (10 mM Tris, 1 mM EDTA, pH 7.4 at room temperature) for control binding or with 0.625 units of DNA–cellulose (in 100 μl of pH 8.8 $TAPS_0$ buffer) and soluble DNA (in 50 μl of TE buffer) for 0.75–2.55 h at 4°C. Samples were processed as described above. Non-specific binding for each condition was determined by parallel incubations of the appropriate amounts of DNA–cellulose and soluble DNA with cytosol labeled in the presence of excess [^{1}H]dexamethasone. Specific binding was calculated as the difference in the amounts of binding to DNA–cellulose by cytosols labeled in the absence and presence (i.e. non-specific binding) of excess [^{1}H]dexamethasone. Specific binding in the presence of soluble DNA was then expressed as a percentage of specific binding to control (i.e. 1 unit of DNA–cellulose), and plotted as in Fig. 6. The theoretical basis of this assay is outlined elsewhere (Simons, 1977).

F. DNA filter binding and footprinting

These assays have not yet been described using affinity labeled glucocorticoid receptors. However, owing to the nearly idential cell-free DNA binding properties of non-covalent dexamethasone, and covalent Dex-Mes, receptor–steroid complexes (Simons *et al.*, 1983; Simons and Miller, 1984, 1986; Miller *et al.*, 1984), it is expected that very similar, if not identical, procedures can be used for non-covalent and affinity labeled complexes. Since the nuclear binding of non-covalent dexameth-

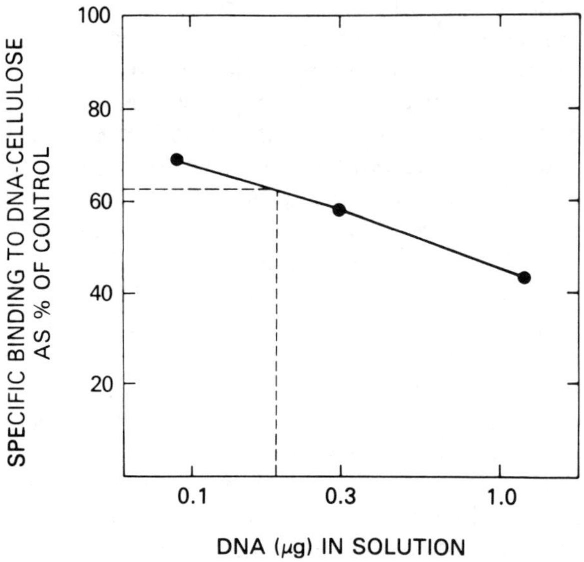

Fig. 6 — Competition of receptor–steroid complex binding of DNA–cellulose by soluble DNA. Activated, [^{3}H]dexamethasone bound HTC cell receptors were incubated with 1 unit of DNA–cellulose (control binding) or 0.625 units of DNA–cellulose plus 0.09–1.2 μg of mouse mammary tumor virus LTR DNA. Binding of receptor–[^{3}H]dexamethasone complexes to DNA–cellulose was measured as described for the DNA pellet assay (Section X.D) and expressed as a percentage of control binding. The amount of LTR DNA with the same binding capacity as 0.375 units of DNA–cellulose is that amount which reduces the binding of receptor–[^{3}H]dexamethasone complexes to 0.625 units of DNA–cellulose to 62.5% of control binding. Interpolation of this value is indicated by the broken lines. (From Miller *et al.*, 1984.)

asone and covalent Dex-Mes complexes are not the same (Simons *et al.*, 1983; Simons and Miller, 1986), different experimental conditions may be necessary when dealing with nuclei.

XI. PROTEOLYSIS OF UNACTIVATED, AND ACTIVATED, AFFINITY LABELED RECEPTORS

A. [^{3}H]Dex-Mes labeled receptors (Reichman *et al.*, 1984; Simons, 1987)

Nearly radiochemically pure samples of activated [^{3}H]Dex-Mes labeled HTC cell receptors were prepared as described in Section IX.A by DNA–cellulose chromatography. Unactivated complexes that were partially purified by $(NH_4)_2SO_4$ precipitation (see Section XI.A) still contained a non-specifically labeled species at ~55 000 daltons. Therefore all digestions of unactivated complexes were performed with a parallel sample of identically prepared, non-specifically labeled (by [^{3}H]Dex-Mes plus 80-fold excess [^{1}H]dexamethasone) cytosols. With these paired samples, one can easily identify those fragments deriving from the labeled 98K receptor. The various receptor preparations were digested both in the native state and after denaturation for 30 min at 22°C with 0.2% SDS–0.5 mM dithiotheitol (Fig. 7).

Native and denatured complexes were treated at 10°C for 20 min with enough stock solution (2.3 mg/ml in water) of trypsin (TPCK treated), chymotrysin (TLCK treated), or *Staphylococcus aureus* V8 protease to give final protease concentrations of 0.06–100 μg/ml. Digestions were stopped by freezing the samples in dry ice. For gel analysis, the samples were thawed just long enough to add 2×SDS sample buffer (0.6 M Tris (pH 8.85), 2% SDS, 0.2 M dithiothreitol, 20% (v/v) glycerol, and bromphenol blue) and refrozen at −78° until heated at 100°C for 5 min and then electrophoresed on 10.8% SDS–polyacrylamide gels.

Limit protease digestion was achieved by treating denatured, [^{3}H]Dex-Mes labeled complexes with 150 μg/ml of protease for up to 48 h at room temperature. These digests were analyzed on 15% SDS–polyacrylamide gels, which resolve bands down to a molecular weight of ~1350 daltons (Simons, 1987).

B. Proteolysis of photoaffinity labeled receptors (Gehring and Hotz, 1983; Gronemeyer *et al.*, 1983)

Several proteases (trypsin, chymotrypsin, endoproteinase Lys-C, endoproteinase Arg-C) have been used to digest photoaffinity labeled receptors. Several different digestion conditions have been used. The main difference with digestion of Dex-Mes labeled receptors is that the concentration of photoaffinity labeled receptors is usually much lower.

XII. LOCALIZATION OF MODIFIED AMINO ACID(S) IN AFFINITY LABELED RECEPTORS

A. [^{3}H]Dex-Mes labeled receptors (Simons *et al.*, 1987)

Trypsin, chymotrypsin, and V8 protease limit digestion of nearly radiochemically pure samples of activated, [^{3}H]Dex-Mes labeled HTC cell receptors in the denatured state was conducted as described in Section XI.A. Efforts to purify a limit digest fragment for direct amino acid sequencing were thwarted by the extreme hydrophobicity of these labeled fragments. Thus the [^{3}H]-labeled, V8 limit digest fragment has an apparent $M_r = 1790 \pm 200$ (SD, $n = 17$) on 15% SDS–polyacrylamide gels (Simons, 1987) and can be separated from most of the other proteins and peptides in the crude digestion mixture on 15% SDS–polyacrylamide gels but not on C_3 HPLC columns. The limit digest fragment could not be eluted from C_{18} HPLC columns. Lyophilization of the digested sample caused most of the [^{3}H]Dex-Mes labeled peptide to stick tightly to the walls of the tube and to resist resolubilization by most solvents. The labeled material was retained equally well by non-specific, and anti-dexamethasone, antibodies and stuck to Amicon 'Centricon 10' ultrafiltration membranes. All the radioactivity in crude V8 protease limit digests appeared in the void volume of Sephadex columns, regardless of whether or not the limit digest solution was first heated at 100°C for 5 min before being chromatographed (a) on LH-20 columns with pH 8.8 $TAPS_0$ buffer containing 10 mM β-mercaptoethanol, (b) on LH-20 columns with $TAPS_0$–β-mercaptoethanol buffer containing 4 M guanadinium hydrochloride, or (c) on either LH-20 or G-25 columns with $TAPS_0$ buffer containing 1% SDS and 0.1 M dithiothreitol.

Good recoveries (60–70%) of the labeled material were obtained from Sep-pak

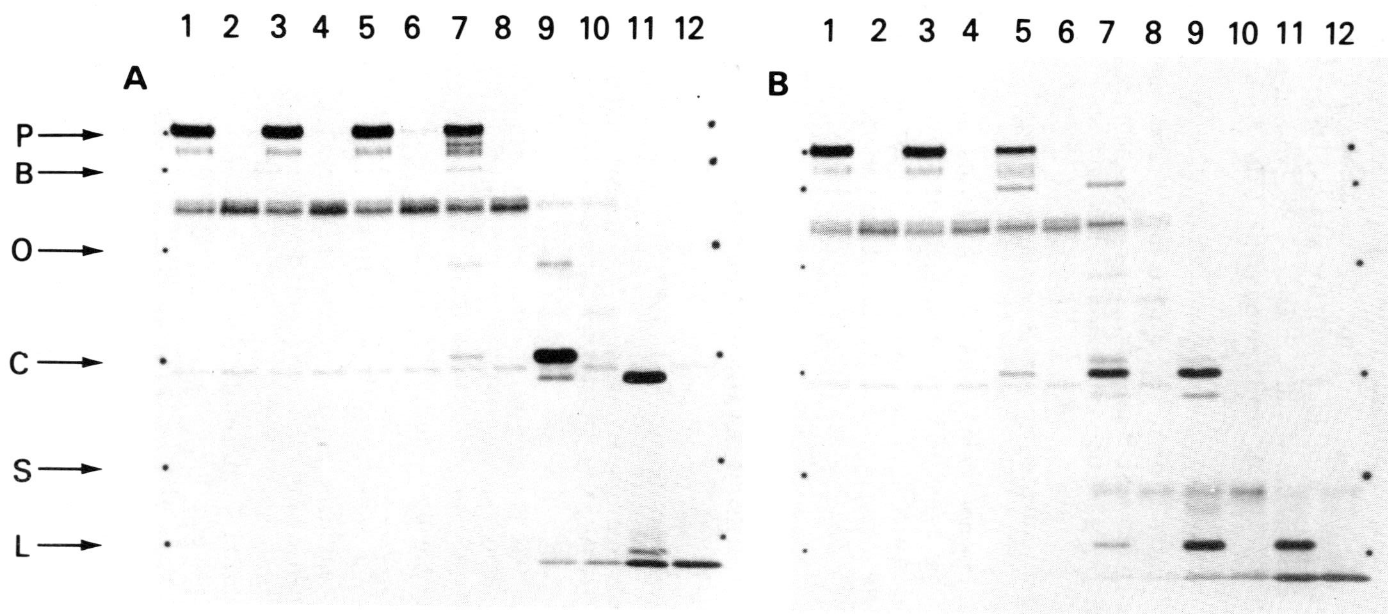

Fig. 7 — Digestion of unactivated, [^{3}H]Dex-Mes labeled HTC cell receptors by trypsin. Paired samples of unactivated receptors that had been covalently labeled by [^{3}H]Dex-Mes without (odd-numbered lanes) or with (even-numbered lanes) an excess of [^{1}H]dexamethasone (see Section IX.A) were kept at 0°C (lanes 1 and 2) or were digested at 10°C for 20 min (lanes 3–12) in the (A) native and (B) denatured (30 min at 22°C in SDS–dithiothreitol) state. Enzyme concentrations (μg/ml) used in the lanes are as follows: (A) lanes 1–4, 0; lanes 5 and 6, 0.063; lanes 7 and 8, 0.5; lanes 9 and 10, 4; lanes 11 and 12, 100; (B) lanes 1–4, 0; lanes 5 and 6, 0.8; lanes 7 and 8, 4; lanes 9 and 10, 20; lanes 11 and 12, 100. (A) Fluorograph of digest of native complexes; (B) fluorograph of digest of denatured complexes. Dots at edge of fluorograph were made by overlaying a phosphorescent marker (UltEmit) on the band of the molecular weight markers on the dried gels. P, phosphorylase b; B, BSA; O, ovalbumin; C, carbonic anhydrase; S, soybean trypsin inhibitor; L, lysozyme (M_r = 14 000; see Fig. 3 legend for other M_r values). (From Reichman *et al.*, 1984.)

C_{18} cartridges (Waters) that were pre-wet with methanol and then water containing 0.1% trifluoroacetic acid (TFA; Pierce). The [^{3}H]Dex-Mes labeled limit digests (100–200 μl) were diluted to 0.5 ml with water, boiled for 5 min, and applied to the cartridges, which were washed with 10 ml of water–0.1% TFA. The sample was eluted with 90% MeOH–10% water–0.1% TFA and 0.5 ml fractions were collected. The peak fractions were determined by scintillation counting. The amount of purification by this procedure was minimal but the sample was now ready for sequential Edman degradation (Fig. 8). If further chromatography on C_3 HPLC

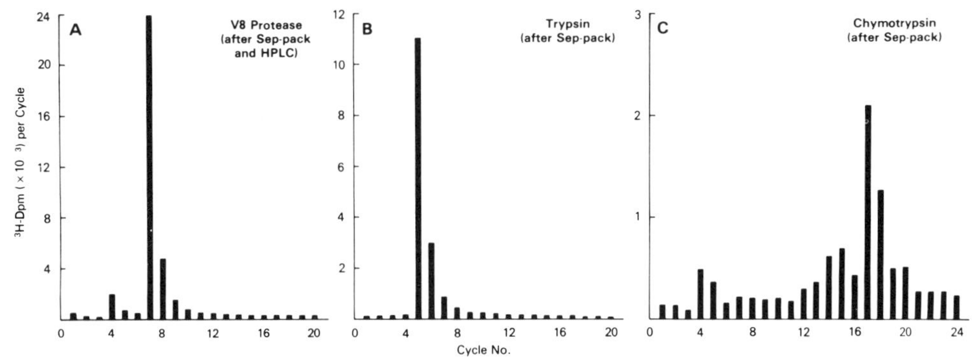

Fig. 8 — Sequential Edman degradation of limit protease digestion fragments of activated, [^{3}H]Dex-Mes labeled HTC cell glucocorticoid receptors. Activated [^{3}H]Dex-Mes labeled receptors were denatured for 30 min at 22°C, digested with V8 protease, trypsin, or chymotrypsin at 150 μg/ml for 16–18 h at 22°C, and chromatographed on C_{18} Sep-pak cartridges. The peak fractions from the Sep-pak cartridges were subjected to sequential Edman degradation on an Applied Biosystems Model 470A protein–peptide sequencer. The dpm per Edman degradation cycle were plotted for (A) the C_{18} Sep-pak chromatographed, and then C_3 HPLC fractionated, V8 protease limit digest, (B) a C_{18} Sep-pak chromatographed, trypsin limit digest, and (C) a C_{18} Sep-pak chromatographed, chymotrypsin limit digest. In each case, the complete conversion of the 98K receptor to the limit protease digestion fragment was demonstrated by electrophoresis on 15% SDS–polyacrylamide gels and detection by fluorography. (From Simons *et al.*, 1987.)

columns (Altex) was desired, the Sep-pak eluted material was diluted with 10 volumes of water–0.1% TFA, applied to columns equilibrated in 10% MeOH–90% water–0.1% TFA, and eluted by a linear gradient of 90% MeOH–10% water–0.1% TFA.

Since the specific cleavage sites of each of the proteases appears to be maintained under the above denaturing conditions (Simons, 1987) (i.e. basic amino acids with trypsin, aromatic amino acids plus leucine for chymotrypsin, and acid amino acids for V8 protease), a partial amino acid sequence could be constructed with members of each class of amino acid positioned as indicated from the Edman degradation data. If the affinity labeling steroid is known to react with specific amino acids (e.g. Dex-Mes specifically reacts with the –SH group of crystaine (Simons *et al.*, 1980a; Simons, 1987)), these amino acids can be added to the above partial sequence.

A comparison of this partial amino acid sequence with the entire sequence of the receptor will give the position(s) of the affinity labeled amino acid(s). If this analysis yields more sites than there are labeled amino acids according to the Edman degradation data, then physical data (e.g. which fragments of partially proteolyzed receptors both contain the affinity labeling steroid and are recognized by monoclonal antibodies) can be used to eliminate the spurious sites.

REFERENCES

Bucala, R., Fishman, J. and Cerami, A. (1982) Formation of covalent adducts between cortisol and 16α-hydroxyesterone and protein: possible role in the pathogenesis of cortisol toxicity and systemic lupus erythematosus. *Proc. Natl. Acad. Sci. USA* **79**, 3320–3324.

Bucala, R., Ulrich, P. C., Chait, B. T., Bencsath, F. A. and Cerami, A. (1986) Structure of lysine adducts with 16α-hydroxystrone and cortisol. *J. Steroid. Biochem.* **25**, 127–133.

Cidlowski, J. A. and Richon, V. (1984) Evidence for microheterogeneity in the structure of human glucocorticoid receptors. *Endocrinology* **115**, 1588–1597.

Danze, P.-M., Formstecher, P., Richard, C. and Dautrevaux, M. (1987) Microheterogeneity of agonist and antagonist glucocorticoid receptor complexes detected by isoelectric focusing and modifications induced by receptor activation. *Biochem. Biophys. Acta* **927**, 129–138.

Eisen, H. J. (1980) An antiserum to the rat liver glucocorticoid receptor. *Proc. Natl. Acad. Sci. USA* **77**, 3893–3897.

Eisen, H. J. and Glinsmann, W. H. (1978) Maximizing the purification of the activated glucocorticoid receptor by DNA–cellulose chromatography. *Biochem J.* **171**, 177–183.

Eisen, H. J., Schleenbaker, R. E. and Simons, S. S. Jr. (1981) Affinity labeling of the rat liver glucocorticoid receptor with dexamethasone 21-mesylate: identification of covalently labeled receptor by immunochemical methods. *J. Biol. Chem.* **256**, 12920–12925.

Foster, C. M., Eisen, H. J. and Bloomfield, C. D. (1983) Covalent labeling of rat thymocyte and human lymphoid glucocorticoid receptor. *Cancer Res.* **43**, 5273–5277.

Gehring, U. and Holtz, A. (1983) Photoaffinity labeling and partial proteolysis of wild-type and variant glucocorticoid receptors. *Biochemistry* **22**, 4013–4018.

Govindan, M. W. and Gronemeyer, H. (1984) Characterization of the rat liver glucocorticoid receptor purified by DNA–cellulose and ligand affinity chromatography. *J. Biol. Chem.* **259**, 12915–12924.

Grandics, P., Miller, A., Schmidt, T. J., Mittman, D. and Litwack, G. (1984) Purification of the unactivated glucocorticoid receptor and its subsequent *in vitro* activation. *J. Biol. Chem.* **259**, 3173–3180.

Gronemeyer, H., Harry, P. and Chambon, P. (1983) Evidence for two structurally related progesterone receptors in chick oviduct cytosol. *FEBS Letters* **156**, 287–292.

Homma, H. and Listowsky, I. (1985) Identification of Y_b-glutathione-*S*-transferase as a major rats liver protein labeled with dexamethasone 21-methanesulfonate. *Proc. Natl. Acad. Sci. USA* **82**, 7165–7169.

Hoschutzky, H. and Pongs, O. (1985) Characterization of glucocorticoid receptor in HeLa-S_3 cells. *Biochemistry* **24**, 7348–7356.

Housley, P. R., Sanchez, E. R., Westphal, H. M., Beato, M. and Pratt, W. B. (1985) The molybdate-stabilized L-cell glucocorticoid receptor isolated by affinity chromatography or with a monoclonal antibody is associated with a 90-92-kDa nonsteroid-binding phosphoprotein. *J. Biol. Chem.*, **260**, 13810–13817.

Kovacic-Milivojevic, B. and Vedeckis, W. V. (1986) Absence of detectable ribonucleic acid in the purified, untransformed mouse glucocorticoid receptor. *Biochemistry* **25**, 8266–8273.

Lamontagne, N., Mercier, L., Pons, M., Thompson, E. B. and Simons, S. S. Jr. (1984) Glucocorticoid *vs* antiglucocorticoid activity: can a single functional group modification of glucocorticoid steroids always convey antiglucocorticoid activity? *Endocrinology* **114**, 2252–2263.

Le Fevre, B., Bailly, A., Sallas, N. and Milgrom, E. (1979) Activated steroid–receptor complex. Comparison of assays using DNA–cellulose or homologous nuclei. *Biochem. Biophys. Acta* **585**, 266–272.

Manz, B. and Govindan, M. V. (1980) Synthesis of 21-diazo-9α-fluoro-16α-methyl-21-deoxy-[1,2(n)-^{3}H]prednisolone. *Hoppe-Seyler's Z. Physiol. Chem.* **361**, 953–957.

Marx, M. and Smith, L. L. (1962) 21-Nitrogen-substituted pregnane derivatives. *Chem. Abs.* **57**, 2289i.

Miller, P. A., Ostrowski, M. C., Hager, G. L. and Simons, S. S. Jr. (1984) Covalent and noncovalent receptor–glucocorticoid complexes preferentially bind to the same regions of the long terminal repeat of murine mammary tumor virus proviral DNA. *Biochemistry* **23**, 6883–6889.

Nordeen, S. K., Lan, N. C., Showers, M. O. and Baxter, J. D. (1981) Photoaffinity labeling of glucocorticoid receptors. *J. Biol. Chem.* **256**, 10503–10508.

O'Farrell, P. Z., Goodman, H. M. and O'Farrell, P. H. (1977) High resolution two-dimensional electrophoresis of basic as well as acidic proteins. *Cell* **12**, 1133–1142.

Payvar, F., DeFranco, D., Firestone, G. L., Edgar., B., Wrange, O., Okret, S., Gustafsson, J.-A. and Yamamoto, K. R. (1983) Sequence-specific binding of glucocorticoid receptor to MTV DNA at sites within and upstream of the transcribed region. *Cell* **35**, 381–392.

Reichman, M. E., Foster, C. M., Eisen, L. P., Eisen, H. J., Torain, B. F. and Simons, S. S. Jr. (1984) Limited proteolysis of covalently labeled glucocorticoid receptors as a probe of receptor structure. *Biochemistry* **23,** 5376–5384.

Sherman, M. R. and Moran, M. C., Tuazon, F. B. and Stevens, Y.-W. (1983) Structure, dissociation and proteolysis of mammalian steroid receptors. Multiplicity of glucocorticoid receptor forms and proteolytic enzymes in rat liver and kidney cytosols. *J. Biol. Chem.* **258**, 10366–10377.

Simons, S. S. Jr. (1977) Glucocorticoid receptor–steroid complex binding to DNA: competition between DNA and DNA–cellulose. *Biochem. Biophys. Acta* **496**, 349–358.

Simons, S. S. Jr. (1987) Selective covalent labeling of cysteines in bovine serum albumin and in HTC cell glucocorticoid receptors by dexamethasone 21-mesylate. *J. Biol. Chem.*, in press.

Simons, S. S. Jr. and Miller, P. A. (1984) Comparison of DNA binding properties of activated, covalent and noncovalent glucocorticoid receptor–steroid complexes from HTC cells. *Biochemistry* **23**, 6876–6882.

Simons, S. S. Jr. and Miller, P. A. (1986) Affinity-labeling steroids as biologically active probes of antiglucocorticoid hormone action. *J. Steroid Biochem.* **24**, 25–32.

Simons, S. S. Jr. and Thompson, E. B. (1981) Dexamethasone 21-mesylate: an affinity label of glucocorticoid receptors from rat hepatoma tissue culture cells. *Proc. Natl. Acad. Sci. USA* **78**, 3541–3545.

Simons, S. S. Jr. and Thompson, E. B. (1982) Affinity labeling of glucocorticoid receptors: New methods in affinity labeling. In: Litwack, G. (ed.) *Biochemical Actions of Hormones*, Vol. 9. Academic Press, New York, pp. 221–254.

Simons, S. S. Jr., Pons, M. and Johnson, D. F. (1980a) α-Keto mesylate: A reactive thiol-specific functional group. *J. Org. Chem.* **45**, 3084–3088.

Simons, S. S. Jr., Thompson, E. B., Merchlinsky, M. J. and Johnson, D. F. (1980b) Synthesis and biological activity of some novel, chemically reactive glucocorticoids. *J. Steroid Biochem.* **13**, 311–322.

Simons, S. S. Jr., Thompson, E. B. and Johnson, D. F. (1979) Anti-inflammatory pyrazolo-steroids: potent glucocorticoids containing bulky A-ring substituents and no C3-carbonyl. *Biochem. Biophys. Res. Commun.* **86**, 793–800.

Simons, S. S. Jr., Schleenbaker, R. E. and Eisen, H. J. (1983) Activation of covalent, affinity labeled, glucocorticoid receptor–steroid complexes. *J. Biol. Chem.* **258**, 2229–2238.

Simons, S. S. Jr., Pumphry, J. G., Rudikoff, S. and Eisen, H. J. (1987) Identification of cysteine-656 as the amino acid of HTC cell glucocorticoid receptor that is covalently labeled by dexamethasone 21-mesylate. *J. Biol. Chem.*, in press.

Smith, A. C. and Harmon, J. M. (1985) Multiple forms of the glucocorticoid receptor steroid binding protein identified by affinity labeling and high-resolution two-dimensional electrophoresis. *Biochemistry* **24**, 4946–4951.

Smith, A. C. and Harmon, J. M. (1987) Structural organization of the human glucocorticoid receptor determined by one- and two-dimensional gel electrophoresis of proteolytic receptor fragments. *Biochemistry* **26**, 646–652.

Thompson, E. B. (1979) Liver cells (HTC). *Methods Enzymol.* **58**, 544–551.

Vogel, C. L., Denham, C., Waalkes, T. P. and DeVita, V. T. (1970) Physiological disposition of the carcinostatic 5(or 4)-[3,3-bis(2-chlorethyl)-1-triazeno]imidazole-4(or 5)-carboxamide (NSC 82196) (imidazole mustard) in mice and dogs. *Cancer Res.* **30**, 1651–1657.

Westphal, H. M., Fleischmann, G. and Beato, M. (1981) Photoaffinity labeling of steroid binding proteins with unmodified ligands. *Eur. J. Biochem.* **119**, 101–106.

Wrange, O., Carlstedt-Duke, J. and Gustafsson, J.-A. (1979) Purification of the glucocorticoid receptor from rat liver cytosol. *J. Biol. Chem.* **254**, 9284–9290.

3

The chicken progesterone receptor

H. Gronemeyer
Institut de Chimie Biologique, Faculté de Médecine, Strasbourg, France

I. PHOTOAFFINITY LABELING

A. Steroids

For the chicken progesterone receptor the only ligand that has been used for crosslinking is the synthetic progestin R5020 (promegestone; 17α, 21-dimethyl-19-nor-4,9-pregnadiene-3,20-dione) which is commercially available both as [17α-methyl-^{3}H]-R5020 and in the non-radioactive form from NEN. The specific activity of the labeled form is typically 70–87 Ci/mmol and the molecular weight is 326.0 Da. Its structure is depicted in Fig. 1. The use of further progestins and progesterone

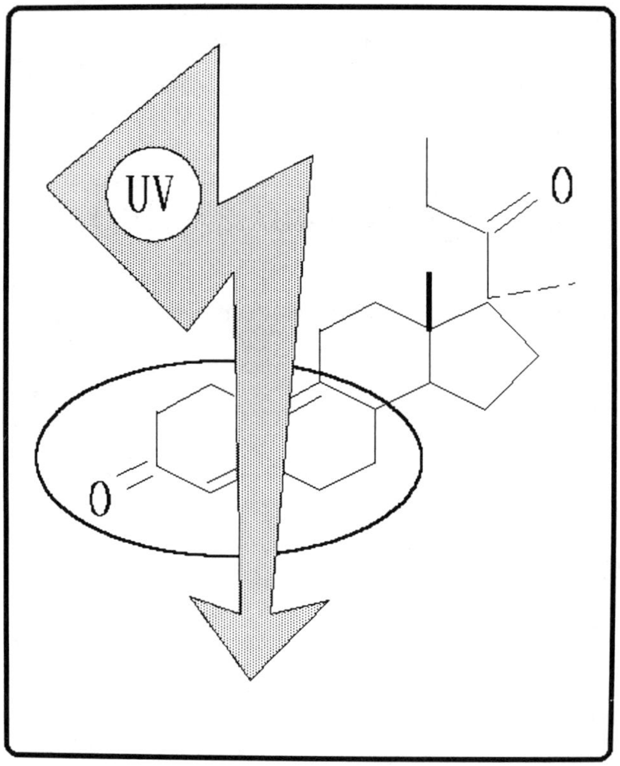

Fig. 1 — Crosslinking with R5020.

derivatives has been reported for other systems (for review and references see Gronemeyer and Govindan, 1986; Katzenellenbogen and Katzenellenbogen, 1984) but only the antiprogestin RU 38.486 has been verified as a crosslinking agent in the human T47D system (see Chapter 4 on the human progesterone receptor). This antihormone does not bind to the chicken receptor.

The mechanism of photoaffinity labeling with unsaturated carbonyl compounds is as yet unknown and no photoaddition products have been isolated and identified from receptor affinity labeling with ligands bearing this type of chromophor. Despite this lack of understanding, the first report on the use of unsaturated ketosteroids for photoaffinity labeling appeared in 1973. It was shown that Δ^5-ketosteroid isomerase from *P. testosteroni* (EC 5.3.3.1) in the presence of 3-oxo-4-estrene-17β-yl acetate is irreversibly inactivated after irradiation with UV light above 300 nm. Later it was demonstrated that the cause of this inactivation was a ligand-dependent photodestruction and the conversion of an aspartic acid residue at position 38 into alanine (Benisek, 1977, and references therein). Similarly, steroid-dependent photoinactivation of the estrogen receptor ligand binding site was reported using 6-oxo-estradiol (Katzenellenbogen *et al.*, 1974). Although this study did not report a photocovalent attachment, the possibility of such a reaction had been demonstrated much earlier by Bellus and coworkers (1969). On irradiation of *o*-acetyltestosterone in toluene the toluene adduct 3-oxo-4-benzyl-17β-acetoxy-andostane was the major reaction product. However, in addition, a variety of side reactions were determined, as depicted in Fig. 2. This figure also shows the putative steps involved in the photoaddition

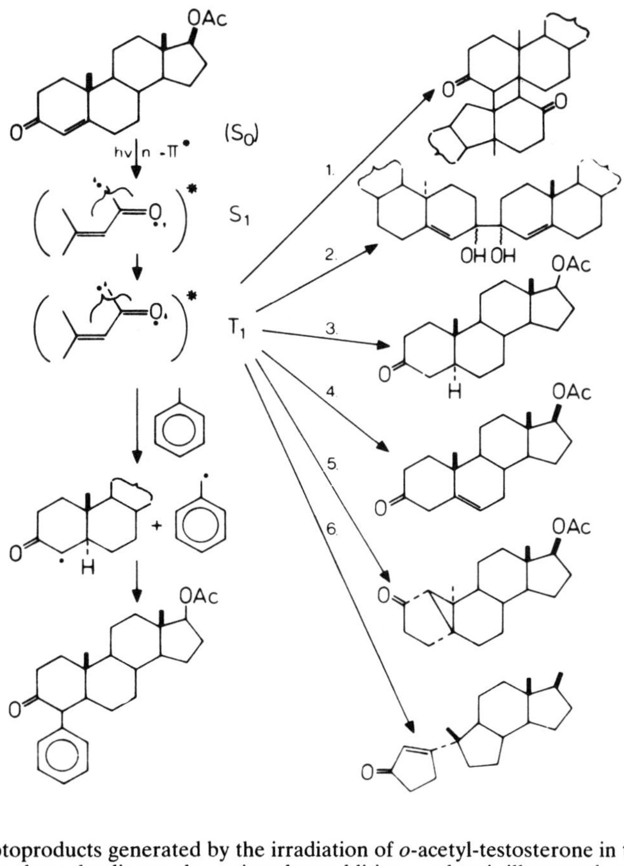

Fig. 2 — Photoproducts generated by the irradiation of *o*-acetyl-testosterone in toluene. The putative pathway leading to the major photoaddition product is illustrated on the left.

reaction. The enone moiety in its electronic ground state (S_0) is photoactivated to a very short-lived singlet (S_1) which subsequently converts to a highly reactive diradical-like triplet state (T_1). The radical pair can combine or undergo additional hydrogen atom transfers or other molecular rearrangements, finally yielding stable products. If the hydrogen atom abstracted initially is part of a binding protein, then photocrosslinking by the excited ketone is, in principle, possible. The chemical specificity of the photoexcited keton is very broad since both nucleophiles and C–H groups are potential reactants. A particular advantage of the excited ketone is that the triplet energy is not sufficient to react with the O–H bonds of water.

B. Equipment for photoaffinity labeling

For the photoactivation of enones a high-power UV source is required which generates a collinear beam that can be filtered and focused onto a cooled cuvette. Simple photochemical apparatus for the irradiation of protein samples at either 254 or >315 nm have been described (Katzenellenbogen *et al.*, 1974). The author of this chapter has satisfactory experience with a 1 kW mercury–xenon lamp connected to a power supply that compensates for the aging of the lamp (Oriel). To exclude UV light below 300 nm, which severely damages protein, we use a toughened 2 mm thick long-pass filter (WG 320, Schott). The spectrum of this filter is characterized by a transmission of 5×10^{-4} at 300 nm, 0.5 at 320 nm and 0.94 at 340 nm, as is shown in Fig. 3. The use of other filter systems, including liquid filters, has been reported also,

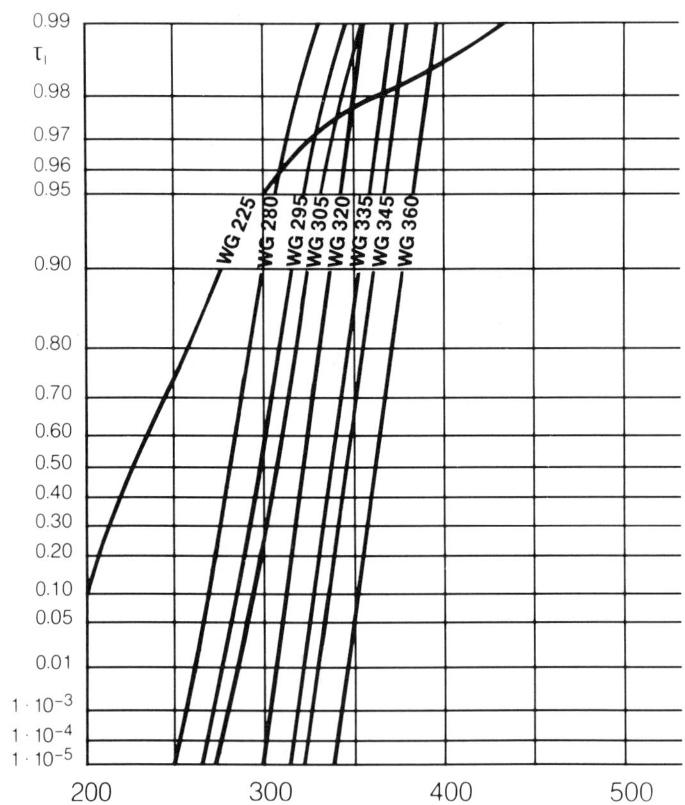

Fig. 3 — Spectrum of a series of long-pass filters (WG series) for photoaffinity labeling commercially available from Schott, FRG.

but these are more expensive or less convenient. To compensate for the heat emission of the lamp, the hormone receptor complex is irradiated in a jacketed cylindrical quartz cuvette, cooled to −10°C. To prevent freezing the irradiation buffer contains 10% glycerol. The crosslinking efficiency is determined by precipitation of the receptor preparation in 10% TCA followed by filtration onto GFC filters (Whatman), and is expressed as the percentage of specifically bound hormone. Because of the variety of possible side reactions (Fig. 2), a crosslinking efficiency of 8–12% of specifically receptor-bound hormone is usually observed.

C. Protocol

A progesterone–receptor preparation in TEBGN50 (10 mM Tris, pH 7.8, 1 mM Na2 EDTA, 1 mM β-mercaptoethanol, 50 mM NaCl, 10% glycerol) is incubated with [^{3}H]-R5020 for at least 2 h at 0–4°C. Typically, for chick oviduct cytosol about 70% of the receptors will be labeled using 20 mM radioactive steroid. For complete labeling, incubation is carried out overnight in an ice bath. To determine non-specific labeling, an aliquot of the preparation is incubated additionally with a 100-fold excess on non-radioactive hormone. If the buffers are not based on Tris, scavengers such as DTT should be included to avoid non-specific crosslinking by long-lived steroid radicals which dissociate from the receptor during irradiation. Excess non-bound steroid is removed, for example by gel filtration (spun column) or by adding 1/10 volume of a solution containing 5% charcoal (Norite A) and 0.5% dextran T70 in buffer and incubation in ice for 30 min to 2 h followed by centrifugation at 4°C (10 000 *g*, 5 min). The steroid hormone–receptor complex is transferred into the precooled cuvette (−10°C), magnetically stirred, and irradiated. The crosslinking kinetics should be determined by TCA precipitation on GFC filters (wash extensively with EtOH). With the UV apparatus described above the photoaddition will be complete after 1 min. The receptor is immediately concentrated by precipitation in 50% saturated ammonium sulfate and further processed or analyzed directly, for example by SDS–polyacrylamide electrophoresis after dialysis against the sample buffer.

II. PURIFICATION

A. Methodologies

There are basically three methodologies which have been used to purify the chicken progesterone receptors forms A and B.

(i) 'Classical' routes employ chromatography on matrices that separate on the basis of size (e.g. Sephadex gel filtration) or charge (e.g. DEAE cellulose chromatography) or general affinity of 'activated' receptors to DNA (e.g. calf thymus DNA cellulose chromatography) in combination with precipitation techniques (using e.g. ammonium sulfate). Several purification strategies based on combinations of these steps have been published in detail by O'Malley's group (for references see Section II on the chicken progesterone receptor).

(ii) Ligand affinity chromatography has only recently become attractive since the problems with ternary binding proteins have been solved. This purification technique has been reviewed recently by Formstecher and Lustenberger (1987), who give the relevant references for the chicken progesterone receptor.

(iii) A combination of a limited number of classical purification steps, controlled by affinity labeling, combined with preparative SDS–gel electrophoresis and devised for the unequivocal purification to homogeneity of large amounts of receptor proteins will be described in detail below.

B. Preparation of cytosol

The basic buffer used for receptor preparation is 10 mM Tris–HCl, 1 mM Na_2 EDTA, 1 mM β-mercaptoethanol, 10% glycerol, and 50 mM NaCl (TEBGN50) at the pH indicated (differing salt concentrations are described as TEBGNX with X denoting the concentration (millimolar) of NaCl). Salt concentrations are determined by conductivity measurement.

20 laying hens are killed, and their oviducts are removed and collected in 1 l of ice-cold TEBGN50, pH 7.4, containing 0.3 mM phenylmethylsulfonyl fluoride and 10 mM Na_2MoO_4. Unless otherwise stated all subsequent manipulations are performed at 0–3°C. All centrifuges and rotors used during purification should be precooled to −5°C. The magnum portion of the oviduct is dissected, sliced open, washed in 2 l of ice-cold buffer, and homogenized in 1.6 l of this buffer by eight 5 s bursts (low speed) in a Waring blender. The homogenate is centrifuged for 15 min at 9500 *g*, and the floating fat layer is aspirated. The supernatant is recentrifuged at 105 000 *g* for 60 min using R45Ti rotors (Beckman). The resulting cytosol (about 1.5 l) is incubated with 20 nM [^{3}H]R5020 (New England Nuclear, 7.71 Ci/mmol), 10^{-6} M cortisol, and 10^{-7} M dexamethasone for 2 h at 4°C. Non-specific binding of R5020 should be below 10% as determined in parallel incubations.

C. Classical chromatography

The basic principle for the purification is differential DNA and phosphocellulose chromatography before and after activation, as developed by Coty and colleagues (1978). Cytosol is prepared in TEBGN50 from laying hen oviducts as this is quantitatively the richest source of progesterone receptor and because no differences have been observed when compared with the progesterone receptor isolated from stimulated chickens. In order to preserve the non-activated status of the receptor during the initial steps, cytosol is prepared in the presence of 10 mM molybdate which is known to inhibit receptor transformation and inactivation. In addition, phenylmethylsulfonyl fluoride is used as protease inhibitor. Incubation of cytosol with radioactively labeled R5020 is performed in the presence of cold dexamethasone and cortisone in order to suppress the labeling of the glucocorticoid receptor which is known to have a weak affinity for R5020, although the number of glucocorticoid receptors in oviduct is about one-tenth that of progesterone receptors. An incubation time of 2 h at 4°C under continuous stirring is usually sufficient to label about 500 fmol/mg of cytosolic protein, representing about 22 nmol of progesterone receptor A and B in the starting phase of a standard purification from 20 laying hen oviducts. It is noteworthy that the extent of labeling is not far below that obtained with estrogenized chicken although endogenous progesterone is present in the oviduct from laying hens. The high affinity of R5020 toward the progesterone receptor may account for this fact, as Schrader *et al.* (1977) could label only a minority of receptors using progesterone. For all calculations only progesterone receptor labeled with R5020 should be considered because the fate of non-labeled receptor is unknown.

Immediately before starting a series of purifications (up to ten) the column matrices are washed according to the manufacturer's instructions. Between subsequent purifications the columns were washed with 1 M NaCl in the equilibration buffer. After incubation, the cytosol is cooled in an ice–water slurry and applied onto a phosphocellulose (Whatman P-11) column I, equilibrated in TEBGN50, pH 7.4 (column diameter, 10 cm; bed volume, 1 l). The flow rate is adjusted by collecting the flowthrough 1.5 m below the column outlet. Owing to accumulation of binding proteins and some residual fat on top of the column the flow rate will decrease during chromatography; the surface of the matrix is, therefore, stirred from time to time with a sterilized glass rod. The appearance of the cytosol filtrate is monitored by its yellow color. When 500 ml of phosphocellulose I flowthrough are obtained, the pH is adjusted to 8.0 by the addition of 1 ml of 1 M Tris base, and the flowthrough is immediately applied onto a DNA–cellulose column I, equilibrated in TEBGN50, pH 8.0 (we use a column having the following characteristics: column diameter, 10 cm; bed volume, 630 ml; DNA content, 150 μg of high molecular weight calf thymus DNA/ml of packed bed volume). The flow rate of this column was adjusted to 8–10 ml/min (average, 7.6 ml/(cm^2 h). Two further 500 ml phosphocellulose I flowthrough fractions as well as a 500 ml wash with TEBGN50, pH 7.4, are adjusted to pH 8.0 and applied onto the DNA–cellulose I column as soon as they are collected. When the yellow-colored DNA–cellulose filtrate appears at the column outlet it is gathered in an ice-cooled reservoir. From there it is immediately pumped in the reverse direction (from bottom to top) onto a DEAE–Sepharose C1-6B (Pharmacia) column equilibrated in TEBGN50, pH 8.0 (guideline: diameter, 5 cm; bed volume, 400 ml) at 14 ml/(cm^2 h). The DNA–cellulose column I is washed with 500 ml of TEBGN50, pH 8.0, until all the yellow–colored material is removed from the column and bound onto the DEAE–Sepharose as a bright yellow band fading about 5 cm inside the matrix. Thereafter, both the phosphocellulose and the DNA–cellulose column are washed with 2 l of the corresponding equilibration buffer containing 1 M NaCl, followed by immediate re-equilibration. Under these conditions the columns will not show any significant decay in binding capacity.

The DEAE–Sepharose column is washed with 1 l of TEBGN50, pH 8.0, from bottom to top. The flow direction is then reversed and the column is washed with an additional 1 l of buffer. Elution of the progesterone receptor from this matrix is performed using TEBGN300 without separating forms A and B. The eluting front can be monitored again by the yellow-colored material that has been absorbed on the DEAE matrix. The 500 ml of eluate usually obtained at this stage are precipitated in a final concentration of 40% ammonium sulfate (calculated for 0°C). For this, 113 g of solid ammonium sulfate are added to the 500 ml of eluate during 1 h in fractions of about 10 g under continuous stirring at 3°C. The pH will fall to 7.6 (pH paper) but need not be readjusted. When all of the ammonium sulfate is dissolved the preparation is stirred for another hour on ice and then centrifuged at 9500 *g* for 30 min. The supernatant is decanted and the tubes are quickly frozen in liquid nitrogen so that the pellets can be quantitatively removed as solid pieces, which are then stored at −70°C. The hormone–receptor complex is very stable under these storage conditions, and preparations have been used up to 6 months after freezing, which, after solubilization, yielded 85% of non-dissociated hormone–receptor complex with normal chromatographic behavior in the succeeding steps.

For subsequent purification the ammonium sulfate pellets of one to two preparations (20 oviducts each) are dissolved in ice-cold TEBG, pH 8.0, to give a salt concentration of 90–100 mM as determined by conductivity measurements. About 400 ml are required for the precipitate of one preparation. After centrifugation (9500 *g*, 15 min) the supernatant is chromatographed on a tandem arrangement of a DNA–cellulose column II (column diameter, 2.5 cm; bed volume, 150 ml; DNA content; see DNA–cellulose column I), equilibrated in TEBGN100, pH 8.0, and operated in the reverse direction, and a phosphocellulose column II (P-11, column diameter; 2.5 cm; bed volume, 40 ml), equilibrated in the same buffer and operated in the usual direction. By using a peristaltic pump in front of the DNA–cellulose column the flow rate is adjusted to approximately 9.5 ml/(cm^2 h). After application of the preparation, the tandem is washed with 1.5 l of TEBGN100, pH 8.0, disconnected, and, with normal flow direction for both columns, each is washed with an additional 500 ml of the same buffer. The phosphocellulose column serves to rescue any receptor molecules which do not bind to the DNA–cellulose. The capacity of the DNA–cellulose column was such that the material prepared from 40 oviducts, which contains at this stage approximately 10 nmol of hormone-binding activity, may be purified in one run. Both columns are eluted with TEBGN500, pH 8.0, and the radioactivity-containing fractions are combined.

Immediately after pooling, the DNA cellulose and phosphocellulose eluates are crosslinked by UV irradiation and precipitated overnight by the addition of solid ammonium sulfate to a final concentration of 50% at 0°C. The precipitates are collected by high speed centrifugation (100 000 *g*, 1 h) in SW 40 polyallomer tubes. After the supernatant is decanted, the walls of the tubes were dried with paper tissues and the sealed tubes are stored at −70°C.

D. Preparative SDS gel electrophoresis

The ammonium sulfate pellet of one tube is solubilized in 0.5 ml of sample buffer (0.0625 M Tris–HCl, pH 6.8, 2% SDS, 10% glycerol, 5% β-mercaptoethanol, 0.001% bromphenol blue) and dialyzed against 250 ml of sample buffer in a collodium bag (Sartorius) for 2 h. The contents of the bag are agitated from time to time with a pipette. After dialysis the sample was heated for 2 min at 95°C and, after the addition of bromphenol blue, loaded onto a precast preparative gel (1100 PG, Bethesda Research Laboratories), assembled as described by the manufacturer. The SDS–PAGE system of Laemmli is used with a 4 ml 7.5% separating and a 2 ml 4% stacking gel. The upper and lower buffer chamber contains 0.025 M Tris, 0.192 M glycine, and 0.1% SDS, while the elution buffer is 0.075 M Tris, 0.576 M glycine, and 0.1% SDS. Electrophoresis is performed under constant current (3.0 mA, 76 V initially). The pumping rate is adjusted to give 1 ml fractions in 5 min. Bromphenol blue is usually eluted in one fraction, designated as fraction 1. The radioactivity is determined in 5–20 μl aliquots of each fraction, and the purity of the peak fractions is analyzed by SDS–polyacrylamide gel electrophoresis of 5 μl aliquots followed by silver staining. Progesterone receptor B is routinely observed as an electrophoretically pure band, whereas form A is contaminated by an upper and a lower protein band. For further purification all fractions containing form A are pooled, dialyzed overnight in Sartorius collodium bags against 10 mM ammonium bicarbonate, and lyophilized. The material is taken up in 0.2 ml of sample buffer, heated for 2 min at 95°C, and rerun on a second preparative gel. For this a 10% polyacrylamide separating gel (4 ml) is used. Stacking gel, buffers, and electrophoretic conditions are as described above. After determination of radioactivity, the corresponding fractions are again analyzed by SDS–polyacrylamide gel electrophoresis and silver staining as is shown in Fig. 4.

E. Generation of antibodies

Polyclonal antibodies against the hen oviduct progesterone receptor forms A and B were raised in rabbits by using the preparative gel-purified receptor proteins as immunogens. Priming was performed by injecting about 16 μg of receptor A or B in complete Freund's adjuvant intradermally, subcutaneously, and intramuscularly at multiple sites. One week later a series of subcutaneous booster injections was initiated, each time with 5–20 μg of receptor proteins A or B in incomplete Freund's adjuvant. After four boosts the rabbits were bled from the ear vein and left to recover for 2 weeks. Thereafter, the animals were reboosted using subcutaneous injections of about 20 μg of receptor–rabbit in incomplete Freund's adjuvant, bled 1 week later, and left to recover for 1 or 2 further weeks. Sera and preimmune sera from coagulated blood of each individual rabbit were obtained, aliquoted, and stored at −70°C.

III. STRUCTURAL ANALYSIS

A. Partial proteolytic maps

Peptide maps of crude or partially purified receptors are generated either by incubating a constant amount of covalently labeled receptor with increasing amounts of a protease or by digesting with a fixed amount of protease and substrate for varying lengths of time. We will describe digestion using α-chymotrypsin below, but other enzymes such as trypsin or staphylococcus V8 protease can be used in a similar way (see also the original paper by Cleveland *et al.* (1977). TLCK-treated α-chymotrypsin and TPCK-treated trypsin are available commercially (Sigma). TLCK (*N*α-4-tosyl-L-lysine chloromethylketone) treatment inactivates the trypsin which is usually present in chymotrypsin, TPCK (L-1-tosylamide-2-phenylethyl chloromethylketone) inactivates the α-chymotrypsin present in the trypsin batches. TLCK-α-chymotrypsin is prepared by adding 1 mg/ml TLCK to a solution of 50 mg/ml α-chymotrypsin in 100 mM Na phosphate, pH 7.2+1 mM Na_2EDTA which is also the buffer for digestion.

Receptor preparations after UV crosslinking are precipitated in 50% ammonium sulfate and the protein pellets are solubilized in digestion buffer containing 8 M freshly deionized urea — if digestions are performed with the denatured protein. For maps of non-denatured receptor the urea is omitted. After dialysis against digest buffer–2 M urea identical aliquots of the $[^3H]$R5020 receptor are digested at 37°C for 30 min by varying the enzyme to protein ratio from 1/20 to 1/500 (cytosolic preparations) or 1/100 to 1/10 000 (partially purified preparations). For complete

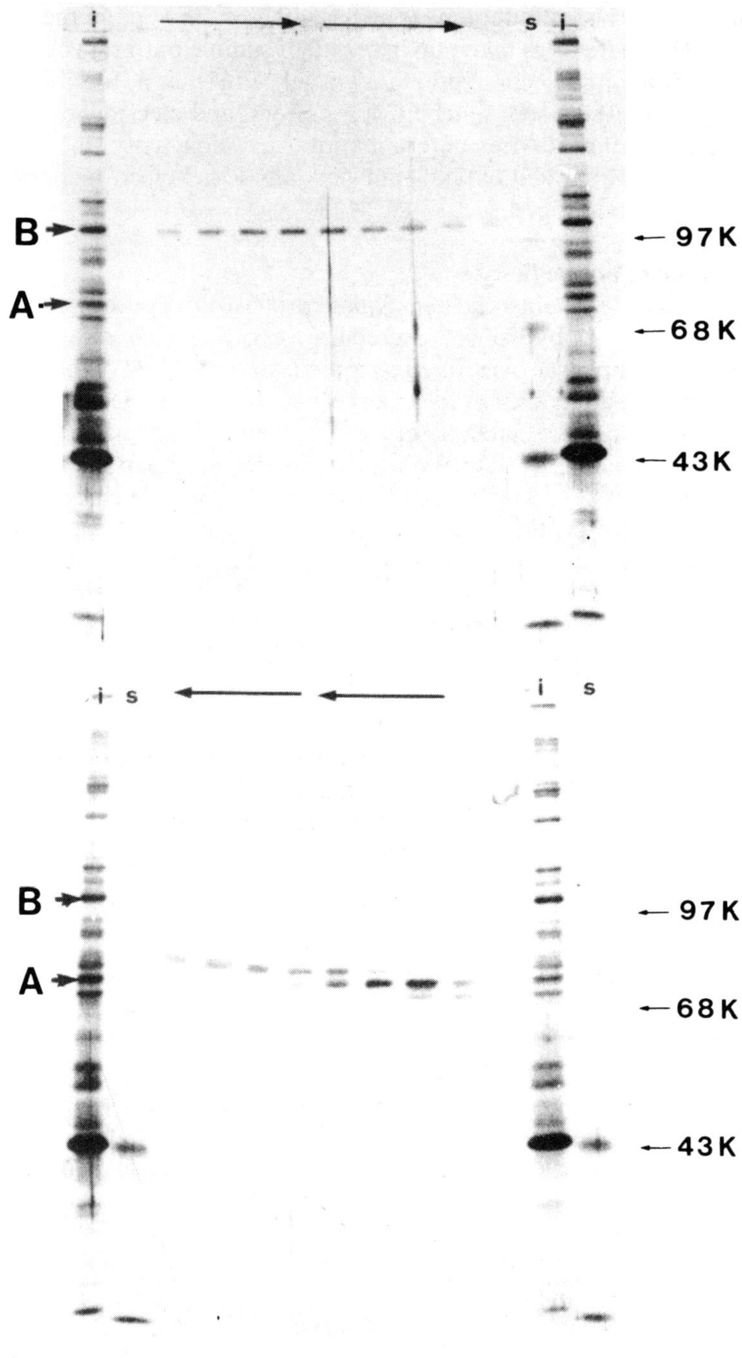

Fig. 4 — Purification of photocrosslinked progesterone receptor by preparative SDS–PAGE (silver-stained SDS–polyacrylamide gel): i, material before separation; s, standards. The positions of the receptor forms A and B are indicated. (From Gronemeyer and Govindan, 1986.)

digestion an enzyme-to-protein ratio of 3 may be used and the digestion should be for 1 h at 37°C.

SDS–polyacrylamide gel electrophoresis of digested samples may be performed on linear or gradient gels (e.g. 7.5–15% polyacrylamide) using [^{14}C]-labeled protein markers (BRL). For fluorography the author has obtained optimal results with EN^{3}HANCE (NEN) used according to the instructions of the manufacturers. After complete digestions good results have been obtained with 8.75–15% SDS polyacrylamide gels containing a gradient of 0–8 M urea. Also 15% polyacrylamide gels in 100 mM Na phosphate, pH 7.2, 0.1% SDS and 6 M urea have been used with success.

If — especially if the amount of crosslinked counts is low — fluorography is too time consuming, the gel is fixed and the lanes are cut into 2 mm slices. These are dissolved in H_2O_2 overnight in closed vials at 55°C and after addition of the scintillation fluid (D22, Roth) the vials are kept overnight at 4°C in the dark to avoid chemoluminescence counting.

B. Peptide isolation and sequencing

1. Receptor purification and concentration

The partial purification and photoaffinity labeling of the laying hen oviduct progesterone receptor is performed as described above. Five subsequent preparations, each starting with cytosol prepared from 20 laying hen oviducts, are stored at −70°C as ammonium sulfate pellets. Only aliquots of each preparation are crosslinked as tracers in the subsequent electrophoresis in order to avoid UV damage of aromatic amino acid residues for the bulk of the preparation. For preparative SDS gel electrophoresis aliquots of 5 mg of total protein of the partially purified receptor preparations are electrophoresed. Progesterone receptor forms A and B are identified by determining the radioactive fractions and the purity of the peak fractions as assayed by analytical SDS gel electrophoresis and silver staining. Only the peak fractions of progesterone receptor form B which do not reveal any impurity on silver-stained gels are pooled, dialyzed against 0.1% SDS and lyophilized. The dried material is taken up in 1 ml H_2O, radialyzed against 0.01% SDS and relyophilized.

2. Preparation of tryptic peptides of progesterone receptor form B

Purified progesterone receptor form B (100 μg in 50 μl 0.5 M Tris HCl buffer, pH 7.9) is reduced by adding dithiothreitol (0.5 μmol) and incubating for 2 h at 37°C. After cooling to room temperature, 2.25 μmol of sodium iodoacetate are added and the sample is kept in the dark for 1 h. The alkylation reaction is halted by the addition of 2.5 μmol of dithiothreitol. *S*-carboxymethylated progesterone receptor is dialyzed against 20 mM ammonium bicarbonate, lyophilized and digested in 100 mM ammonium bicarbonate with 10 μg of TPCK–trypsin for 2 h at 37°C.

3. Purification of polypeptides by high performance liquid chromatography

The initial fractionation of tryptic peptides on a conventional analytical HPLC column (4.0 mm internal diameter) is performed under gradient elution conditions using a Waters liquid chromatograph model 680 fitted with model F6000A pumps. Peptides are detected by monitoring the absorbance of the column eluent simultaneously at 206 and 280 nm using a serial arrangement of two LC spectrophotometers (Uvicord S model 2138, LKB; Lambda Max model 481, Waters Associates, Milford, MA, USA). Fractions are collected manually in siliconized Eppendorf tubes and stored at −70°C.

Separation of peptides on microbore columns (2.1 or 1.0 mm internal diameter) is performed on a Hewlet–Packard model 1090 liquid chromatograph fitted with a model 1040 diode-array detector. This configuration allows the column eluent to be monitored simultaneously at 210, 254 and 280 nm and real-time spectral data (200–400 nm at 2 nm band width) to be obtained on eluting peaks. Chromatographic and spectral data are stored on disk (Hewlett–Packard model 9121) for subsequent analysis (e.g. second-order derivative spectroscopy) using the EVALU 1 and 2 software packages (Hewlett–Packard, Waldbronn, FRG). Samples are injected using a Rheodyne model 7125 injector equipped with a 2 ml injection loop installed in the column oven compartment. Samples (typically, 500–600 μl) from the conventional column (4.0 mm internal diameter) are applied repetitively onto the microbore columns in volumes of 50 μl after which the solvent gradient is started. Eluent fractions from microbore columns (2.1 and 1.0 mm internal diameter) are recovered manually after correcting for the dead volume of tubing between the detector flow cell and the collection port. Purified peptides from the microbore columns are either stored at −10°C or sequenced directly.

4. *Column supports*

The following columns were used for the purification of tryptic peptides from chicken progesterone receptor form B: (a) LiChrospher 500 CH-8 (50 nm pore size, 10 μm particle diameter support packed into a stainless steel glass-lined column 250 mm×4.0 mm internal diameter; E. Merck Co., Darmstadt, FRG); (b) MOS-Hypersil (12 nm pore size, 5 μm octylsilica packed into a stainless steel column 100 mm×2.1 mm internal diameter; Hewlett–Packard Co., Waldbronn, FRG); (c) Beckman Ultrasphere (6 nm pore size, 5 μm particle diameter octylsilica packed into a stainless steel column 250 mm×1.0 mm internal diameter; Beckman Instrument Co., CA, USA).

5. *Amino acid sequence analysis*

Automated amino acid sequence analyses of tryptic peptides were performed using an Applied Biosystems sequencer model 470A. Polybrene was used as a carrier. Amino acid phenylthiodydantoins were analyzed on a Hewlett–Packard liquid chromatograph (model 1090) equipped with a filter detector (254 nm filter) according to the procedure described elsewhere (Grego *et al.*, 1985; Simpson *et al.*, 1985).

REFERENCES

Bellus, D., Kearns, D. R. and Schaffner, W. (1969) Zur Photochemie von α,β-ungesättigten cyclischen Ketonen: spezifische Reaktionen der nπ*- und ππ*-Triplettzustände von *o*-Acetyl-testosteron und 10-Methyl-$\Delta^{1,9}$-octalon-(2). *Helv. Chim. Acta* **52**, 971–1009.

Benisek, W. F. (1977) Labeling of Δ^5-3-ketosteroid isomerase by photoexcited steroid ketones. *Methods Enzymol.* **46**, 469–479.

Coty, W. A., Schrader, W. T. and O'Malley, B. W. (1978) Purification and characterization of the chick oviduct progesterone receptor A subunit. *J. Steroid Biochem.* **10**, 1–12.

Formstecher, P. and Lustenberger, P. (1987) Purification of various steroid hormone receptors (glucocorticoid, progesterone, mineralocorticoid) by affinity chromatography. Design of suitable adsorbents and their applications. In: Moudgil, V. K. (ed.) *Molecular Mechanism of Steroid Hormone Action.* Walter de Gruyter, Berlin.

Grego, B., Van Driel, I. R., Stearne, P. A., Goding, J. W., Nice, E. C. and Simpson, R. J. (1985) A microbore high-performance liquid chromatography strategy for the purification of polypeptides for gas-phase sequence analysis. Structural studies on the murine transferrin receptor. *Eur. J. Biochem.* **148**, 485–491.

Gronemeyer, H. and Govindan, M. V. (1986) Affinity labelling of steroid hormone receptors. *Mol. Cell. Endocrinol.* **46**, 1–19.

Katzenellenbogen, J. A. and Katzenellenbogen, B. S. (1984) Affinity labeling of receptors for steroid and thyroid hormones. *Vitam. Horm. (N.Y.)* **41**, 213–274.

Katzenellenbogen, J. A., Johnson, H. J., Carlson, K. E. and Myers, H. N. (1974) The photoreactivity of some light-sensitive estrogen derivatives. The use of an exchange assay to determine their photointeraction with the rat uterine estrogen binding protein. *Biochemistry* **13**, 2986–2994.

Schrader, W. T., Kuhn, P. W. and O'Malley, B. W. (1977) Progesterone-binding components of chick oviduct: receptor B subunit protein purified to apparent homogeneity from laying hen oviducts. *J. Biol. Chem.* **252**, 299–307.

Simpson, R. J., Smith, J. A., Moritz, R. L., O'Hare, M. J., Rudland, P. S., Morrison, J. R., Lloyd, C. J., Grego, B., Burgess, A. W. and Nice, E. C. (1985) Rat epidermal growth factor: complete amino acid sequence. Homology with the corresponding murine and human proteins; isolation of a form truncated at both ends with full *in vitro* biological activity. *Eur. J. Biochem.* **153**, 629–637.

4

Human progesterone receptor

K. B. Horwitz and M. D. Francis
Department of Medicine, University of Colorado, Denver, USA

I. OPTIMIZATION OF PHOTOLYSIS CONDITIONS

In the study reported by Dure *et al.* (1980), [^{3}H]R5020 labeled chick oviduct cytosol progesterone receptors (PR) were concentrated by ammonium sulfate precipitation, and the redissolved complexes were photolysed on ice using an 85 W low-pressure mercury vapor lamp (GE H85A3) and irradiation times of up to 2 h. Borosilicate glass was used to absorb UV below 300 nm to protect the proteins. As reported by General Electric, this lamp emits more than 75% of its spectral energy at wavelengths of greater than 350 nm; less than 15% of the energy output is emitted between 270 and 330 nm. However, when we performed a spectrophotometric scan of R5020, we found that it maximally absorbs UV at 304 nm (Fig. 1). We therefore decided to use a lamp that had a more effective energy distribution range. It turned out that DNA transilluminators, present in most laboratories for visualization of ethidium bromide–DNA complexes, had the desired emission spectrum, with a peak at 302 nm. With a transilluminator (we use the Foto-UV 300 from Fotodyne Inc., New Berlin, WI) we increased the yield of specifically photolabeled receptors 10-fold

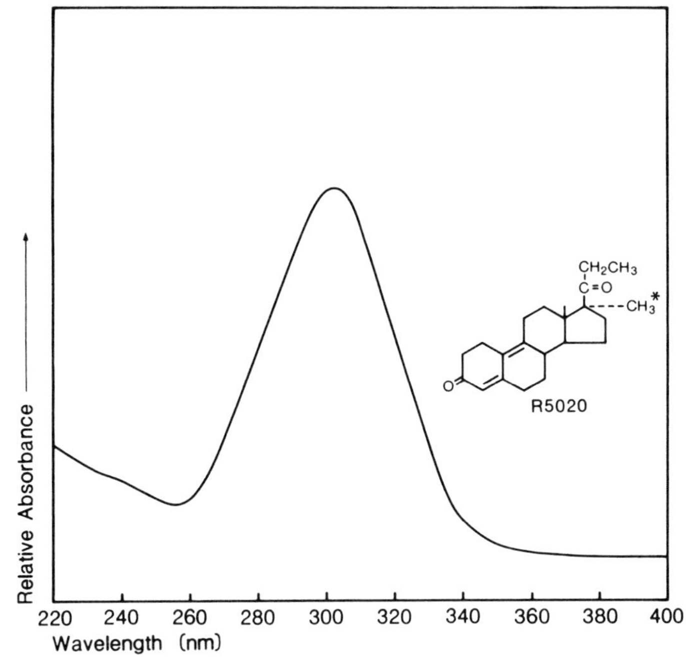

Fig. 1 — Absorption spectrum of R5020 showing a peak at 304 nm.

(from 200 to 2000 cpm per peak extractable from electrophoretic gels) and decreased the irradiation time at least 15-fold (from 30–120 min to 2 min). A time-course of the photolabeling rate is shown in Fig. 2. Photoaffinity labeling was judged to be 10–15% efficient by comparison of the total specific hormone binding activity loaded on gels with the amount of specifically bound counts in A- plus B-receptors eluted from gels. This method slightly underestimates efficiency.

II. *IN VITRO* PHOTOAFFINITY LABELING OF PR FROM HUMAN BREAST CANCER CELLS

The conditions described below apply to the photocrosslinking of [^{3}H]R5020 to PR in cell-free conditions. The PR can be in a variety of states: hormone free as obtained in cytosols of untreated cells; progesterone (but not R5020) filled after 0.4 M KCl extraction of chromatin-bound PR (while progesterone can be exchanged for [^{3}H]R5020 under the conditions described below, R5020 exchange is poor (Mockus *et al.*, 1982)); PR concentrated by ammonium sulfate precipitation and redissolved.

Prepare stock buffers and solutions as follows. TEDG: 10 mM Tris (pH 7.4 at 4°C); 1 mM Na_2EDTA; 1 mM dithiothreitol (DTT); 10% glycerol. TEDG Mo+I: TEDG containing 20 mM Na_2MoO_4; 47.5 μg/ml leupeptin, 100 μg/ml bacitracin; 77 μg/ml aprotinin; 10 μg/ml pepstatin A. Hank's–EDTA: 1 mM Na_2EDTA in Hank's balanced salts solution. TEK: 10 mM Tris (pH 8.5 at 4°C), 1.5 mM

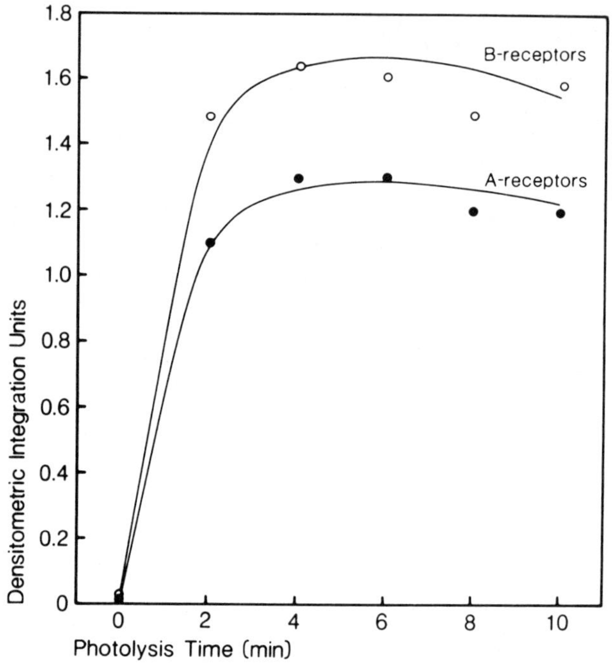

Fig. 2 — Rate of *in situ* photoaffinity labeling of PR by 300 nm in UV. Cells growing in Petri dishes were incubated with serum-free medium containing 80 nM [^{3}H]R5020 for 5 min at 37°C. The medium was aspirated. The cell surface was washed with ice-cold buffer and the dishes were inverted on the surface of a DNA transilluminator and irradiated for 2–10 min. The cells were then harvested and homogenized, receptors were extracted from nuclei with buffer containing 0.4 M KCl, and the nuclear extract was concentrated by ammonium sulfate precipitation and analyzed by SDS–PAGE and fluorography. Band intensity of the B- and A-receptors was quantitated densitometrically.

Na_2EDTA; 0.4 M KCl. SDS–PMSF–NEM: 1% SDS; 2 mM phenylmethylsulfonylfluoride; 1 mM *N*-ethylmaleimide. SDS–DTT–BPB: 1 part 10% SDS, 1 part 0.1 M DTT, 2 parts 0.02% (by weight) bromophenol blue in 50% (by volume) glycerol.

(1) Harvest ~10^8 cells, growing in two T175 culture flasks, by incubating the flasks with 15–20 ml Hank's–EDTA, for 10 min at 37°C. Collect cells in a 50 ml conical polypropylene tube (Falcon) and hold in ice. All subsequent work is done with cold reagents, in an ice bucket and a 4°C cold-room.

(2) Wash cells once in 10 ml TEDG Mo+I, transfer to the homogenizer, wash again in 4 ml TEDG Mo+I, pellet and resuspend in 1.2 ml of the same buffer for homogenization. The homogenizer used is a 3 ml (with pestle) capacity Duall® tissue grinder, size 0021, having a ground-glass base (Kontes, K-885452) with a Teflon pestle (Kontes, K-885481). Homogenize with 20–80 strokes. The number is determined empirically for each cell type and buffer volume by sequential light microscopic examination of Trypan Blue stained cells. Look for >98% broken cells with minimal nuclear swelling.

(3a) If cells were untreated, then PR will be found predominantly in cytosols. In that case, transfer the homogenate to a 10 ml polycarbonate tube (Nalgene), and centrifuge for 30 min at 100 000*g* at 4°C. Collect the supernatant and either discard the pellet or save it for DNA determination if it is desirable to normalize PR values among several experiments.

(3b) If cells were treated with progesterone, then PR will be predominantly associated with nuclei. In that case, transfer the homogenates to polycarbonate tubes and centrifuge for 15 min at 800*g*. Remove the supernatant cytoplasm and discard, or transfer it to a second polycarbonate tube for cytosol preparation as described above. Resuspend the nuclear pellet in 5 ml TEMo, mix and repellet. Discard the supernatant wash. Add 500–1000 μl of TEK to the pellet, mix vigorously, and incubate for 1 h at 4°C, remixing every 15 min. At the end of the incubation place the polycarbonate tube in a rotor, and centifuge the salt extracted suspension at 100 000*g* for 30 min at 4°C. Save the supernatant salt extract containing PR; discard the residual nuclear pellet or freeze it for later DNA analysis.

(4) Dilute the cytosol to 1.2 ml with TEDG Mo+I. Use the nuclear extract in undiluted form. Add 1 part 10× [^{3}H]R5020 stock in buffer to 9 parts cytosol or nuclear extract to a final concentration of 20 nM [^{3}H]R5020. If unlabeled competitor is added, combine 8 parts cytosol, 1 part [^{3}H]R5020 and 1 part 10× unlabeled R5020 or buffer. Mix and incubate at 4°C for 3–4 h in a 12 mm × 75 mm polypropylene tube (Falcon).

(5) In the cold-room lay two sheets of Saran Wrap on the surface of a cooled 300 nm transilluminator, raising the edges of the plastic film with a 10 mm × 10 cm frame made from plastic straws. Pour the labeled cytosol into the centre, turn the transilluminator on, and irradiate for 2 min (wear UV-protective glasses). Turn the UV off, and return the cytosol to the polypropylene tube using a plastic transfer pipet, being careful not to tear the Saran Wrap or to scratch the surface of the transilluminator.

(6) Cytosols can be used immediately for SDS–polyacrylamide gel electrophoresis; they can be frozen at −20°C for later use or they can be concentrated. Nuclear 0.4 M KCl extracts are always ammonium sulfate precipitated and redissolved. This accomplishes two ends: it concentrates the sample while at the same time reducing the total salt content ~65% (as measured by conductivity). Reducing the salt content of the extract prevents its crystallization in the BPB sample buffer, and avoids spreading and thickening of lanes during electrophoresis. For ammonium sulfate precipitation, combine equal volumes of labeled cytosol or nuclear extract with saturated ammonium sulfate, incubate for 30 min at 4°C with intermittent mixing, then centrifuge at 35 000*g* for 30 min at 4°C. Discard the supernatant and redissolve the pellet with TEMo buffer in 1/4–1/3 the original volume of cytosol or nuclear extract. Usually the pellet from 500 μl cytosol or nuclear extract is redissolved in 150 μl TEMo. Either the ammonium sulfate pellet or the redissolved suspension can be stored frozen at −20°C in 80–160 μl aliquots using 1.5 ml capped microcentrifuge tubes.

III. *IN SITU* PHOTOAFFINITY LABELING

With this method, R5020 is covalently linked to PR at any intracellular location using whole cells. Briefly, cells either attached to plastic, or harvested but still intact, are incubated with [^{3}H]R5020 for 2–4 h at 0°C to keep PR in the untransformed cytosolic state, or for 5–10 min at 37°C to transform PR to the tight nuclear binding state. The intact cells are then irradiated with UV for 2 min. The cells are then solubilized directly, or first homogenized, and then various subcellular compartments including cytosols, nuclear extracts, microsomal pellets and/or residual nuclear pellets are solubilized and/or denatured.

(1a) Cells are plated and grown in 100 mm Petri dishes designated for cell culture (Falcon). At confluence, remove growth medium from two plates ((15–25)$\times 10^6$ cells) and wash the cell surface with 5.0 ml/plate serum-free growth medium. Aspirate the medium. Add 2 ml serum-free growth medium containing 80 nM [^{3}H]R5020 alone or together with 8000 μM unlabelled R5020. Incubate the plates for 2–4 h at 0°C (in the cold-room on a bed of ice) or for 5–10 min at 37°C. Aspirate the hormone solution, and wash the cell surface twice with 5 ml/plate ice-cold TEMo. Aspirate the wash buffer. In a 4°C cold-room, cover the surface of a 300 nanometer UV light source with two sheets of Saran Wrap. Invert the Petri dishes on the surface and irradiate for 2 min. Add 5 ml ice-cold TEMo buffer to each plate and harvest the cells by scraping with a rubber policeman. Cells from two plates are pooled for homogenization, or they can be solubilized directly.

(1b) Alternatively, *in situ* photoaffinity labeling can be carried out with suspended cells by first harvesting $\sim 10^8$ cells. These are washed in serum-free medium at 4°C, then suspended in 4–5 ml of serum-free medium containing 80 nM [^{3}H]R5020 and incubated in an ice–water bath with gentle shaking for 3–4 h. At the end of incubation the cells are pelleted, the supernatant hormone is discarded and the cells are washed twice in 5 ml TEDG Mo+I. To link [^{3}H]R5020 covalently to PR, the suspended cells (5 ml are poured onto two layers of Saran Wrap placed directly over the surface of the DNA transilluminator and irradiated for 2 min as described above. Collect the cells with a transfer pipet and recover remaining cells adhering to the plastic with an additional 5 ml TEDG Mo+I wash. Pool the cells, transfer to the homogenizer.

(2) Washed cells prepared as described in (1) above can be solubilized directly by incubation in 500 μl SDS–PMSF–NEM for 30 min at room temperature. Otherwise, cells can be homogenized as described above and fractionated into cytosol, nuclear extract, microsome and salt-resistant nuclear pellets. The cytosols and nuclear extracts solutions are concentrated by ammonium sulfate precipitation, redissolved and stored frozen as described above. The nuclear pellet is solubilized in 500 μl, and the microsomes are solubilized in 200 μl SDS–PMSF–NEM for 30 min at room temperature. Insoluble debris is removed by a 35 000g, 30 min centrifugation and the cleared supernatant is used directly or stored for later use in 80–160 μl aliquots in 1.5 ml capped microcentrifuge tubes at −20°C.

IV. SDS–POLYACRYLAMIDE GEL ELECTROPHORESIS

Cytosols, redissolved ammonium sulfate pellets, supernatant from solubilized whole cells, microsome or nuclei in volumes of 80–160 μl are mixed with 40 μl SDS–DTT–BPB sample buffer, denatured by boiling for 2 min, then cooled.

Electrophoresis is performed according to standard procedures, which we will not detail. Briefly, 120–200 μl of the bromophenol-containing solutions are analyzed. Electrophoresis is performed on discontinuous 1.5 mm thick polyacrylamide slab gels (PROTEAN cell, Bio-Rad Laboratories). Running gels consist either of linear 7.5–19% acrylamide gradients or 7.5% gels as described by Laemmli (1970) and 0.08–0.2% *N,N'*-methylene-bis-acrylamide (0.8–7.5% gels); stacking gels contain 3% acrylamide and 0.08% *N,N'*-methylene-bis-acrylamide. Samples are run for 4 h at 40 mA/gel or until the dye front reaches the end of the gel. ^{14}C-labeled molecular weight standards purchased from Bethesda Research Laboratories (Gaithersburg, MD) are as follows (in daltons): myosin (H-chain), 200 000; phosphorylase B, 97 400; bovine serum albumin, 68 000; ovalbumin, 43 000; α-chymotrypsinogen, 25 700; β-lactoglobulin, 18 400; cytochrome C, 12 300. These are run in a parallel lane. Prestained molecular weight standards are also available from BRL. These are particularly useful for immunoblotting. Note, however, that their migration on SDS–PAGE is slowed compared with that of [^{14}C]-labeled standards, leading to overestimation of molecular weights by 2000–5000 daltons. Gels can then be sliced and counted or used for fluorography, for electroelution or for immunoblotting.

A. Gel slicing and counting

(1) Remove the gel sandwich from the electrophoresis support frame. Lay the glass sandwich flat on a bench and gently pry the upper plate away from the gel, keeping the gel, including the stacking gel, intact.

(2) Measure Rf values of prelabeled standards, if desired. If the standards are radiolabeled, the lane is also sliced and counted.

(3) Cut the gel between lanes with a sharp knife (by pressing rather than dragging the knife across the gel to prevent tearing) using the stacking gel and the dye front as a guide. Tease the stacking gel away from the running gel and discard it. Cover the gel slices with a thin layer of electrophoresis buffer followd by a sheet of parafilm to prevent drying and shrinking.

(4) Adjust the razor blades in a commercial slicer (Bio-Rad or Hoefer) to obtain slices of desired thickness (usually 2–4 mm). Lay each gel lane on the blades of the slicer, cover the gel with a small amount of electrophoresis buffer, followed by four layers of parafilm and four folded paper towels. Firmly press the gel into the slicer, Remove paper towels and parafilm. If the gel has not been cut all the way through, press it into the blades with a wet spatula, and wet the gel with a small amount of electrophoresis buffer if it seems to be drying out.

(5) Using wet forceps, put each gel slice into a separate scintillation vial. Add 5.0 ml of fresh PPO–POPOP–PROTOSOL scintillation cocktail to each vial. Cap and shake vigorously. Incubate overnight at 37–40°C. Shake vigorously again and place in a refrigerated liquid scintillation counter. Allow the samples to cool and dark adapt for at least 1 h in the counter, then count for 5 min per vial.

PPO–POPOP–PROTOSOL. Use 8.0 g PPO (2,5-diphenyloxazole), 0.1 g POPOP (*p*-bis-(2-(5-phenyloxazolyl))benzene) and 60 ml PROTOSOL (New England Nuclear); make up to 2.0 l with toluene. Stir at room temperature until all components are dissolved. Store in an amber bottle for not more than 1 week; best if used the same day that it is made.

B. Fluorography

(1) Remove the gel from the glass sandwich, and place it in a polyethylene tray with 100–200 ml of fixative. Cover, and rotate using a flat-surfaced clinical rotator for at least 1 h at room temperature (may go overnight).

(2) Remove the fixative, add ~100 ml of En^{3}Hance (New England Nuclear) to the tray, and rotate the gel at room temperature for 60 min.

(3) Remove the En^{3}Hance and wash the gel in ~100 ml of cold distilled water to precipitate the floor, rotating at room temperature for 1 h.

(4) Remove the gel from water, place it on filter paper, and dry the gel using a gel dryer at a low temperature setting (60°C). Drying time is 2–3 h depending on gel thickness and acrylamide concentration. In a dark room place the dried gel over Kodak XAR 5 X-ray film held in a metal cassette, and expose the film for 1–2 weeks at −70°C. The use of pre-flashed film and intensifying screens reduces exposure time somewhat.

(5) If a stained gel is desired, remove the gel from the electrophoresis apparatus, place it in 100 ml of Coomassie Blue stain and rotate at room temperature for 5–15 min. Remove the strain and replace it with 100–200 ml of fixative solution to destain. Destain in several changes of fixative until bands are visible or overnight. Photograph the wet gel, then continue with steps (2)–(4) for fluorography.

(6) If desired, a single electrophoretic gel containing photoaffinity labeled receptors can be analyzed both for the presence of tritium labeled receptors and for the presence of immunoreactive protein. Briefly, this is done by electrophoretically transferring the proteins to a nitrocellulose sheet. The receptor proteins are first visualized by sequential incubations of the nitrocellulose with the A- and B-specific anti-human PR monoclonal antibody AB-52, followed by horseradish-conjugated goat anti-mouse IgG, followed by substrate 4-chloro-1-naphthol–methanol and H_2O_2 (Wei *et al.*, 1987). After the desired band intensity is achieved, the filter is photographed to record the immunoblot results. The filter is then air dried, sprayed with En^{3}Hance Spray (New England Nuclear) and placed against X-ray film, according to the manufacturer's instructions, for fluorography of the tritium-labeled receptors. This is photographed at the same scale as the immunoblot and results from the two methods directly compared. A mixture of prestained and [^{14}C]-labeled molecular weight standards is visualized in both data sets.

Fixative. Use 1200 ml distilled H_2O, 600 methanol (30% final) and 200 ml glacial acetic acid (10% final).

Coomassie Blue Stain. Use 1.25 g Coomassie Blue powder, 250 ml methanol (50% final) and 37.5 ml glacial acetic acid (7.5% final); make up to 500 ml with distilled water.

REFERENCES

Dure, L. S., Schrader, W. T. and O'Malley, B. W. (1980) Covalent attachment of a progestational steroid to chick oviduct progesterone receptor by photoaffinity labeling. *Nature* **283**, 784–786.

Laemmli, K. (1970) Cleavage of structural proteins during the assembly of the head of bacteriophage T4. *Nature* **227**, 680–684.

Mokus, M. B., Lessey, B. A., Bower, M. A. and Horwitz, K. B. (1982) Estrogen insensitive progesterone receptors in a human breast cancer cell line: characterization of receptors and of a ligand exchange assay. *Endocrinology* **110**, 1564–1571.

Wei, L. L., Sheridan, M. L., Krett, N. L., Francis, M. D., Toft, D. O., Edwards, D. P. and Horwitz, K. B. (1987) Immunologic analysis of human breast cancer progesterone receptors, II. Structure, phosphorylation and processing. *Biochemistry* **26**, 6262–6272.

5

Androgen receptor

D. J. Tindall
Baylor College, Department of Cell Biology, Houston, USA

I. SYNTHESIS OF AFFINITY LABELING STEROIDS

A. Protocol for synthesis of [^{3}H]DHT-bromoacetate

(1) Transfer 1.9 μg (4.5 nmol) of [1,2,4,5,6,7,16,17-$^3H_8$]dihydrotestosterone to a test tube and evaporate the solvent under a stream of N_2.

(2) Make stock solution A by diluting 10 μl of bromoacetylbromide in 10 ml of toluene (2.3 ng/μl final concentration). Make stock solution B by diluting 20 μl of diisopropylethylamine in 10 ml of toluene (1.5 ng/μl final concentration). Add 50 μl of each stock solution to the test tube containing [^{3}H]dihydrotestosterone.

(3) Incubate the solution for 4.5 h and evaporate the solvent under N_2.

(4) Add 50 μl each of stock solutions A and B.

(5) Incubate the solution for 18 h and evaporate the solvent under N_2.

(6) Chromatograph the residue on a microsilica gel column (400 mg of absorbent, 70–230 mesh) using hexane–ethyl acetate (6:4) solvent and collect 0.5 ml fractions. Pure affinity label should appear in fractions 2 and 3.

B. Protocol for synthesis of non-radioactive DHT-bromoacetate

(1) A solution of bromoacetyl bromide (3.8 g, 18.9 mmol) in 25 ml of toluene is added dropwise over a period of 1 h to a solution of dihydrotestosterone (5 g, 17.2 mmol) in 50 ml of toluene which is cooled in an ice bath.

(2) Add diisopropylethylamine dropwise (3.4 ml, 10.5 mmol) and allow the mixture to stand at room temperature for 8 h, then filter.

(3) The mixture is extracted with 5% sodium bicarbonate in water and dried with Na_2SO_4, and the solvent removed *in vacuo*. This will produce approximately 7.5 g of residue.

(4) The residue is chromatographed on silica gel, eluting with a linear gradient of 8:2 to 6:4 hexane–ethyl acetate.

(5) The desired material is then crystallized from acetone–hexane to give analytically pure DHT–bromoacetate.

II. PROTOCOL FOR PURIFICATION OF ANDROGEN RECEPTOR

(1) Prepare cytosol by homogenizing the tissue in 4 volumes (w/v) of TEDG buffer (50 mM Tris–HCl, 1.5 mM EDTA, 1.5 mM dithiothreitol, 20% glycerol; pH 7.4 at 22°C) and centrifugation at 105 000*g* for 1 h at 4°C. All subsequent steps are performed at 4°C unless indicated otherwise.

(2) Apply the cytosol to a DNA–Sepharose column (denatured calf thymus DNA–Sepharose 4B; 3 ml sample/ml bed volume) and collect the flow-through fraction.

(3) Precipitate androgen receptor at 40% saturation by slowly adding solid ammonium sulfate while stirring over a 30 min period. Allow the sample to precipitate for an additional 45 min at 4°C.

(4) Collect the precipitate by centrifugation at 25 000*g* for 15 min. Decant the supernate and freeze the pellets at −90°C overnight.

(5) On day 2, thaw the precipitated pellets at 22°C and resuspend in 0.16 ml TEDG buffer/ml of flow-through sample from step (2).

(6) Add 0.6 g of affinity resin (testosterone 17β-hemisuccinyl-3,3′-diamino-dipropylamine–Sepharose 4B) per milliliter of sample in step (5). Incubate with gentle stirring for 75 min.

(7) Wash the affinity resin in a sintered glass funnel with 3 bed volumes of TED buffer containing 0.5 M NaCl.

(8) Elute androgen receptor from the affinity resin by addition of 2 ml of testosterone solution (100 μg/ml TEDG buffer) per gram of resin. Stir gently for 6 h.

(9) Separate the eluted sample from the resin by using a sintered glass funnel. Wash the resin with an additional 3 ml TEDG buffer per gram of resin.

(10) Apply the combined eluate to a second DNA–Sepharose 4B column (bed volume = $\frac{1}{4}$ that of DNA column I from step (2)). Discard the flow-through fraction.

(11) Wash the column with TED buffer (4 ml/ml bed volume) and elute the receptor with TED buffer containing 0.5 M NaCl (4 ml/ml bed volume).

This protocol has been used to purify androgen receptor from rat ventral prostate, steer seminal vesicle, and Dunning R3327 tumor (see references in Chapter 12).

III. PROTOCOL FOR AFFINITY LABELING OF PARTIALLY PURIFIED ANDROGEN RECEPTOR USING DIHYDROTESTOSTERONE 17β-BROMOACETATE

(1) A fresh, pure batch of dihydrotestosterone 17β-bromoacetate (17β-[(bromoacetyl)oxy]-[1,2,4,5,6,7,16,17-$^{3}H_{8}$]-5α-androstan-3-one; abbreviated [^{3}H]DHT-BrAc; 147 Ci/mmol; molecular weight, 411.38) is dried under nitrogen gas and resuspended in 100% ethanol to make a stock solution of 0.5 mCi/ml. The concentration of the stock non-radioactive DHT-BrAc is 1.0 μg/ml in 100% ethanol. Both stock solutions should be protected from light by using foil and can be stored at −20°C for up to six months.

(2) Aliquots of partially purified samples, such as eluates from the steroid affinity resin or the second DNA–cellulose column, are precipitated by addition of ammonium sulfate to 45% saturation.

(3) Centrifuge at 5000*g* for 20 min, and resuspend the pellet in 10 mM phosphate buffer (pH 7.4). Note: buffers such as Tris, which contain primary amines, must not be used during affinity labeling because these functional groups will react with bromine.

(4) Labeling of receptor samples is performed in a glass 12×75 mm tube using 10 nM [^{3}H]DHT-BrAc, with or without 10 μM non-radioactive DHT-BrAc in a volume of 0.5 ml. This requires addition of 1.5 μl of the stock [^{3}H]DHT-BrAc ± 2.0 μl of the non-radioactive DHT-BrAc stock to the 0.5 ml sample. Mix well.

(5) The labeled samples are incubated for 3 h at 0°C and then for 30 min at 22°C to effect exchange of ligands and covalent attachment.

(6) Samples are treated with an equal volume of 20% trichloroacetate for 1 h at 4°C and centrifuged at 5000*g* for 20 min.

(7) The precipitate is redissolved in Laemmli sample buffer and applied to SDS polyacrylamide gels.

(8) Following electrophoresis, gels are washed extensively with 40% methanol–7% acetate to remove unbound steroid. The gels are then treated with EN^{3}HANCE (New England Nuclear), dried, and exposed to Kodak X-oMat film for fluorography.

6

Thyroid receptor

Z. D. Horowitz and H. H. Samuels
Department of Medicine, NY University Medical Center, New York, USA

I. BACKGROUND AND METHODS

The synthesis and purification of the iodothyronine photoaffinity label (PAL) derivatives have been reported in detail (Casanova *et al.*, 1984; Pascual *et al.*, 1982) and have been used to modify covalently thyroid hormone receptor in intact cells and in nuclear extract (see Chapter 13). We have further simplified this methodology, as described below (Horowitz *et al.*, 1987). The yield of each of the different PAL derivatives is between 60% and 80%.

On exposure to UV light (254 nm), a 2-diazo-3,3,3-trifluoropropionyl moiety, which is coupled to [^{125}I]iodothyronine by peptide bond linkage, generates a highly reactive carbene intermediate which covalently modifies residues which are sterically favored in the hormone binding site. During subsequent steps of receptor isolation the radioactive label remains covalently bound. Any unreacted

[^{125}I]iodothyronine–PAL or free [^{125}I]iodide which may have been generated by UV exposure is easily removed by precipitation and washing in 70% ethanol (v/v).

Photoaffinity labeled thyroid hormone receptor can be quantitated by counting in a γ spectrophotometer or visualized by SDS–PAGE followed by autoradiography. The radioactive bands containing receptor may then be excised and counted. Alternatively, photoaffinity labeled receptor may be transferred from an acrylamide gel to nitrocellulose by electroblotting, where its exact position on the nitrocellulose can be directly visualized and quantitated by autoradiography.

II. TECHNICAL PROTOCOLS

A. Buffers

(1) STM: 20 mM Tris, pH 7.85 at 25°C; 0.25 M sucrose; 1 mM $MgCl_2$; STM+Triton X-100 0.2% (v/v).

(2) *Nuclease buffer*: 10 mM Tris, pH 7.45 at 25°C; 10 mM KCl; 1 mM $CaCl_2$; 1 mM $MgCl_2$; 1 μl/ml Trasylol; 5 mM 2-mercaptoethanol.

(3) *Extraction buffer*: STM buffer containing 0.4 M KCl; 1 μl/ml Trasylol; 5 mM 2-mercaptoethanol.

B. Cell culture

GH_1 cells, GC cells (two rat pituitary cell lines), and other cell types (e.g. human fibroblasts) are grown in monolayer in 25 cm^2 flasks, or in roller bottle flasks, in 95% air–5% CO_2. GH_1 cells are grown in Ham's F-10 medium supplemented to 15% (v/v) with horse serum and to 2.5% (v/v) with fetal calf serum. GC cells and human fibroblasts are grown in Dulbecco's Modified Eagle Medium supplemented with 10% fetal calf serum. Thyroid hormone concentrations in the serum component of these media are physiologic (2 nM L–T3, 100 nM L-T4). For hormone-binding experiments and for photoaffinity labeling experiments, the cells are maintained for 24 h in media lacking thyroid hormone, as previously described (Samuels *et al.*, 1979), in order to deplete cells of thyroid hormone. Hormone-binding experiments and photolabeling of intact cells are carried out in serum-free media when cells are in late logarithmic phase of growth.

C. Preparation of nuclear extracts

(1) Following incubation with hormone and/or the PAL derivatives, remove the medium by aspiration and gently wash the cell monolayer three times with ice cold normal saline. All subsequent steps are carried out at 4–7°C.

(2) Harvest cells in 1 ml of ice-cold STM+0.2% (v/v) Triton X-100 buffer per 25 cm^2 flask using a rubber policeman. Identical samples may be pooled. When cells are harvested from roller bottle cultures, because much more material is obtained, cells are first scraped into 8–10 ml STM per bottle, then homogenized (15 strokes at 2000 rpm) in a glass–Teflon homogenizer. Centrifuge at 600*g* for 10 min and resuspend the pellet in STM+Triton X-100 as described above. If the nuclear extract is to be prepared using micrococcal nuclease digestion (step (6)), an aliquot of the initial cell suspension in STM + 0.2% Triton X-100 is saved to determine the DNA content ($A_{260\,nm}$ units).

(3) Vortex the cell suspension five times at setting 5–6 on a Vortex Genie and centrifuge at 600*g* for 10 min.

(4) Carefully aspirate the supernatent and resuspend the pellet in 2 ml of STM + 0.2% Triton X-100. Centrifuge at 600*g* for 10 min.

(5) Resuspend the pellet in 2 ml of STM. Centrifuge at 600*g* for 10 min. This step is repeated twice.

(6) Resuspend the washed pellet in extraction buffer and incubate for 20 min at 4°C. Centrifuge at 2230*g* for 15 min. The nuclear extract containing thyroid hormone receptor (supernatant) may be saved for further processing, or precipitated. Alternatively, the washed pellet generated at step (5) may be resuspended in 0.5 ml nuclease buffer and incubated with micrococcal nuclease (15 units/$A_{260\ nm}$ unit) at 4°C for 45 min, after which EDTA (final: 10 mM EDTA, pH 7.4) and KCl (final: 0.4 M) is added. The samples are then incubated for an additional 20 min, followed by centrifugation.

D. Synthesis and purification of PAL derivatives

All steps may be carried out in the presence of low levels of incandescent light. Glass tubes should be used, as thyroid hormone may adsorb to plastic.

(1) Dry an appropriate amount of radioactive hormone (L-[^{125}I]3,5,3′-triiodo-L-thyronine (L-[^{125}I]T3), L[^{125}I]3,5,3′,5′-tetraiodo-L-thyronine (L-[^{125}I]T4) or reverse L-T3, L-[^{125}I]3,5′,3′-triiodo-L-thyronine (L-[^{125}I]rT3) in a 12 mm × 75 mm tube.

(2) Add aproximately 1 mg of *p*-nitrophenyl-2-diazo-3,3,3-trifluoropropionate (Pierce Chemical Company).

(3) Dissolve in 0.1–0.2 ml of anhydrous dimethyl formamide (DMF) by gentle swirling. Incubate in the dark for approximately 18 h at room temperature.

(4) In order to purify any unreacted hormone from the derivatized material, chromatograph the derivatized hormone preparation through two K1F, K1, or K5 silica gel columns (0.5 cm × 7 cm Pasteur pipet). The silica gel may be obtained from Whatman TLC plates by scraping off the silica gel coating.

(a) Pre-equilibrate the columns with glacial acetic acid–ethyl acetate (1:9 v/v) for L-[^{125}I]T3–PAL, or 2:8 v/v for L-[^{125}I]T4–PAL and L-[^{125}I]rT3–PAL.

(b) Layer the sample on the column and allow it to soak in completely.

(c) Rinse the tube with approximately 0.1 ml of the glacial acetic acid–ethyl acetate mixture and layer on the column, allowing it to soak in completely.

(d) Elute with the same mixture, collecting fractions of approximately 0.3 ml.

(e) Count 1 μl aliquots of each fraction in a γ counter and pool the fractions containing the peak which usually elutes between fractions 4–6 inclusive.

(f) Dry the pooled material under nitrogen to a volume of approximately 0.1 ml. Avoid splashing.

(g) Layer the concentrated material on the second silica gel column and repeat steps (c)–(f). Dry completely.

(5) Resuspend the dried material in a small volume of anhydrous DMF, so that the final concentration of DMF in the photolabeling reaction is less than 1%. The purified material may be stored in the dark at room temperature if used within 24 h, or at 4°C for up to a week.

E. Photoaffinity labeling

1. In intact cells

(a) After 24 h in thyroid hormone depleted media (in 25 cm^2 flasks), cells are incubated in 2 ml of serum-free medium with L-[^{125}I]T3–PAL, L-[^{125}I]T4–PAL, or L-[^{125}I]rT3–PAL for 1.5 h at 37°C in the dark. In some experiments underivatized [^{125}I]iodothyronine may be used but results in a lower coupling efficiency than the PAL derivatives. The concentration of the PAL derivatives or underivatized hormone is between 1 and 5 nM. Experiments are in parallel with 1000-fold molar excess non-radioactive underivatized hormone as competitor. Exposure to ambient light is minimized until UV irradiation, at which time photocoupling occurs. Since a covalent bond develops between the photoaffinity label derivative and proteins on activation by UV light, all subsequent steps may be carried out in the presence of incandescent or fluorescent light without significant increase in non-specific photolabeling.

(b) Aspirate the radioactive medium and gently wash the cell monolayer three times with ice-cold normal saline.

(c) Aspirate dry and crack and remove the flat tops of the flasks using a hemostat.

(d) Place the flask on a flat surface of ice and irradiate the cell monolayer for 2 min at a distance of 2 cm with a lamp (Ultra-Violet products, model R-52) having a wavelength peak of 254 nm.

(e) Immediately following irradiation add 1 ml ice cold STM + 0.2% (v/v) Triton X-100 and proceed with the preparation of nuclear extract as described above. Any unbound hormone or PAL derivative is removed by precipitation with 80% ethanol (−20°C) and washing with 70% (v/v) ethanol (room temperature).

2. *In nuclear extracts*

(a) Nuclear extract prepared with or without micrococcal nuclease digestion is made from cells grown in flasks or in roller bottle cultures, as described above. The extract is used fresh for optimal photoaffinity labeling of receptor.

(b) Pre-adjust the pH of the photoaffinity label to pH 7.5 (in DMF) by adding 1 M Tris, pH 8.2 (final concentration approximately 20 mM Tris). Incubate at room temperature in the dark for 15 min.

(c) Add the pH adjusted hormone and/or PAL to the nuclear extract so that the final DMF concentration is no more than 1% and incubate in the dark for 1.5–3 h at 0°C. The concentration of the PAL derivative or underivatized hormone is between 1 and 5 nM. Experiments are performed in parallel with 1000-fold molar excess non-radioactive hormone as competitor.

(d) Layer 0.5 ml of sample on an ice-cold 2.5 ml Sephadex G-25 (fine) column (a 3 cm^3 syringe is a convenient column vessel), pre-equilibrated with extraction buffer. Allow to absorb completely.

(e) Elute with 1.5 ml of extraction buffer. Collect the whole eluent as a single fraction. This step removes unbound hormone or PAL derivative.

(f) Apply 0.75 ml of eluent per well of a flat-bottom (9.62 cm^2) six-well Linbro tissue culture plate on ice.

(g) Irradiate at 254 nm for 2 min at a distance of 2 cm.

(h) Following irradiation, samples may be pooled and precipitated with 80% ethanol (−20°C) as described previously. Ethanol precipitation and washing remove any non-covalently bound hormone of PAL derivative.

3. Notes

We found that L-[^{125}I]T3 (Dupont–NEN) (>3000 μCi/μg) appears to be chemically stable for many weeks and L-[^{125}I]T3–PAL derivative displays a consistent efficiency of coupling. However, L-[^{125}I]T4 (>5000 μCi/μg) can deiodinate and, therefore, the L-[^{125}I]T4–PAL coupling efficiency may vary greatly. Therefore, L-[^{125}I]T4 should be used as fresh as possible (within two weeks). We have not determined the chemical stability of L[^{125}I]rT3 in storage.

ACKNOWLEDGMENTS

The research from the authors' laboratory described in this chapter was supported by Grants AM16636 and AM32352 (H.H.S.) and AM01372 (Z.D.H.) from the National Institutes of Health.

REFERENCES

Casanova, J., Horowitz, Z. D., Copp, R. P., McIntyre, W. R., Pascual, A. and Samuels, H. H. (1984) Photoaffinity labeling of thyroid hormone nuclear receptors: influence of *n*-butyrate and analysis of the half-lives of the 47,000 and 57,000 molecular weight receptor forms. *J. Biol. Chem.* **259**, 12084–12091.

Horowitz, Z. D., Casanova, J., Pascual, A., Sahnoun, N. and Samuels, H. H. (1987) Photoaffinity labeled iodothyronine derivatives identify receptor molecular weight forms with coupling efficiency inversely proportional to binding affinity. Manuscript submitted.

Pascual, A., Casanova, J. and Samuels, H. H. (1982) Photoaffinity labeling of thyroid hormone nuclear receptors in intact cells. *J. Biol. Chem.* **257**, 930–938.

Samuels, H. H., Stanley, F. and Casanova, J. (1979) Depletion of L-3′,5,5′-triiodothyronine and L-thyroxine in euthyroid calf serum for use in cell culture studies of the action of thyroid hormone. *Endocrinology* **105**, 80–85.

Part II
Application of affinity labeling techniques to study steroid and thyroid hormone receptors

7

Affinity labeling of estrogen receptors

B. S. Katzenellenbogen[a] and J. A. Katzenellenbogen[b]
[a]Department of Physiology and Biophysics and [b]Department of Chemistry,
University of Illinois, Urbana, USA

I. INTRODUCTION: OVERVIEW OF THE ESTROGEN RECEPTOR SYSTEM AND BIOLOGICAL AND MEDICAL IMPORTANCE OF ESTROGENS AND ANTIESTROGENS

Although estrogen affects the functioning of many tissues to some degree, clearly its principal effect is to stimulate the growth and maturation of the female reproductive system and to regulate its function. Involved in these estrogen stimulated processes are increased syntheses of RNA and protein, findings which early on suggested gene expression as a site of estrogen action (Katzenellenbogen and Gorski, 1975; Katzenellenbogen, 1980).

Since the identification of high affinity estrogen receptor proteins some 25 years ago, a variety of experiments have documented that the estrogen receptor is essential for mediating the changes effected by estrogens at the cellular and molecular level. Recent studies employing cell autoradiography (Martin and Sheridan, 1982), immunohistochemistry with monoclonal antibodies to the estrogen receptor (King and Greene, 1984), and cell enucleation techniques (Welshons *et al.*, 1984) have indicated that estrogen receptors are probably predominatly intranuclear in location, even in the absence of ligand, and that the main effect of the hormone is to increase the affinity of the receptor for chromatin binding sites. Hence, it is now generally believed that the enhanced nuclear localization of receptors on hormone binding does not involve translocation of receptor from an extranuclear (cytoplasmic) to an intranuclear location, but rather involves an increase in the affinity of its association with chromatin.

In addition to the dramatic and important effects of estrogens on the reproductive tract, including the oviduct, uterus and vagina, estrogens also influence the growth and properties of the mammary gland and of tumors that develop in the mammary gland. It is well established now that the estrogen receptor is found in about two-thirds of human breast cancer specimens and that its presence correlates well with the response of breast cancer to endocrine therapy. Thus, quantitation of the estrogen receptor has proven very important clinically in predicting the value of endocrine therapy in breast cancer treatment (McGuire, 1979).

Normally, the binding of natural and synthetic estrogens to receptor in target tissues is reversible. On binding of ligand to receptor, the avidity of receptor for chromatin increases (Katzenellenbogen, 1980). In estrogen-receptor-positive human breast cancer cells, such as MCF-7 cells, estrogen stimulation is associated with an increase in cell proliferation and the induction of several proteins such as the progestrone receptor and plasminogen activator, increases in secreted glycoproteins and polypeptide growth factors, and increases in selective RNA species, such as pS2 RNA (reviewed in Chambon *et al.*, 1984; Katzenellenbogen *et al.*, 1985, Dickson and Lippman, 1987). In rat uterus and chick oviduct, two other model systems used frequently for studying the mechanisms of estrogen action, estrogen also stimulates growth and cell differentiation; also stimulated is the synthesis of specific proteins. These include the rat uterine estrogen induced protein (IP), shown to be the BB isozyme of creatine kinase, and secreted oviduct proteins incorporated into the egg such as ovalbumin and conalbumin. Progesterone receptor levels in both rat uterus and chick oviduct are increased as well. The precise mechanisms by which estrogen evokes increased gene transcription as well as selective effects on RNA stability still remain challenging and important aspects to be detailed. Similarly, the mechanisms by which antiestrogens antagonize estrogen action remain of great interest and medical importance. Affinity labeling antiestrogens have proven useful in addressing aspects of the receptor and gene interactions of antiestrogens (reviewed in Katzenellenbogen *et al.*, 1985).

Although antiestrogens were developed initially by pharmaceutical companies as fertility regulating agents (Jordan, 1984), they are of particular interest and importance today because of their clinical efficacy in controlling the growth and spread of hormone-responsive breast and uterine tumors. Clinical trials employing antiestrogens have documented the effectiveness of antiestrogen treatment as endocrine therapy for breast cancer, with far fewer side effects than are associated with pharmacological hormone therapy (McGuire, 1979). Antiestrogens selectively inhibit the proliferation of estrogen-receptor-containing breast cancer cells and they inhibit estrogen stimulation of several specific RNA and protein synthetic activities in breast cancer cells. There is considerable evidence that the antagonistic action of antiestrogens is mediated, at least in large measure, via the estrogen receptor. These compounds bind to the estrogen receptor but presumably inhibit the events that promote cell growth (reviewed in Katzenellenbogen *et al.*, 1985). Hence, antiestrogens remain important compounds for elucidating aspects of receptor-antihormonal interactions resulting in estrogen antagonism, growth antagonism and antitumor activity.

Although the presence of the estrogen receptor in estrogen-responsive tissues has been known for over 20 years, the very low concentration of the receptor in cells and its lability have made it difficult to perform detailed studies of the physicochemical properties of the estrogen receptor and its dynamics within target cells. Almost all studies until recently used reversibly binding radiolabeled ligands, such as estradiol, which has precluded detailed analyses of receptors under more vigorous disaggregating conditions, since ligand dissociation would render the results invalid. The availability of efficient and selective affinity labels for the estrogen receptor (Katzenellenbogen *et al.*, 1983; Elliston *et al.*, 1987; Salituro *et al.*, 1987; Simpson *et al.*, 1987; Zablocki *et al.*, 1987), has now enabled investigators to address several important aspects of estrogen action, namely the physicochemical characteristics of the estrogen receptor under strongly disaggregating conditions and the kinetics of receptor turnover and dynamics in a variety of target cells. The results of some of these studies and future directions still being pursued are described in the sections that follow.

II. REVIEW OF PREVIOUS WORK

Prior to the development of the aziridine-based affinity labels, described herein, there were a number of attempts to label the estrogen receptor covalently. Most of the early work involved the preparation of steroidal and non-steroidal estrogens derivatized with alkylating functions with little regard to the consequences for receptor binding affinity, and often they were assayed only in terms of the duration of their biological activity, the hypothesis being that prolonged activity would be indicative of covalent attachment. Few compounds were radiolabeled and studied directly, and these without success. These studies have been reviewed in detail elsewhere (Katzenellenbogen, 1977).

Our initial work on affinity labeling of the estrogen receptor utilized a series of photoreactive derivatives of steroidal and non-steroidal estrogens (Katzenellenbogen *et al.*, 1973a). These compounds were characterized in terms of their receptor binding affinity under reversible conditions (in the dark) (Katzenellenbogen *et al.*, 1973b) and their capacity for specific, irreversible binding to the receptor on irradiation (Katzenellenbogen *et al.*, 1974). The most promising compound from these studies was hexestrol azide, which has an affinity for estrogen receptor 70% that of estradiol, and a 10–15% efficiency for light-activated irreversible binding to receptor. This compound was prepared in high specific activity tritium-labeled form for direct labeling studies (Katzenellenbogen *et al.*, 1977a). Its labeling, however, proved to be quite non-selective, so specific covalent attachment could only be demonstrated in receptor preparations that were already quite pure (Katzenellenbogen *et al.*, 1977b). This work has also been reviewed in detail elsewhere (Katzenellenbogen and Katzenellenbogen, 1984).

There have been some reports suggesting that 11β-chloromethylestradiol, a potent estrogen receptor that binds to the estrogen receptor and does not undergo exchange, may be undergoing covalent attachment to the receptor (Van den Broek *et al.*, 1983; Reiner and Katzenellenbogen, 1983; Reiner 1984; LeClercq *et al.*, 1980; Ratajczak *et al.*, 1981). Recent studies in which [^{3}H]11β-chloromethylestradiol has been synthesized and studied directly have shown, however, that the binding of this compound, albeit of very high affinity, is fully reversible (Bindal *et al.*, 1987).

The efficiency, selectivity, and convenience of the aziridine-based covalent labels for the estrogen receptor have made them the reagents of choice for affinity labeling studies.

III. DETAILS OF THE AFFINITY LABELING SYSTEMS

A. The reaction characteristics of aziridines

The high ring strain of aziridines makes them electrophilic, and their reaction with nucleophiles is well documented (Dermer and Ham, 1969). The electrophilicity of aziridines is increased by the attachment of electron withdrawing groups on the ring nitrogen or by ring nitrogen protonation to form aziridinium ions. Aziridines, however, are significantly less basic than amines; so, they protonate only under more acidic conditions. These patterns of reactivity are summarized in Scheme 1.

B. The structure of aziridine-based affinity labels for the estrogen receptor

We have incorporated *N*-aziridine substituents into four non-steroidal estrogens and utilized these agents as affinity labels for the estrogen receptor. The structures of these aziridine-based affinity labels are shown in Scheme 2.

The first compound prepared, tamoxifen aziridine (Tam-Az), is a close structural analog of the antiestrogen tamoxifen, in which the dimethylamino group is replaced by the aziridine (Robertson *et al.*, 1981). Desmethylnafoxidine aziridine (Naf-Az) is an analog of the antiestrogen nafoxidine; it has somewhat improved labeling characteristics, and the cyclic B-ring avoids problems with geometric isomerism

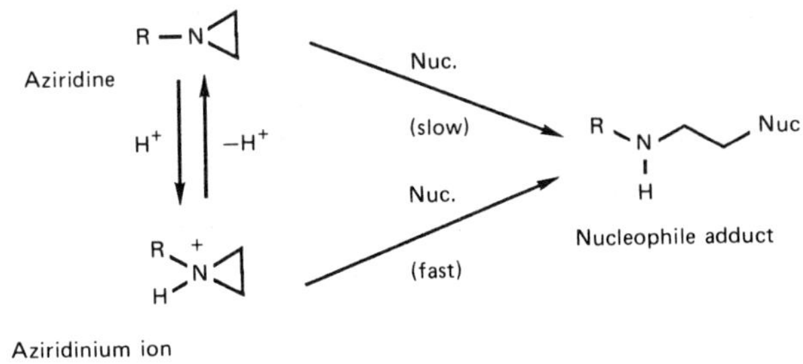

Scheme 1 — Electrophilic reactivity of aziridines.

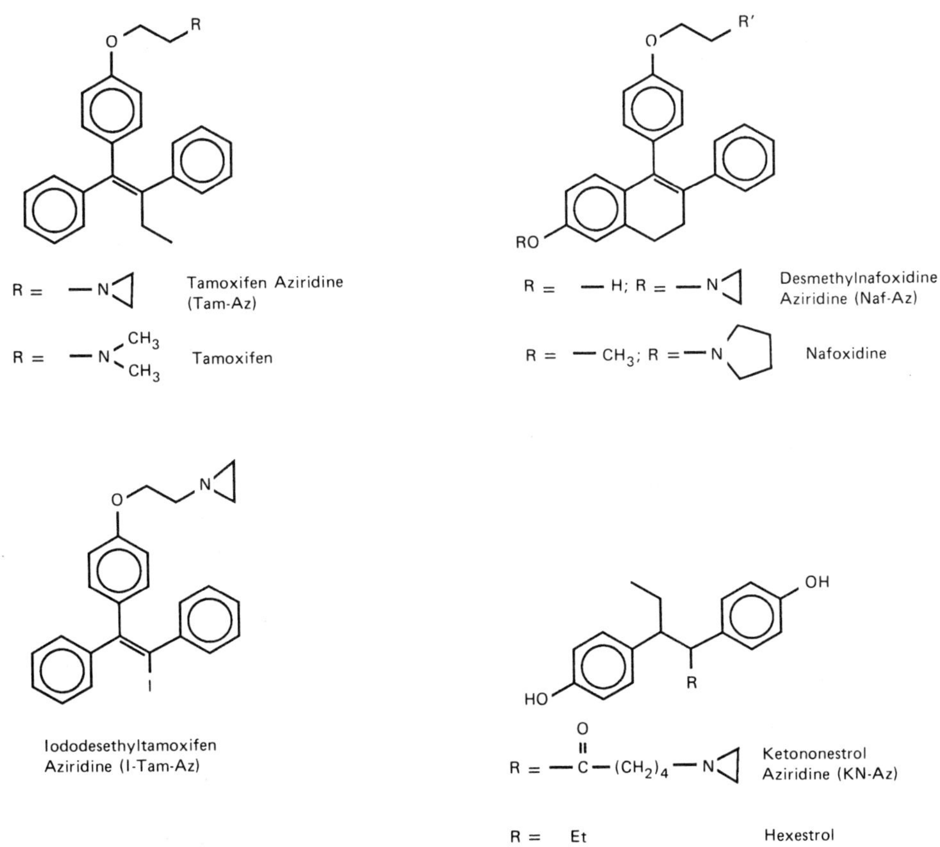

Scheme 2 — Aziridine-based affinity labels for the estrogen receptor.

encountered with Tam-Az (Simpson *et al.*, 1987). An iodinated analog of Tam-Az, iododesethyl tamoxifen aziridine (I-Tam-Az), was prepared to provide an affinity label that could be synthesized at very high specific activity (Salituro *et al.*, 1987). Finally, the derivative ketononesterol aziridine (KN-Az) is based on the potent non-steroidal estrogen hexestrol (Zablocki *et al.*, 1987). In contrast to the antagonist-based agents (Tam-Az, Naf-Az and I-Tam-Az), KN-Az is an estrogen agonist (Elliston *et al.*, 1987).

C Synthesis of aziridine-based affinity labels

All the aziridine derivatives are prepared by a common route: displacement of an immediate methanesulfonate precursor with ethylenimine (Scheme 3). The high

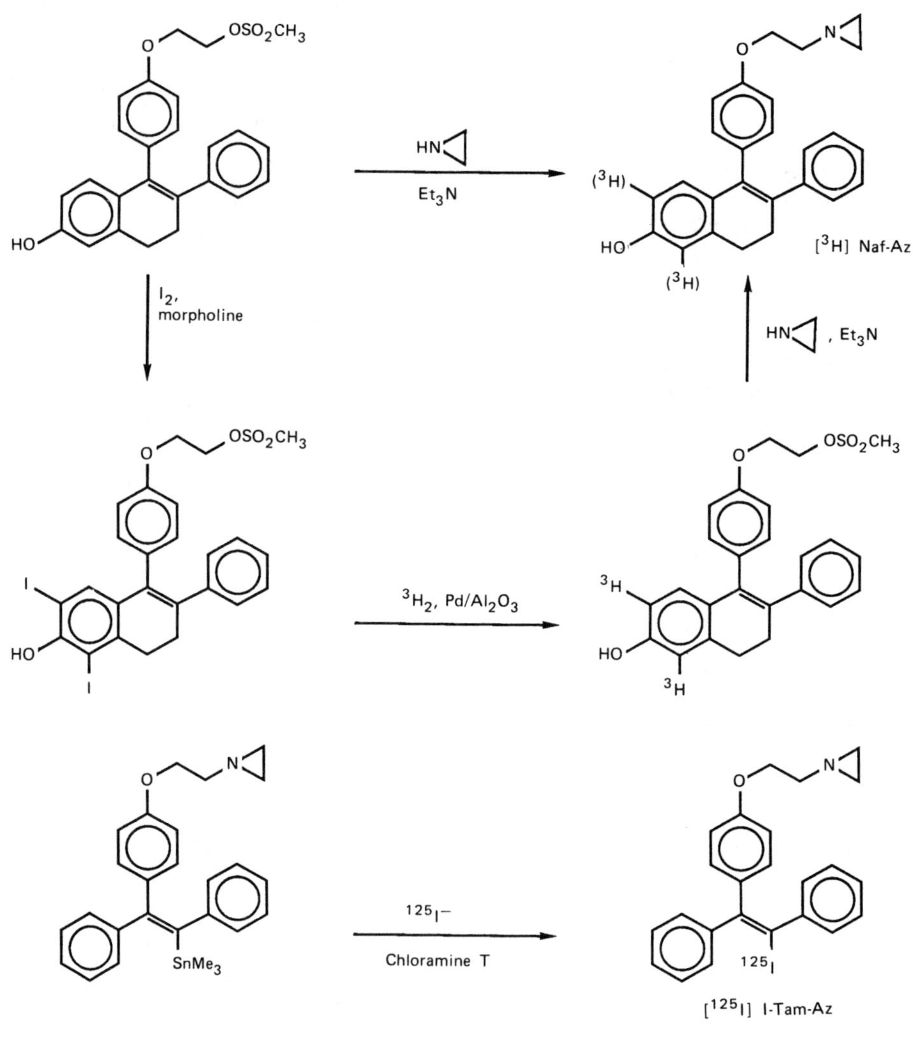

Scheme 3 — Radiolabeling of aziridine-based affinity labels for the estrogen receptor.

troxicity and suspected carcinogenicity of ethylenimine, coupled with its volatility and chemical instability, demand that great care be exercised in its preparation and use (see the cautionary note in Chapter 1, Section I.A; Dermer and Ham, 1969). As no further chemical transformations are required, the final product needs only to be purified by HPLC. In fact, the aziridine is relatively sensitive, and further chemical manipulations can only be done with low efficiency.

D. Radiolabeling of aziridine-based affinity labels

The three tritium-based aziridines (Tam-Az, Naf-Az and KN-Az) were radiolabeled by tritium-iodine exchange on an iodinated analog of the methanesulfonate precursor, followed by ethylenimine displacement (Scheme 3) (Katzenellenbogen *et al.*, 1983: Simpson *et al.*, 1987; Elliston *et al.*, 1987). Labeling by this method cannot be carried out at the aziridine stage because the aziridine function is unstable to the reduction conditions and is reactive towards iodine ion, a reaction by-product. After tritium–iodine exchange, the radiolabeled methanesulfonate, which is stable, is purified carefully. Then, as a final step, it is subjected to the ethylenimine displacement, with a final quick purification by HPLC. I-Tam-Az can be labeled at the final stage by radioiododestannylation of a vinyl tin precursor already bearing the aziridine function (Scheme 3); the mildness and rapidity of this reaction is consistent with aziridine stability (Salituro *et al.*, 1987).

E. Purity, specific activity and storage of aziridine-based affinity labels

Aziridines are compounds of moderate stability (Dermer and Ham, 1969); under the appropriate conditions, they can be stored and used successfully for years. As described above, the most extensive purification after radiolabeling is done at the stage of the methanesulfonate, which is very stable. Since this purification involves extensive HPLC separations, it is also convenient to determine the specific activity at the same time by quantitative HPLC analysis. The final conversion to the aziridine does not alter the specific activity and, when pure methansulfonate is used, the aziridine product can be brought to greater than 95% radiochemical purity by a quick HPLC run.

Trace levels of aziridines (such as the radiolabeled material) are not stable when stored in aqueous solutions; they are also destroyed when organic solutions are evaporated to dryness in either glass or plastic vessels (Katzenellenbogen *et al.*, 1983). The origin of this instability is not known, but may involve destruction of the sensitive aziridine function by active sites on these surfaces. The aziridines may be stored in organic solvents and manipulated successfully without drying down. We have found ethanol to be the most suitable solvent; stability in acetonitrile is relatively poor. Using these precautions, we have stored samples of Tam-Az in ethanol at −20°C for several years, with only a 5–10% loss in radiochemical purity. The other aziridines appear to have comparable stability.

F. Self-absorption in liquid scintillation counting

It may be necessary to use some special procedures to perform accurate radiometric determinations of proteins covalently labeled with the tritium-labeled aziridine agents: under conditions where reversibly-bound tritium-labeled ligands would be extracted into the organic, scintillant-rich phase, protein-bound tritium activity will partition into the aqueous, scintillant-poor phase and may, therefore, experience severe absorption quenching that would not be detected by the external standard quench correction method (Katzenellenbogen *et al.*, 1983). These problems are particularly severe under conditions where separate phases or micellar structures persist, such as counting of sucrose gradients. They can usually be minimized, however, by increasing the concentration of detergent or solubilizing agent in the scintillation fluid. It is important, therefore, to verify counting efficiency of covalently labeled samples by the samples channel ratio and to add progressive amounts of aqueous solubilizer until a stable, maximum radiometric value is established.

G. Conditions for affinity labeling

The conditions for covalent labeling of the estrogen receptor with the aziridine-based affinity labels are very simple.

1. Soluble receptor preparations

Soluble preparations of receptor are incubated with the aziridine at concentrations of 5–30 nM, for 30 min to 3 h, at 0–25°C. The time-course of reaction is more rapid at the higher concentrations and temperatures but, at temperatures above 25°C, receptor degradation may compete with labeling and thereby lower labeling efficiency.

In cytosol preparations that are relatively rich in receptor and free from endogenous ligands, such as immature rat uterine cytosol, labeling efficiency can approach 100%, and labeling selectivity (i.e. receptor labeling as a percentage of total labeling) can be as high as 70–90% (Katzenellenbogen *et al.*, 1983). Cytosol preparations less rich in receptor and containing endogenous ligand are labeled with lower efficiency and selectivity. In extreme cases, where the level of specific covalent attachment may be very low relative to total labeling, it may be difficult to measure accurately without additional fractionation. However, even in receptor-poor cases (e.g. cytosol from rat DMBA-induced mammary tumors), specific labeling of receptor is clearly evident after fractionation on SDS–PAGE (Wei, 1984).

The time-course of receptor labeling can be followed conveniently by an ethanol disk assay, in which aliquots from the labeling incubation are spotted onto disks of Whatman 3 paper, which are then extracted with boiling ethanol and counted. The hot solvent precipitates the protein in the pores of the paper and extracts any radioactivity that is not attached to the protein covalently (Katzenellenbogen *et al.*, 1977b). Examples of the time course of covalent attachment of Tam-Az to rat uterine cytosol are shown in Fig. 1.

Although the aziridines are electrophilic, their reaction with receptor appears to be quite resistant to competing nucleophiles, such as 50–100 mM mercaptoethanol, and high labeling efficiencies can be achieved in buffers than contain low concentration of other thiols (Katzenellenbogen *et al.*, 1983). In this respect they are quite different from the affinity labels based on α-keto halides and methanesulfonates (Katzenellenbogen and Katzenellenbogen, 1984).

The aziridine-based labeling agents are quite lipophilic; so, in unfractionated receptor preparations, they can show considerable non-specific binding. The inclusion of low concentrations of a dipolar aprotic organic solvent (e.g. 3–5% dimethylformamide) that has little effect on the integrity of the receptor markedly reduces

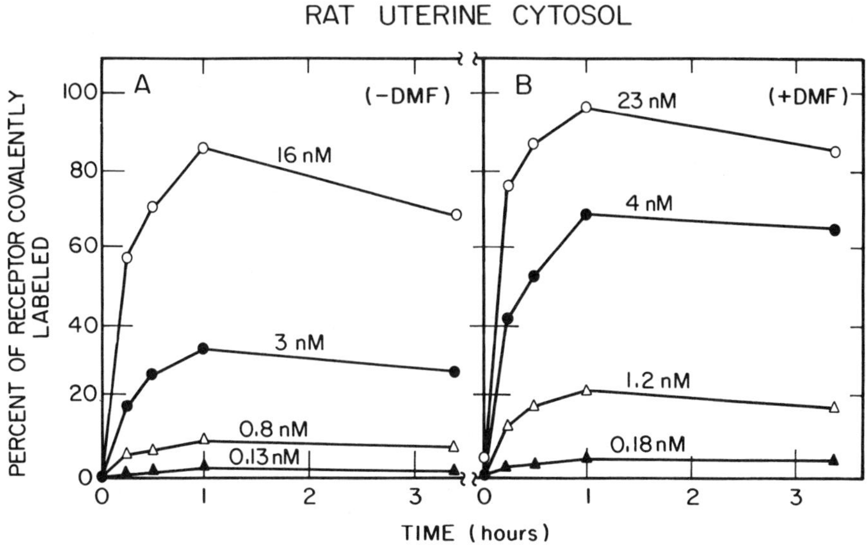

Fig. 1 — Time-courses of covalent labeling of estrogen receptor in rat uterine cytosol. A rat uterine cytosol preparation containing 3.0 nM of estrogen receptor was treated with the indicated concentrations of [^{3}H]Tam-Az in the absence or presence of 3000 nM unlabeled estradiol, both (B) with 7% dimethylformamide (DMF) and (A) without DMF. The extent of covalent labeling was determined by ethanol extraction of aliquots spotted on filter disks. The data plotted represent estrogen-specific covalent labeling (the difference between labeling in the absence and presence of unlabeled estradiol). (From Katzenellenbogen *et al.*, 1983.)

this non-specific binding and improves the rate and efficiency of receptor covalent labeling, presumably by increasing the concentration of free reagent (Katzenellenbogen *et al.*, 1980, 1983) (cf. Fig. 1).

In some situations where it is necessary to label covalently estrogen receptor that is already saturated with a ligand (such as material that has been eluted from an estrogen affinity column), receptor can be labeled under exchange conditions. For this, exposure for longer periods of time or at the higher temperatures (25°C) is required, and labeling efficiency is generally lower (Young *et al.*, 1986; G. L. Greene, personal communication).

2. Labeling of estrogen receptors in intact cells

The estrogen receptor in intact cells can be covalently labeled with good efficiency. Such studies have been carried out in MCF-7 human breast cancer cells (Katzenellenbogen *et al.*, 1983; Monsma *et al.*, 1984) and in primary cultures of rat uterine cells (Nardulli and Katzenellenbogen, 1986; Katzenellenbogen *et al.*, 1983), where exposure of cells (in tissue culture medium containing 5% charcoal–dextran treated serum) to 20 nM [^{3}H]Tam-Az for 1 h at 37°C results in covalent labeling of 80–100% of the estrogen receptors. When cells are incubated with [^{3}H]Tam-Az alone or [^{3}H]Tam-Az plus excess unlabeled estradiol for 1 h at 37°C and cells are then homogenized and fractionated, most of the [^{3}H]Tam-Az-labeled receptor (estradiol-competable binding) is found in the salt extract of the nuclear fraction (Fig. 2). In

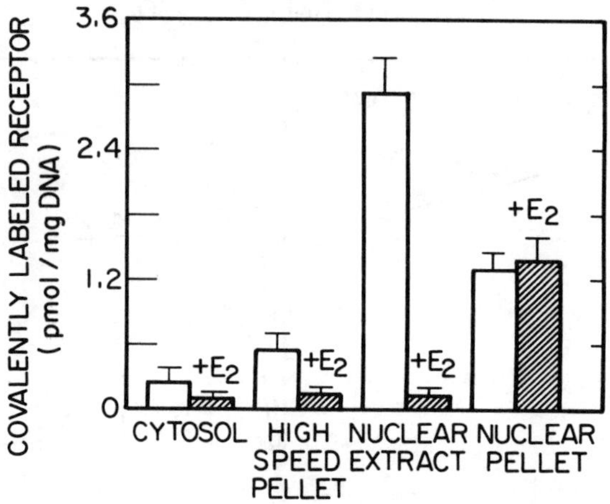

Fig. 2 — Covalent labeling of estrogen receptors in MCF-7 cells with [^{3}H]Tam-Az. Separate flasks of MCF-7 cells were exposed to 10 nM [^{3}H]Tam-Az in the absence (open bars) or presence (shaded bars) of 1 μM estradiol for 1 h at 37°C, and the indicated cell fractions were prepared. The extent of covalent binding was determined by the ethanol filter disk assay. The error bars indicate the range of two experiments. (From Katzenellenbogen *et al.*, 1983.)

addition, the extent of non-specific binding of tritiated Tam-Az in this nuclear receptor extract is similar to that seen with tritiated estradiol and is quite low, making it possible to use tritiated Tam-Az to examine the properties of covalently labeled estrogen receptors. Most of the non-specific binding of Tam-Az is found in the nuclear (non-salt-extractable) pellet fraction and is believed probably to represent Tam-Az associated with the nuclear membrane (Monsma *et al.*, 1984).

In time-course studies on the covalent labeling of receptors in whole cells at 37°C, we have found that about 40% of the receptors are labeled by [^{3}H]Tam-Az in 20 min; maximal labeling is achieved by 1 h, and there is no further increase in labeling by 3 h. 1 h was also found to be optimal for maximal nuclear receptor labeling by tritiated estradiol, as reported previously (Eckert and Katzenellenbogen, 1982).

The labeling by Tam-Az was found to be estrogen specific in that it was blocked only by estrogens and antiestrogens Monsma *et al.*, 1984). Full competition with [^{3}H]Tam-Az (20 nM) binding to receptor was seen with a 20- or 200-fold excess of the estrogen diethylstilbestrol and estradiol or the antiestrogens trans-hydroxytamoxifen or tamoxifen aziridine. Androgens, progestins, and glucocorticoids had no effect on receptor labeling by Tam–Az (Monsma *et al.*, 1984). In addition, the estrogen specificity of the Tam-Az labeling in cells was further confirmed by labeling experiments with MDA-MB-231 human breast cancer cells; these cells lack estrogen receptors, and no specific (estradiol-displaceable) covalent attachment of Tam-Az was observed in these cells.

3. *Labeling of estrogen receptors in target tissues* in vivo

In certain cases, it is possible to affinity label successfully estrogen receptors in target tissues *in vivo*. This has been particularly useful in studying the properties and dynamics of estrogen receptors in the rat uterus, since small amounts of [^{3}H]Tam-Az can be directly injected into the lumen of the uterus of rats using fine-gauge Hamilton syringes. With a 1 h labeling period, intralumenally applied [^{3}H]Tam-Az covalently labeled 92%±6% of the total uterine estrogen receptors and, in the nuclear salt extracts which contained essentially all the estrogen receptors, there was very little non-specifically labeled proteins. Non-specific (estradiol non-competable) labeling represented less than 5% of the total radiolabeling in this extract (Nardulli and Katzenellenbogen, 1986).

4. Bioactivity of affinity labeling antiestrogens and estrogens

It is important to determine whether affinity labeled receptors mediate biological responses in a manner analogous to that evoked by reversibly binding ligands, since hormone action normally involves only reversible interactions between the hormonal agent and the receptor. With Tam–Az aziridine and with KN-Az, we have found that these compounds have bioactivities much like those of their reversibly binding counterparts.

Tam-Az behaves as an antiestrogen and has a bioactivity much like that of its parent compound, tamoxifen. It inhibits the growth of MCF-7 human breast cancer cells *in vitro*, suppresses estradiol-stimulated uterine growth in rats, and elicits the regression of hormone-dependent dimethylbenzanthracene-induced mammary tumors in rats (Wei *et al.*, 1985). Similarly, the agonostic activity of KN-Az is evident in MCF-7 cells in culture, where it increases cell proliferation rate and elevates the level of progesterone receptor (Zablocki *et al.*, 1987; Elliston *et al.*, 1987). Hence, these receptor affinity labeling probes appear to have appropriate bioactivity for studies on receptor structure and function.

IV. APPLICATIONS OF ESTROGEN RECEPTOR AFFINITY LABELING

A. Physicochemical characterization of receptor properties and structure and relatedness of receptors in different species and target cells

With a covalently labeled receptor, one can apply quite rigorous methods for receptor purification and characterization, such as polyacrylamide gel electrophoresis in the presence of sodium dodecyl sulfate, isoelectric focusing in the presence of 8 M urea, and reversed-phase high performance liquid chromatography on denatured proteins. Since affinity labeling of estrogen receptors can be performed on unpurified or only partially purified preparations, this has allowed facile comparisons of the physicochemical properties of receptors from different species, different target tissues, or from mutant or variant cells having altered responsiveness to hormones.

By SDS–polyacrylamide gel electrophoresis, the M_r of the Tam-Az labeled estrogen receptor appears to be *ca* 66000 (Fig. 3, lanes 1). This is the case for receptor from rat or lamb uterus, from MCF-7 human breast cancer cells, and from rat pituitary and DMBA-induced mammary tumurs (Katzenellenbogen *et al.*, 1983; Monsma *et al.*, 1984; Nardulli and Katzenellenbogen, 1986; Wei, 1984). Both cytosolic estrogen receptor labeled by treating cytosol with [^{3}H]Tam-Az *in vitro* and nuclear receptor obtained from intact cells or tissues incubated with [^{3}H]Tam-Az or

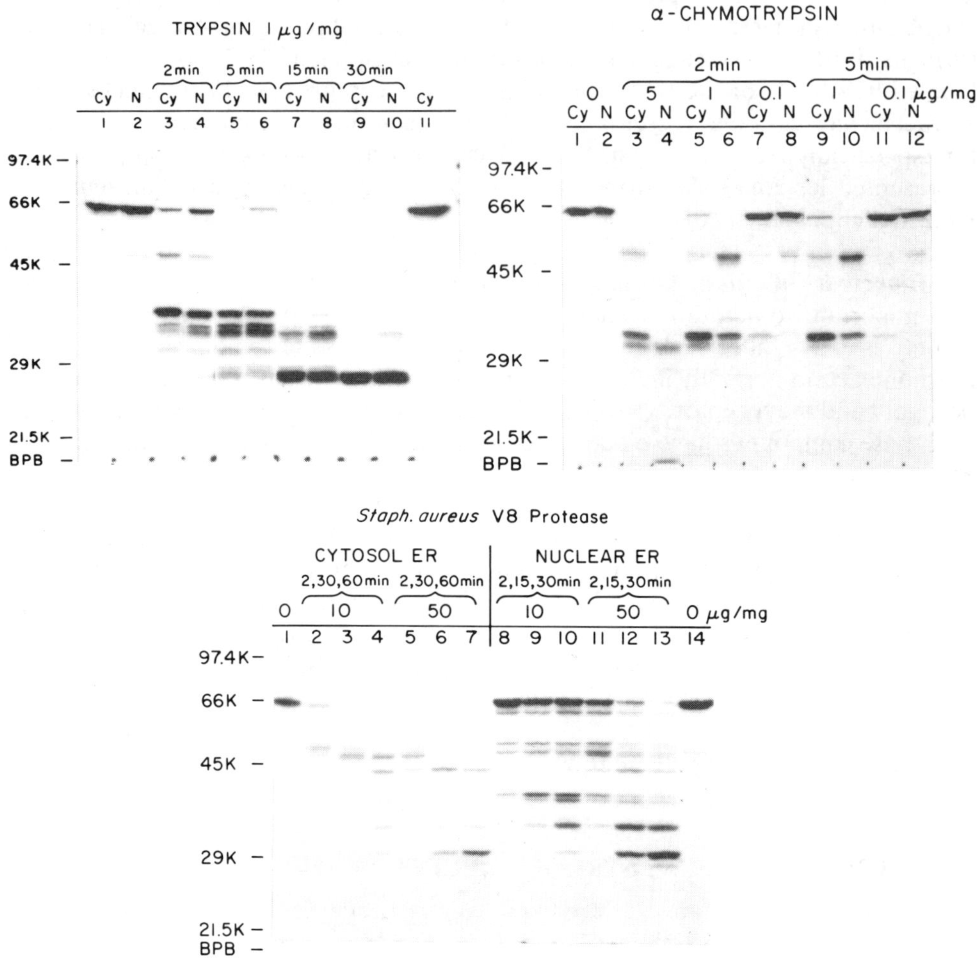

Fig. 3 — Comparison of the proteolytic digestion with trypsin, α-chymotrypsin or *Staphylococcus aureus* V8 protease of MCF-7 cytosol and nuclear receptors labeled with [^{3}H]Tam-Az. MCF-7 cytosol receptors and nuclear salt extracted receptors covalently labeled with [^{3}H]Tam-Az were precipitated with ammonium sulfate and the resuspended precipitate was treated with 1 μg trypsin/mg protein for 2-30 min at 22°C, or with various amounts of α-chymotrypsin for 2 and 5 min, 22°C, or 10 or 50 μg V8 protease/mg protein for 2–60 min at 22°C and were analyzed side-by-side on SDS–PAGE followed by fluorography. Treatments are of cytosol (Cy) and nuclear receptors (N). For trypsin: lanes 1 and 2, untreated; lanes 3 and 4, 2 min, 22°C; lanes 5 and 6, 5 min, 22°C; lanes 7 and 8, 15 min, 22°C; lanes 9 and 10, 30 min, 22°C. For chymotrypsin: lanes 1 and 2, untreated; lanes 3 and 4, 5 μg α-chymotrypsin/mg protein, 2 min; lanes 5 and 6, 1 μg α-chymotrypsin/mg protein, 2 min; lanes 7 and 8, 0.1 μg α-chymotrypsin/mg protein, 2 min; lanes 9–12, same concentrations as lanes 5–8 respectively for 5 min, 22°C. For V8 protease: lanes 1 and 14; untreated cytosol and nuclear receptors; lanes 2, 3 and 4, cytosol treated with 10 μg V8 protease/mg protein for 2 min, 30 and 60 min respectively; lanes 5, 6 and 7, cytosol treated with 50 μg V8 protease/mg protein for 2 min, 30 min and 60 min respectively; lanes 8, 9 and 10, nuclear receptor treated with 10 μg V8 protease/mg protein for 2 min, 15 min and 30 min respectively; lanes 11, 12 and 13; nuclear receptor treated with 50 μg V8 protease/mg protein for 2 min, 15 min and 30 min respectively. (From Katzenellenbogen *et al.*, 1987.)

[^{3}H]KN-Az show identical molecular weights (Fig. 3, Monsma *et al.*, 1984; Katzenellenbogen *et al.*, 1987; Elliston *et al.*, 1987). *In situ* labeling of nuclear receptor, as can be done with the MCF-7 cells, together with direct lysis of cells denaturing medium, can be used to ascertain that the molecular weight of receptor in intact cells is also 66 000 under conditions where proteolytic degradation of receptor is unlikely (Katzenellenbogen *et al.*, 1983). Indeed, this 66 000 molecular weight receptor size is the same as calculated for the MCF-7 ER by analysis of cDNA for the estrogen receptor (Green *et al.*, Greene *et al.*, 1986).

Studies with the [^{3}H]Tam-Az labeled nuclear receptor indicate that the M_r 66 000 receptor form is stable for over 24 h at 25°C in the nuclear salt extracts; cytosolic receptor from MCF-7 cells, however, is rapidly degraded to a 52 000 and then eventually to a *ca* 30 000 molecular weight species, this conversion being blocked completely by the presence of leupeptin of molybdate (Monsma *et al.*, 1984). Affinity receptor interacts with several monoclonal antibodies to MCF-7 estrogen receptor (prepared by Geoffrey L. Greene, University of Chicago), and it can be purified extensively by immunoadsorbent chromatography on antireceptor antibody columns (Monsma *et al.*, 1984).

The labeling of whole cells with [^{3}H]Tam-Az also indicates that there are at most only minor quantities of estrogen binding species in the particulate fractions (microsomal and membrane–nuclear pellet fractions) remaining after salt-extraction that are labeled covalently by this agent (Katzenellenbogen *et al.*, 1983; Monsma *et al.*, 1984). However, these findings do not rule out the possible presence of estrogen receptors in these subcellular fractions; these sites may simply not be labeled covalently with Tam-Az or their low levels could be masked by the non-specific background binding that is observed in the membrane fractions.

It is notable that Tam-Az labeled estrogen receptor from R3 and R27 variants of the MCF-7 cell line that are resistant to the growth inhibitory effect of antiestrogens have molecular weights and isoelectric points indistinguishable from those of the wild-type MCF-7 cell receptor (Miller *et al.*, 1984). These and other findings (Nawata *et al.*, 1981) suggest that the defect in antiestrogen responsiveness of these variant cells may lie at steps beyond the initial binding of ligand. However, it is important to note that changes in small numbers of amino acids in the receptor would probably not be observable as alterations with these methods, even though they might be of great biological significance.

It is difficult to determine the isoelectric point of estrogen receptor labeled with reversible ligands, because aggregation with other proteins prevents the receptor from entering the focusing gel and distorts the focusing pattern. With Tam–Az labeled receptor, isoelectric focusing can be carried out in the presence of 8 M urea. Under these disaggregating conditions, both nuclear and cytosolic MCF-7 receptor are found to have pI values of *ca* 5.7 (Monsma *et al.*, 1984). The focused bands in the 1-dimensional isoelectric focusing disk gels are rather broad, and 2-dimensional (IEF–SDS) gel analysis (Fig. 4) reveal charge heterogeneity (2–3 spots between $pI =$ 5.5–6.2) of the 66 000 M_r estrogen receptor (Katzenellenbogen *et al.*, 1987) that might arise from different degrees of receptor phosphorylation, acetylation or other modification.

We have used limited proteolysis of affinity labeled estrogen receptors, coupled with antireceptor antibody immunoreactivity, to assess structural features of recep-

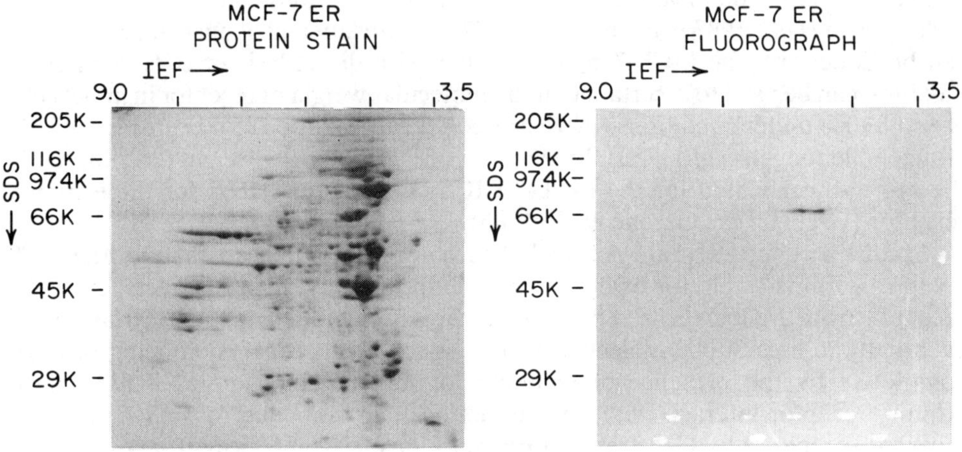

Fig. 4 — 2-dimensional gel analysis of MCF-7 nuclear receptor. MCF-7 estrogen receptor covalently labeled with [^{3}H]Tam-Az was run on an isoelectric focusing gel (pH 3.5–10) in the presence of 8 M urea of the first dimension. This was then applied to the top of an SDS–10% polyacrylamide slab gel and electrophoresed in the vertical direction. Several IEF gels were run in parallel. One was sliced for determination of pH and radioactivity and another was electrophoresed on an SDS gel. (A) Coomassie Blue stain of 2-D gel used for fluorography. (B) Fluorogram of gel in (A). (From Katzenellenbogen *et al.*, 1987.)

tor and the relatedness of estrogen receptor from two different target cells, namely MCF-7 human breast cancer cells and rat uterine cells (Katzenellenbogen *et al.*, 1987). MCF-7 estrogen receptor preparations covalently labeled with [^{3}H]Tam-Az were treated with trypsin (T), α-chymotrypsin (C) or *S. aureus* V8 protease prior to electrophoresis on SDS gels. Fluorography (Fig. 3) revealed a distinctive ladder of receptor fragments containing Tam–Az for each protease generated from the M_r 66 000 ER: for T, fragments of 50, 38, 36, 31, 29, and 28K that with longer exposure generated a 6 K fragment; for C, fragments of 50, 38, 35, 33, 31, 19 and 18 K that with longer exposure generated 14 and 6 K fragments, and for V8, *ca* 10 fragments 62 and 28 K.

Immunoblot detection with the primate specific antibody D75P3γ revealed that all immunoreactive fragments correspond to Tam-Az labeled fragments, but that some small Tam–Az labeled fragments (V8-generated forms <47 K and T-generated forms <31 K) were no longer immunoreactive. In contrast, use of the antibody H222Spγ revealed a correspondence between Tam–Az labeled and immunoreactive fragments down to the smallest fragments generated, *ca* 6 K for T and C and 28 K for V8. MCF-7 nuclear and cytosol estrogen receptors showed very similar digest patterns, and there was a remarkable similarity in the Tam–Az labeled and H222 immunoreactive fragments generated by proteolysis of both MCF-7 and rat uterine estrogen receptor (Katzenellenbogen *et al.*, 1987). These findings reveal great structural similarities between the human (breast cancer) and rat (uterine) estrogen receptor and between nuclear and cytosol estrogen receptor, and allow a comparison of the immunoreactive and hormone attachment site domains of the receptor. These protease digest studies, in conjunction with antibody mapping (Greene, 1984) and receptor mutational analysis (Kumar *et al.*, 1986), clearly confirm that the hormone binding domain is near the carboxy-terminal end of the receptor. The observation that T and C generate a *ca* 6 K Tam-Az labeled fragment that is also detectable with the H222 antibody will be useful in further studies on estrogen receptor structure and particularly in analysis of the amino acids in the hormone binding site of the receptor.

Related studies have been performed with [^{3}H]KN-Az, an aziridine derivative of hexestrol, which is an estrogenic affinity label for the estrogen receptor (Elliston *et al.*, 1987). As with [^{3}H]Tam-Az and with Naf-Az which are antiestrogens, the covalent labeling with [^{3}H]KN-Az proceeds in a time-, concentration-, and temperature-dependent manner; labeling is efficient and selective and, by competition studies, can be shown to be estrogen specific. Receptor, covalently labeled with [^{3}H]Kn-Az or [^{3}H]Tam-Az, sediments as a 4 S species on high salt sucrose gradients, and its sedimentation profile is shifted by treatment with monoclonal antireceptor antibodies and, on SDS–polyacrylamide gels, the principal [^{3}H]Kn-Az labeled species migrates with an M_r of 66 000 (Elliston *et al.*, 1987). Studies on the sensitivity of receptors to proteolytic degradation by exogenous and endogeneous proteases when labeled with KN-Az vs. Tam-Az, characterization of the receptor fragments formed, and analysis of the amino acids in the binding domain of receptor covalently labeled by KN-Az and Tam-Az are in progress. Needless to say, the availability of covalently attaching ligands will enable a determination of the amino acids in the binding site with which these compounds directly interact. In addition, the preparation of these affinity labeling ligands in ^{125}I-labeled form, with high specific activities (up to *ca* 2000 Ci/mmol), has enabled detection of receptors on SDS gel fluorograms with only overnight exposure (Salituro *et al.*, 1987), and should be particularly helpful in studies probing for receptor–gene interaction sites in chromatin.

Finally, it is of note that although triphenylethylene antiestrogens such as tamoxifen interact with an additional class of antiestrogen binding sites that are distinct from the receptor (Sutherland *et al.*, 1980; Sudo *et al.*, 1983; Miller and Katzenellenbogen, 1983; Katzenellenbogen *et al.*, 1985), the interaction of Tam-Az with these sites is reversible and shows no evidence of covalent interaction (Monsma *et al.*, 1984).

B. Studies on estrogen receptor dynamics, turnover and dimerization

Because of its high labeling selectivity, Tam-Az has been used to label estrogen receptors in MCF-7 breast cancer or rat uterine cells in a pulse-chase manner (Monsma *e al.*, 1984; Nardulli and Katzenellenbogen, 1986). After the labeling period with [^{3}H]Tam-Az, excess estradiol is added as a chaser, and the fate of the covalently labeled receptor can be followed with time to determine the half-life of the receptor. As is seen in Figs. 5 and 6, it is evident that the 66 000 molecular weight receptor component is lost quite quickly, with a half-life of 4 h, consistent with the rate of receptor turnover in these cells determined in other experiments by a heavy amino acid density shift technique (Eckert and Katzenellenbogen, 1982; Eckert 1984; Nardulli and Katzenellenbogen, 1986).

It is of note that smaller molecular weight fragments are not observed during the degradation of receptor in cells, either in the nuclear salt-extractable fraction or in the cytosol or the microsomal or nuclear pellet fractions (Monsma *et al.*, 1984).

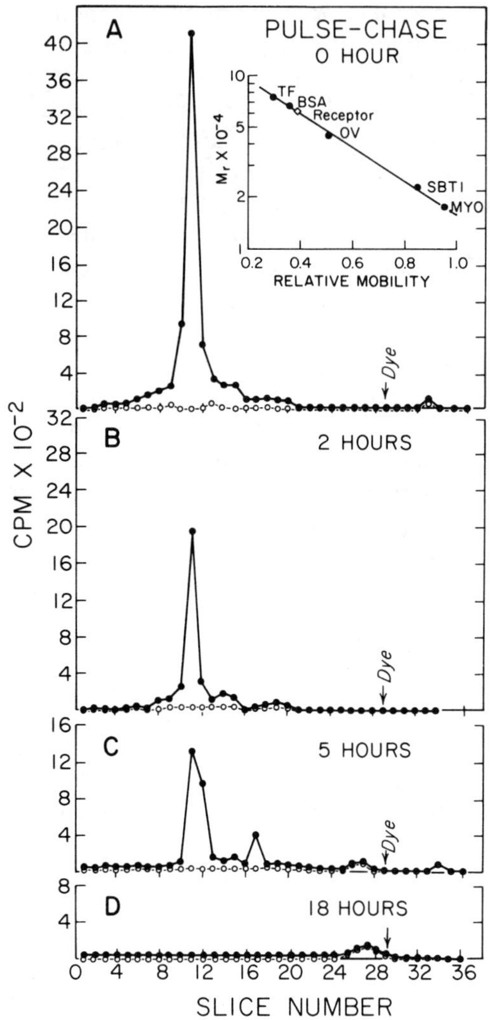

Fig. 5 — SDS–polyacrylamide gel electrophoretic analysis of the MCF-7 nuclear receptor. Pulse-chase experiment. T-150 flasks of MCF-7 cells were labeled with 20 nM [^{3}H]Tam-Az±2 μM estradiol for 1 h at 37°C and then chased with 10^{-6} M E_2 for various time periods ranging from 0 h (no chase) to 18 h. (A) 1 h [^{3}H]Tam-Az label, no chase; (B)–(D) 1 h [^{3}H]Tam-Az label and then 2, 5 or 18 h chase with E_2. After the indicated chase period, nuclear salt extracts were prepared and analyzed on SDS–polyacrylamide gels. The inset of (A) shows the mobility of protein standards and of receptor relative to the bromophenol blue dye band. The protein standards were transferrin (TF, M_r = 74 000), bovine serum albumin (BSA, M_r = 67 000), ovalbumin (OV, M_r = 43 000), soybean trypsin inhibitor (SBTI, M_r = 20 000) and myoglobin (MYO, M_r = 17 000). (From Monsma *et al.*, 1984.)

Thus, in MCF-7 and uterine cells, the Tam-Az labeled receptor must be broken down in such a manner that the portion retaining the label is quickly degraded to small molecular weight fragments that are no longer detectable by SDS–gel electro-

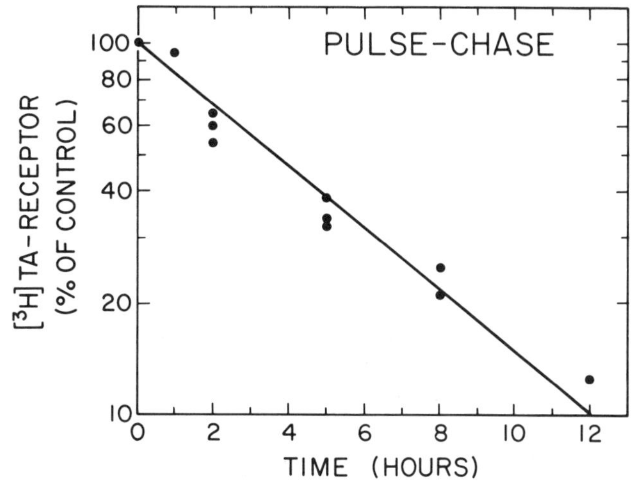

Fig. 6 — Rate of degradation of nuclear receptor in MCF-7 cells covalently labeled with [^{3}H]Tam-Az (TA) determined from pulse-chase experiments and SDS gel electrophoretic analysis. Pulse-chase experiments were performed exactly as described in the legend to Fig. 5. Cells were harvested after the indicated time of chase with 10^{-6} M estradiol, and data are presented as the precentage of [^{3}H]Tam-Az labeled receptor present after 1 h of [^{3}H]Tam-Az exposure and no chase (i.e. 0 h chase). Each point represents data obtained in duplicate from one T-150 flask of cells. All values are corrected for the low, *ca* 10% of total, hot plus cold binding measurement. (From Monsma *et al.*, 1984.)

phoretic techniques (Katzenellenbogen *et al.*, 1983; Monsma *et al.*, 1984).

Immunoblotting methods also failed to detect the presence of defined receptor fragments during receptor turnover (Nardulli and Katzenellenbogen, 1986). Since all samples showed only one immunoreactive species of molecular weight 66 000, it seems probable that degradation of the receptor in intact cells occurs very rapidly, through short-lived intermediate forms, making it difficult to isolate defined degradation products (Nardulli and Katzenellenbogen, 1986).

Affinity labeling of estrogen receptors, combined with chemical crosslinking agents, has been used to characterize the subunit nature of nuclear estrogen receptors (Miller *et al.*, 1985). When MCF-7 cells are incubated with tritiated estradiol or tritiated antiestrogen and nuclear estrogen receptor complexes are extracted and chemically crosslinked with the crosslinker 2-iminothiolane, the crosslinked-receptor complexes sediment as a 5.4 S species on 3 M urea-containing sucrose gradients, while the non-crosslinked species are 4 S. SDS–polyacrylamide gel analyses of these crosslinked nuclear receptor complexes labeled with the covalently attaching ligand [^{3}H]Tam-Az revealed a species of about 130 000 molecular weight, while the non-crosslinked or the crosslinked, but mercaptan-cleaved, receptor is 66 000 molecular weight. These studies, plus additional dense amino acid incorporation studies indicating that the nuclear 5.4 S receptor complex consists of two similarly sized units which turn over with similar half-lives, provide strong evidence that this nuclear receptor complex is a homodimer of two 4 S, 66 000 molecular weight monomers (Miller *et al.*, 1985).

C. Gene interaction studies

The availability of estrogen receptor affinity labels that have antiestrogenic character (Tam-Az and Naf-Az) and estrogenic character (KN-Az) should be of particular utility in probing for chromatin interaction sites of the estrogen receptor that are related to hormone-stimulated gene expression, since isolation of receptor with associated DNA fragments requires isolation procedures in which reversibly binding ligands would dissociate. Affinity labels provide a permanent tag and, when combined with the use of DNA–protein crosslinking methods, should enable isolations under relatively harsh disaggregating and denaturing conditions of receptor associated with relevant DNA. The use of ^{125}I-labeled covalently attaching ligands of very high specific activity should allow relatively refined mapping of the region of the responsive gene and/or flanking regions with which the receptor is associated. Such studies also allow determination of whether the 'efficiency' of association of receptor with primary genes known to be activated by hormone, but not antihormone, differs for receptors labeled in cells with estrogen vs. antiestrogen. Hence, these receptor affinity labeling probes should be very useful in continuing studies of receptor structure, function and chromatin interaction.

ACKNOWLEDGMENTS

We are very grateful to many of our associates for their important contributions to the studies described in this chapter. We wish to acknowledge David W. Robertson and Daniel F. Heiman for synthesis of Tam-Az, David M. Simpson for synthesis of Naf-Az, Jeffrey A. Zablocki for synthesis of hexestrol aziridines and Francesco G. Salituro for synthesis of I-Tam-Az. We also very much appreciate the contributions of Katherine E. Carlson, Jonathon F. Elliston, Frederick J. Monsma, Jr., Ann M. Nardulli, Mary Jane Norman, Penelope A. Springer and Lisa L. Wei in the biological and biochemical characterization of the interaction of the affinity labels with estrogen receptors in target cells.

REFERENCES

Bindal, R. D., Carlson, K. E., Reiner, G. C. A. and Katzenellenbogen, J. A. (1987) 11β-Chloromethyl-[^{3}H]estradiol-17β: A very high affinity, reversible ligand for the estrogen receptor. *J. Steroid Biochem.* Submitted.

Chambon, P., Dierich, A., Gaub, M. P., Jakowlew, S., Jongstra, J., Krust, A., LePennec, J. B., Oudet, P. and Reudelhuber, T. (1984) Promoter elements of genes coding for proteins and modulation of transcription by estrogens and progesterone. *Recent Prog. Hormone Research* **40**, 1–42.

Dermer, O. D. and Ham, G. E. (eds.) (1969) *Ethyleneimine and Other Aziridines.* Academic Press, New York.

Dickson, R. B. and Lippman, M. E. (1987) Estrogenic regulation of growth and polypeptide growth factor secretion in human breast carcinoma. *Endocrine Reviews* **8**, 29–43.

Eckert, R. L. and Katzenellenbogen, B. S. (1982) Physical properties of estrogen receptor complexes in MCF-7 human breast cancer cells: Differences with antiestrogen and estrogen. *J. Biol. Chem.* **257**, 8840–8846.

Eckert, R. L., Mullick, A., Rorke, E. A. and Katzenellenbogen, B. S. (1984) Estrogen receptor synthesis and turnover in MCF-7 breast cancer cells measured by a density shift technique. *Endocrinoloy* **114**, 629–637.

Elliston, J. F., Zablocki, J. A., Katzenellenbogen, B. S. and Katzenellenbogen, J. A. (1987) Ketononestrol aziridine, an agonistic estrogen receptor affinity label: Study of its bioactivity and estrogen receptor covalent labeling. *Endocrinology*, in press.

Green, S., Walter, P., Kumar, V., Krust, A., Bornert, J.-M., Argos, P. and Chambon, P. (1986) Human oestrogen receptor cDNA: Sequence, expression and homology to v-erb-A. *Nature* **320**, 134–139.

Greene, G. L. (1984) Application of immunochemical techniques to the analysis of estrogen receptor structure and function. *Biochem. Actions Hormones* **11**, 207–239.

Greene, G. L., Gilna, P., Waterfield, M., Baker, A., Hort, Y. and Shine, J. (1986) Sequence and expression of human estrogen receptor complementary DNA. *Science* **231**, 1150–1154.

Jordan, V. C. (1984) Biochemical pharmacology of antiestrogen action. *Pharmacological Reviews* **36**, 245–276.

Katzenellenbogen, B. S. (1980) Dynamics of steroid hormone receptor action. *Ann. Review Physiol.* **42**, 17–35.

Katzenellenbogen, B. S. and Gorski, J. (1975) Estrogen actions on syntheses of macromolecules in target cells. *Biochem. Actions Hormones* **3**, 187–243.

Katzenellenbogen, B. S., Miller, M. A., Mullick, A. and Sheen, Y. Y. (1985) Antiestrogen action in breasts cancer cells: Modulation of proliferation and protein synthesis, and interaction with estrogen receptors and additional antiestrogen binding sites. *Breast Cancer Res. Treatment* **5**, 231–243.

Katzenellenbogen, B. S., Elliston, J. F., Monsma, F. J. Jr., Springer, P. A., Zeigler, Y. S. and Greene, G. L. (1987) Structural analysis of covalently labeled estrogen receptors by limited proteolysis and monoclonal antibody reactivity. *Biochemistry*, **26** 2364–2373.

Katzenellenbogen, J. A. (1977) Affinity labeling as a technique in determining hormone mechanisms. *Biochem. Actions Hormones* **4**, 1–84.

Katzenellenbogen, J. A. and Katzenellenbogen, B. S. (1984) Affinity labeling of receptors for steroid and thyroid hormones. *Vitamins and Hormones*, **41**, 213–274.

Katzenellenbogen, J. A., Myers, H. N. and Johnson, H. J. Jr. (1973a) Reagents for photoaffinity labeling of estrogen binding proteins. Synthesis of some diazo and azide derivatives of estradiol, estrone, and hexestrol. *J. Org. Chem.* **38**, 3525–3533.

Katzenellenbogen, J. A., Johnson, H. J. Jr. and Myers, H. M. (1973b) Reagents for photoaffinity labeling of estrogen binding protein. The binding affinity of some azide and diazo derivatives of estradiol, estrone, and hexestrol for the estrogen binding protein of rat uterus *Biochemistry* **12**, 4085–4092.

Katzenellenbogen, J. A., Johnson, H. J. Jr., Carlson, K. E. and Myers, H. M. (1974) The photoreactivity of some light-sensitive estrogen derivatives. The use of an exchange assay to detemine their photointeraction with the rat uterine estrogen binding protein. *Biochemistry* **13**, 2986–2994.

Katzenellenbogen, J. A., Myers, H. M., Johnson, H. J. Jr., Kempton, R. J. and Carlson, K. E. (1977a) Estrogen photoaffinity labels I: Chemical and radiochemical synthesis of hexestrol diazoketone and azide derivatives: Photochemical studies in solution. *Biochemistry* **16**, 1964–1970.

Katzenellenbogen, J. A., Carlson, K. E., Johnson, H. J. Jr. and Myers, H. N. (1977b) Estrogen photoaffinity labels II: Reversible binding and covalent attachment of photosensitive hexestrol derivatives to the uterine estrogen receptor. *Biochemistry* **16**, 1970–1976.

Katzenellenbogen, J. A., Heiman, D. F., Carlson, K. E., Payne, D. W. and Lloyd, J. W. (1980) Optimization of the binding selectivity of estrogens. In: Raus, J., Martens, H. and Leclerq, G. (eds.) *Cytotoxic Agents in Hormone Receptor Tumors.* Academic Press, London.

Katzenellenbogen, J. A., Carlson, K. E., Heiman, D. F., Robertson, D. W. and Katzenellenbogen, B. S. (1983) Efficient and highly selective covalent labeling of the estrogen receptors with ^{3}H-tamoxifen aziridine. *J. Biol. Chem,.* **258**, 3487–3495.

King, W. J. and Greene, G. L. (1984) Monoclonal antibodies localize oestrogen receptor in nuclei of target cells. *Nature* **307**, 745–747.

Kumar, V., Greene, S., Staub, A. and Chambon, P. (1986) Localization of the oestradiol-binding and putative DNA-binding domains of the human oestrogen receptor. *EMBO Journal* **5**, 2231–2237.

Laemmli, K. (1970) Cleavage of structural proteins during the assembly of the head of bacteriophage T4. *Nature* **227**, 680–684.

LeClercq, G., De Vleeschouwer, N., Legros, N. and Heuson, J. C. (1980) *In vitro* screening of cytotoxic estrogens of potential therapeutics activity. In: (Raus, J., Martens, H. and LeClercq, G. (eds.) *Cytotoxic Estrogens in Hormone Receptor Tumors.* Academic Press, London, pp. 165–181.

Martin, P. M. and Sheridan, P. J. (1982) Towards a new model for the mechanism of action of steroids. *J. Steroid Biochem.* **16**, 2155–2229.

McGuire, W. L. (1979) Steroid receptor sites in cancer therapy. *Adv. Intern. Med.* **24**, 127–140.

Miller, M. A. and Katzenellenbogen, B. S. (1983) Characterization and quantitation of antiestrogen binding sites in estrogen receptor-positive and -negative human breast cancer cell lines. *Cancer Research* **43**, 3094–3101.

Miller, M. A., Lippman, M. E. and Katzenellenbogen, B. S. (1984) Antiestrogen binding in antiestrogen growth resistance estrogen-responsive clonal variants of MCF-7 human breast cancer cells. *Cancer Research* **44**, 5038–5045.

Miller, M. A., Mullick, A., Greene, G. L. and Katzenellenbogen, B. S. (1985) Characterization of the subunit nature of nuclear estrogen receptors by chemical crosslinking and dense amino acid labeling. *Endocrinology* **117**, 515–522.

Monsma, F. J. Jr., Katzenellenbogen, B. S., Miller, M. A., Ziegler, Y. S. and Katzenellenbogen, J. A. (1984) Characterization of the estrogen receptor and its dynamics in MCF-7 human breast cancer cells using a covalently-attaching antiestrogen. *Endocrinology* **115**, 143–153.

Nardulli, A. M. and Katzenellenbogen, B. S. (1986) Dynamics of estrogen receptor turnover in uterine cells *in vitro* and in uteri *in vivo. Endocrinology* **119**, 2038–2046.

Nawata, H., Bronzeit, D. and Lippman, M. (1981) Isolation and characterization of a tamoxifen-resistant cell line derived from MCF-7 human breast cancer cells. *J. Biol, Chem.* **256**, 5016–5021.

Ratajczak, T., Sheppart, P. N., Capon, R. J. and Hahnel, R. (1981) The synthesis and study of some potential affinity labeling reagents for estrogen receptors. *Steroids* **38**, 537–555.

Reiner, G. A. and Katzenellenbogen, B. S. (1983) A potential estradiol-based affinity label for studying estrogen receptors in human breast cancer. *Breast Cancer Res. Treat.* **3**, 304.

Reiner, G. A., Katzenellenbogen, B. S., Bindal, R. D. and Katzenellenbogen, J. A. (1984) Biological activity and receptor binding of a strongly interacting estrogen in human breast cancer cells. *Cancer Research* **44**, 2302–2308.

Robertson, D. W., Wei, L. L., Hayes, J. R., Carlson, K. E., Katzenellenbogen, J. A. and Katzenellenbogen, B. S. (1981) Tamoxifen aziridines: Effective inactivators of the estrogen receptor. *Endocrinology* **109**, 1298–1300.

Salituro, F. G., Elliston, J. F., Carlson, K. E., Katzenellenbogen, B. S. and Katzenellenbogen, J. A. (1987) [^{125}I]Iododesethyl tamoxifen aziridine: Synthesis and covalent labeling of the estrogen receptor with an iodine-labeled affinity label. *Steroids*, submitted.

Sheen, Y. Y. and Katzenellenbogen, B. S. (1987) Antiestrogen stimulation of the production of a 37,000 molecular weight secreted protein and estrogen stimulation of the production of a 32,000 molecular weight secreted protein in MCF-7 human breast cancer cells, *Endocrinology* **120**, 1140–1151.

Simpson, D. M., Elliston, J. F. and Katzenellenbogen, J. A. (1987) Desmethylnafoxidine aziridine: An electrophilic affinity label for the estrogen receptor with high efficiency and selectivity. *J. Steroid. Biochem.*, **28**, 233–245.

Sudo, K., Monsma, F. J. Jr. and Katzenellenbogen, B. S. (1983) Antiestrogen-binding sites distinct from the estrogen receptor: Subcellular localization, ligand specificity, and distribution in tissues of the rat. *Endocrinology* **112**, 425–434.

Sutherland, R. L., Murphy, L. C., Foo, M. S., Green, M. D., Waybourne, A. M. and Krozowski, S. Z. (1980) High affinity antiestrogen binding site distinct from the estrogen receptor. *Nature* **288**, 273–275.

Van den Broek, A. J., Leemhuis, J., de Winter, M. S. and Zeelen, F. J. (1983) Org 4333, a potent, irreversibly binding estrogen agonist. *Weekblad Sci. Edit.* **5**, 182–183.

Wei, L. L. (1984) Potential antiestrogen-based cytotoxic agents and an affinity label for the estrogen receptor. *Ph.D. Thesis.* University of Illinois at Urbana-Champaign.

Wei, L. L., Mangel, W. F. and Katzenellenbogen, B. S. (1985) Biological activities of tamoxifen aziridine, an antiestrogen-based affinity label for the estrogen receptor *in vivo* and *in vitro*. *J. Steroid Biochem.* **232**, 875–881.

Welshons, W. V., Lieberman, M. E. and Gorski, J. (1984) Nuclear localization of unoccupied oestrogen receptors. *Nature* **307**, 747–749.

Young, R. L., Harrison, R. W. and Parl, F. F. (1986) Analysis of estrogen receptor purified by a novel affinity chromatography and denaturation method. *Endocrinology* **118** (Supplement), 212 (Abstract 724).

Zablocki, J. A., Katzenellenbogen, J. A., Carlson, K. E., Norman, M. J. and

Katzenellenbogen, B. S. (1987) Estrogenic affinity labels: Synthesis, irreversible receptor binding and bioactivity of aziridine-substituted hexestrol derivatives. *J. Med. Chem.*, in press.

8

Affinity labeling of glucocorticoid receptors: applications

S. S. Simons, Jr.
Steroid Hormones Section, National Institutes of Health, Bethesda, USA

I. INTRODUCTION

Studies of glucocorticoid receptor action have been hampered by the extremely low levels of receptor molecules in target cells. These studies have been further impeded by the dissociation of steroid ligands, and of anti-receptor antibodies, under many of the conditions that would otherwise be employed to study receptors. Affinity labeling of receptors, to give covalent, non-dissociable receptor–steroid complexes which can be studied and purified under denaturing conditions, offers a solution to the problem of dissociation. Classically, affinity labels have been used to identify the macromolecule(s) to which they specifically bind and to gain information about the individual amino acids present in the ligand binding site (Jakoby and Wilchek, 1977; Chowdhry and Westheimer, 1979). However, affinity labels can be even more useful if the covalent ligand–macromolecule complex retains biological activity.

The purpose of this chapter is two-fold. First, the evidence that affinity labeled glucocorticoid receptors can be biologically active will be reviewed. How these studies have advanced our understanding of the mechanism of action of glucocorticoid receptors will also be discussed. Second, the uses of affinity labels in probing receptor structure will be summarized. This chapter does not cover affinity labeling studies of mutant glucocorticoid receptors; this material is contained in Chapter 9. The present chapter will concentrate on those studies with the electrophilic affinity label dexamethasone 21-mesylate (Dex-Mes) since it is the only affinity label for glucocorticoid receptors, other than cortisol 21-mesylate (CM; Fig. 1), that has been

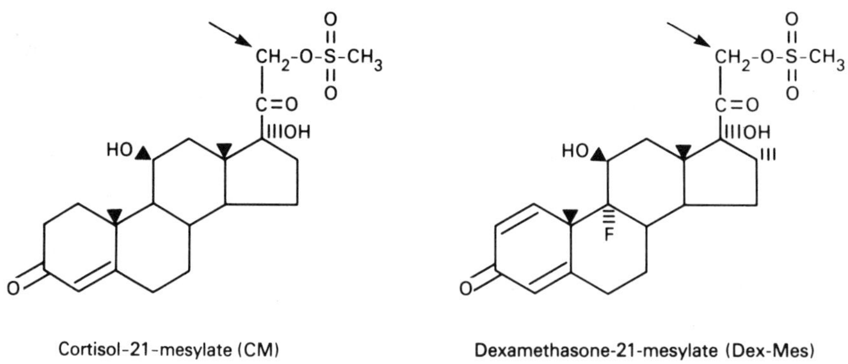

Fig. 1 — Structures of CM and Dex-Mes. The arrows indicate the reactive position in each molecule.

documented to give biologically active covalent complexes. The reader is referred to other reviews (Katzenellenbogen, 1977; Simons and Thompson, 1982; Simons, 1985) for discussions of choice of the chemical class of affinity label and of identification of prospective affinity labeling steroids.

II. BIOLOGICAL ACTIVITY OF AFFINITY LABELED GLUCOCORTICOID RECEPTORS

Two facts initially suggested that it might not be possible to obtain affinity labeled receptors that were biologically active. First, electrophilic affinity labeling functional groups are generally quite reactive and might be expected to be consumed by the numerous nucleophiles present in tissue culture medium, in cells, and in the cell-free extracts containing crude receptors. Photoaffinity labels can be chemically more inert but the yields of photoaffinity labeled macromolecules are usually so low ($\leqslant$10%) that any biological activity could be difficult to detect. Second, virtually all affinity labeled enzymes are inactive; in fact, loss of catalytic activity is one criterion for the affinity labeling of enzymes (Groman *et al.*, 1975). However, since regions of the steroid receptor other than the steroid binding cavity appear to be involved in the expression of biological activity (Carlstedt-Duke *et al.*, 1982), a covalent receptor–steroid complex was not *a priori* incompatible with some, albeit altered, activity.

A. Affinity labels as irreversible antiglucocorticoids

CM does not induce the glucocorticoid regulated enzyme tyrosine amino transferase (TAT: EC 2.6.1.5) in rat hepatoma tissues culture (HTC) cells (Simons *et al.*, 1980b). Dex-Mes gives suboptimal TAT induction (Simons and Thompson, 1981). Both steroids are active as antiglucocorticoids. In each case, the antagonist activity occurs at lower steroid concentrations than would be expected from their affinity for glucocorticoid receptors. This behavior is not due to whole cell toxicity. Instead, it appears to be due to the formulation of covalent steroid–macromolecule adducts since the inhibitory effects of pre-incubating HTC cells with the antiglucocorticoid progesterone, but not CM or Dex-Mes, could be rapidly reversed simply by washing the cells before adding the potent synthetic glucocorticoid dexamethasone (Fig. 2) (Simons et al., 1980b; Simons and Thompson, 1981).

No data have yet appeared on the biological activity of photoaffinity labeled receptors.

B. Irreversible antiglucocorticoid activity is mediated by covalent receptor–Dex-Mes complexes

Several lines of evidence indicate that affinity labeled receptors are responsible for the irreversible anticocorticoid activity of Dex-Mes (and CM) in HTC cells.

(1) The antiglucocorticoid effect of Dex-Mes is prevented by excess dexamethasone (Simons and Miller, 1986).

(2) The potency of the mesylates as antiglucocorticoids (Dex-Mes > CM) follows the same structure–activity order as that of the glucocorticoid activity of the parent steroids (dexamethasone > cortisol) (Simons and Thompson, 1981).

(3) The apparent affinity of Dex-Mes for HTC cell receptors in a standard cell-free competition assay is higher than that of CM (Simons and Thompson, 1981).

(4) The kinetics of Dex-Mes competition of dexamethasone binding to cell-free receptors are those expected for the initial reversible formation of a non-covalent Dex-Mes complex followed by the irreversible formation of a covalent Dex-Mes complex (Lamontagne *et al.*, 1985).

(5) Pre-incubation of cell-free receptors with Dex-Mes, or CM, blocks the

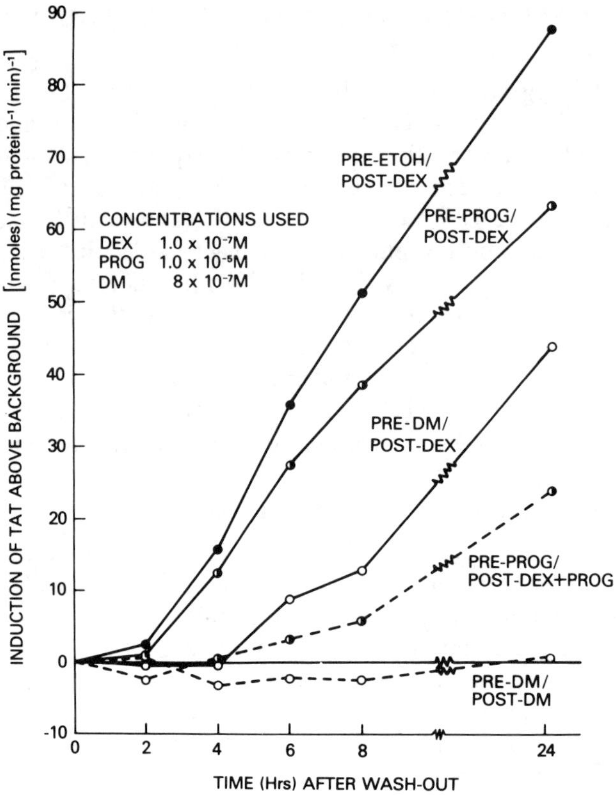

Fig. 2 — TAT induction after EtOH, progesterone, or Dex-mes (DM) pretreatment of HTC cells. Spinner cultures containing 200 ml of cells at 2×10^5/ml were treated with 1.0 ml of EtOH with or without 2.0 mM preogesterone or 160 μM Dex-Mes. After incubation at 37°C for $14\frac{3}{4}$ h, 7×10^7 cells were centrifuged at 600*g* for 10 min at 20°C and resuspended in 200 ml of 'conditioned' medium that also had been 'incubated' at 37°C for $14\frac{3}{4}$ h. After a second centrifugation, each cell pellet was again resuspended in 215 ml of conditioned medium. Duplicate aliquots (3 ml) were removed for zero time protein and TAT determinations, and duplicate 50 ml aliquots were placed in 100 ml Wheaton bottles containing 300 μl of EtOH solutions of (i) 0 or 16.7 μM dexamethasone for the EtOH-preincubated cells, (ii) 16.7 μM dexamethasone with or without 1.67 mM progesterone for the progesterone-pretreated cells, or (iii) 16.7 μM dexamethasone or 133 μM Dex-Mes for the Dex-Mes-pretreated cells. The cells were incubated in a rotating water bath (37°C at 175 rpm) with 3 ml aliquots being removed at various time points for protein and TAT determinations. The specific enzyme activity of TAT above the zero time point was then plotted against the length of incubation after the addition of steroid to washed cells for EtOH-pretreated cells treated after washout with dexamethasone (●—●), for progesterone-pretreated cells treated after washout with dexamethasone (◐—◐) or dexamethasone plus progesterone (◐---◐), and for Dex-Mes-pretreated after washout with dexamethasone (○—○) or Dex-Mes (○--○). At $T=0$, the level of TAT specific activity was 15 nmol/(mg protein min), 18 nmol/(mg protein min), and 27 nmol/(mg protein min) for EtOH, progesterone, and Dex-Mes pretreated cells, respectively. (Data from Simons and Thompson, 1981.)

subsequent exchange binding of [^{3}H]dexamethasone, and Dex-Mes is a more potent blocker than is CM. This inhibition of exchange binding can be prevented by excess cortisol during the pre-incubation with Dex-Mes or CM (Simons *et al.*, 1980b; Simons and Thompson, 1981).

(6) SDS–polyacrylamide gel electrophoresis of [^{3}H]Dex-Mes treated, crude HTC cell receptor preparations revealed only one major, specifically labeled species at 98 000 daltons† (Simons and Thompson, 1981; Simons *et al.*, 1983; Reichman *et al.*, 1984; Simons and Miller, 1984).

(7) This species at 98K is the only [^{3}H]Dex-Mes labeled macromolecule that is also recognized by polyclonal antibodies against the rat glucocorticoid receptor (Fig. 3) (Eisen *et al.*, 1981).

(8) Optimal cell-free affinity labeling conditions (see Chapter 2) afford yields of the [^{3}H]Dex-Mes labeled, 98K receptor that equal 98% ± 11% (SD, $n = 8$) of the available receptor as determined by [^{3}H]dexamethasone binding (Simons and Miller, 1984).

(9) The chromatographic properties of covalent [^{3}H]Dex-Mes labeled and non-covalent [^{3}H]dexamethasone bound, unactivated‡. HTC cell receptors of DEAE–cellulose columns are the same (Simons, 1985).

(10) The ability of [^{3}H]Dex-Mes labeled and [^{3}H]dexamethasone bound receptors to be activated to DNA binding complexes is almost identical (Simons *et al.*, 1983; Simons and Miller, 1984). The relative binding affinities of covalent and non-covalent complexes to biologically active vs. non-specific DNA sequences are virtually the same (Miller *et al.*, 1984).

(11) [^{3}H]Dex-Mes specifically labels just one species at 98K in intact HTC cells (Fig. 4) (Simons and Miller, 1986) and in the related rat HTC cells Fu5-5 (P. A. Miller and S. S. Simons Jr., in preparation).

(12) The amount of antagonist activity seen with Dex-Mes is greater in HTC cells than in HTC cells (Mercier *et al.*, 1983, 1986). At the same time, the amount of covalently labeled 98K receptors that are formed in whole-cell incubations with [^{3}H]Dex-Mes is more in HTC cells than in Fu5-5 cells (Simons and Miller, 1986). Furthermore, Dex-Mes is a nearly pure antagonist in two mouse fibroblast cell lines where almost all the available whole cell receptors are covalently labeled (F. D. Sistare, G. L. Hager, and S. S. Simons Jr., in preparation).

(13) The cloned rate glucocorticoid receptor cDNA restores glucocorticoid regulation of TAT when transfected into receptor-negative, unresponsive HTC cells. Furthermore, these transfected cells synthesize a 98K protein that is specifically labeled by [^{3}H]Dex-Mes and immunoprecipitated by the monoclonal antireceptor antibody BUGR2 (Miesfeld *et al.*, 1986).

Collectively, these data argue that the irreversible antiglucocorticoid activity of Dex-Mes is expressed by the covalent attachment of Dex-Mes in the steroid binding site of the glucocorticoid receptor.

† The reported molecular weights for affinity labeled glucocorticoid receptors vary from ~85K to ~105K. Part of this variation is due to the problems associated with obtaining accurate M_r values from sliced-and-counted gels; fluorographic analysis of the intact gel is inherently more accurate. Another major cause for these differences arises from using values other than 97 400 (Titani *et al.*, 1977) for the M_r of phosphorylase b. Since most of the affinity labeled receptors migrate with, or slightly more slowly than, phosphorylase b, we have used 98K (instead of the various published values) as the M_r of the receptor throughout this article in order to avoid confusion.

‡ The terms 'activate' and 'activation' are used in this paper to describe the currently unknown mechanism by which initially formed receptor–steroid complexes, with little or no affinity for DNA or nuclei, are converted by manipulations such as heat, dilution, increased salt, or increased pH to complexes with relatively high affinity for DNA or nuclei.

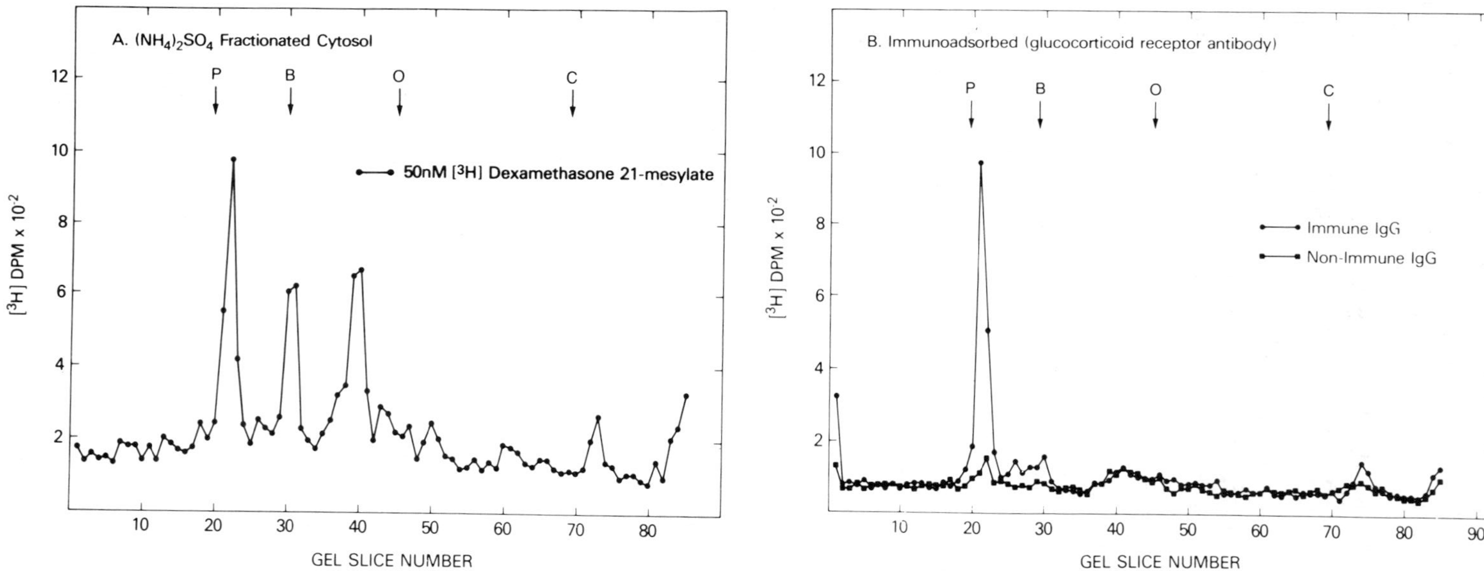

Fig. 3 — Affinity labeling and immunoadsorption of glucocorticoid receptor following $(NH_4)_2SO_4$ fractionation. (A) $(NH_4)_2SO_4$ fractionation of rat liver cytosol, followed by treatment with [^{3}H]Dex-Mes and analyses of [^{3}H]Dex-Mes binding moieties by SDS–polyacrylamide gel electrophoresis. Rat liver cytosol containing 20 mM Na_2MoO_4 wa precipitated by 40% $(NH_4)_2SO_4$, resuspended pH 8.2 TAPS buffer (25 mM TAPS, 1 mM EDTA, 10% (v/v) glycerol) containing 20 mM sodium molybdate, and labeled with 50 mM [^{3}H]Dex-Mes for 2 h at 4°C. An aliquot (100 μl) was treated with 1 ml of 10% trichloroacetic acid and the precipitate analyzed by SDS–polyacrylamide gel electrophoresis (8% acrylamide, 0.03% bis-acrylamide). Gel lane contains protein from 10 μl of original sample. (B) Adsorption of [^{3}H]Dex-Mes labeled receptors by immune and control (non-immune) IgG–protein A–Sepharose. The [^{3}H]Dex-Mes samples (2.5 ml) of (A) were treated with immobilized IgG–protein A–Sepharose preparations for 2 h at 4°C. The absorbed proteins were eluted by 0.1 M acetic acid and analyzed by SDS–polyacrylamide gel electrophoresis. Each gel lane contains 10% of the total protein eluted from the protein A–Sepharose. ●—●, immune IgG; ■—■; non-immune IgG. Protein standards in this figure: phosphorylase b, 94 000; albumin, 67 000; ovalbumin, 43 000; carbonic anhydrase, 30 000 (from Eisen *et al.*, 1981). (Note: the correct molecular weight of phosphorylase b is 97 400 (Titani *et al.*, 1977).)

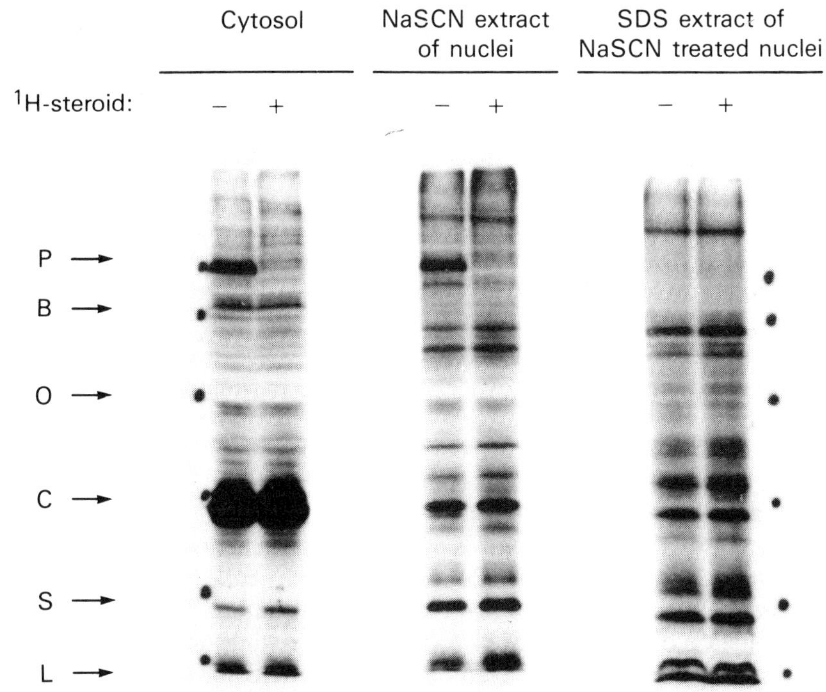

Fig. 4 — Subcellular distribution of receptor labeled in whole cells by [^{3}H]Dex-Mes. HTC cells (5.5×10^6) in 1 ml of medium containing 10% serum plus 1.4% EtOH and 5.0×10^{-8} M [^{3}H]dexamethasone $\pm3.0\times10^{-5}$ M [^{1}H]dexamethasone were incubated at 37°C for 30 min. The various subcellular fractions were prepared, chromatographed on SDS–polyacrylamide (10.8%) slab gels, and analyzed by fluorography as described in Chapter 2, Section VII.A. The film was deliberately overexposed in an effort to reveal low levels of labeled receptor in the SDS extract. Dots at edge of fluorograph were made by overlaying a fluorescent marker (UltEmit from New England Nuclear) on the band of the molecular weight markers in the dried gels. P, phosphorylase, b, 97 000; B, bovine serum albumin, 66 300; 0, ovalbumin, 45 000; C, carbonic anhydrase, 30 600; S, soybean trypsin inhibitor, 21 500; L, lysozyme, 14 400. (From Simons and Miller, 1986.)

III. MECHANISM OF ANTIGLUCOCORTICOID ACTION OF COVALENT RECEPTOR DEX-MES COMPLEXES

A. Variation in amount of antagonist activity of Dex-Mes

As was mentioned in item (12) of Section I.B, the amount of antagonist activity (i.e. induction of TAT) seen with Dex-Mes is different in the rat hepatoma cell lines Fu5-5 and HTC. A similar variability of Dex-Mes and CM antagonist (and agonist) activity has been documented in several other systems. The electrophilic label CM usually is a full antagonist. (Simons *et al.*, 1980b; Bohn *et al.*, 1984; Lanoux *et al.*, 1985; Lewis *et al.*, 1986; Mathis *et al.*, 1986; Richardson *et al.*, 1986) but is a full agonist for the synthesis of milk protein in cultured mammary gland cells (Terada and Oka, 1983). Dex-Mes almost always displays some agonist activity (Simons and Thompson, 1981; Mercier *et al.*, 1983; Bohn *et al.*, 1984). However, Dex-Mes is reported to be a full agonist in inhibiting the growth of hamster melanoma cells (Weisz *et al.*, 1983) but is a full antagonist of dexamethasone induction of the hybrid MMTV LTR/v-ras gene in infected mouse fibroblasts (F.D. Sistare, G. L. Hager, and S. S. Simons, Jr., in preparation).

This difference in amount of agonist activity has also been observed in Fu5-5 and HTC cells with all the reversible antiglucocorticoids (except for cyproterone acetate (Mercier *et al.*, 1986)) that we have examined (Mercier *et al.*, 1986; Simons and Yen, 1987). Thus this property is not unique for affinity labeling antiglucocorticoids. In fact, it appears to be a general and poorly understood phenomenon that the amount of agonist activity of all antisteroids varies between test systems (see references in Simons and Yen, 1987).

The accumulating evidence from cultured cells suggests that this variation in agonist activity of antisteroids may depend on the particular gene that is being regulated. Thus, in MCF-7 cells, the antiestrogen tamoxifen inhibits estrogen induction of p-S2 mRNA (Westley *et al.*, 1984) but acts as an estrogen in the synthesis of progesterone receptors (Horwitz *et al.*, 1978; Westley *et al.*, 1984). Likewise, the antiprogestin RU 486 in $T47D_{co}$ cells gives an agonist-like response in preventing cell growth but antagonizes progesterone induction of insulin receptors (Horwitz, 1985). HTC and Fu5-5 cells constitute a particularly attractive system in which to examine this phenomenon. Not only does the amount of agonist activity for Dex-Mes induction of TAT differ between the two cell lines but also the amount of Dex-Mes agonist within the *same* cell line varies depending on the gene product that is monitored (Fig. 5) (Mercier *et al.*, 1986; S. S. Simons Jr. *et al.*, submitted). This result eliminates preferential metabolism of the antisteroid and other cell-to-cell differences as possible causes for unequal agonist activity.

The above results in HTC and Fu5-5 cells were obtained by monitoring gene induction at the level of enzyme activity. However, identical results have been obtained when TAT mRNA levels were examined, either in HTC vs. Fu5-5 cells (S. S. Simons, Jr. *et al.*, submitted) or in comparison with mouse mammary tumor virus (MMTV) RNA levels in MMTV infected Fu5-5 cells (G. Wasner *et al.*, in preparation). Thus, at least in HTC and Fu5-5 cells, the variation in Dex-Mes agonist activity is not seen for all glucocorticoid inducible genes and may be a property of the TAT gene. Support for this hypothesis comes from the observation that the dose–response curves for TAT enzyme induction by 8-CPT-cAMP (a metabolism-resistant derivative of c-AMP) in HTC and Fu5-5 cells are different (Wasner and Simons, 1987).

Despite the above complications, many properties of the covalent Dex-Mes labeled vs. the non-covalent dexamethasone bound glucocorticoid receptors have been defined in an effort to explain the antiglucocorticoid activity of the covalent Dex-Mes labeled complexes. At this point, it should be noted that attempts to prepare an irreversible, affinity labeling, glucocorticoid agonist have been unsuccessful (Lamontagne *et al.*, 1985; Simons and Miller, 1986). Nevertheless, such steroids appear to be theoretically possible; Katzenellenbogen *et al.* (1987) have recently described an affinity labeling estrogen which is an agonist.

B. Mechanism of irreversible antiglucocorticoid action of Dex-Mes

Whole-cell nuclear binding experiments provided an early clue concerning the mechanism of action of Dex-Mes as an irreversible antiglucocorticoid. The distribu-

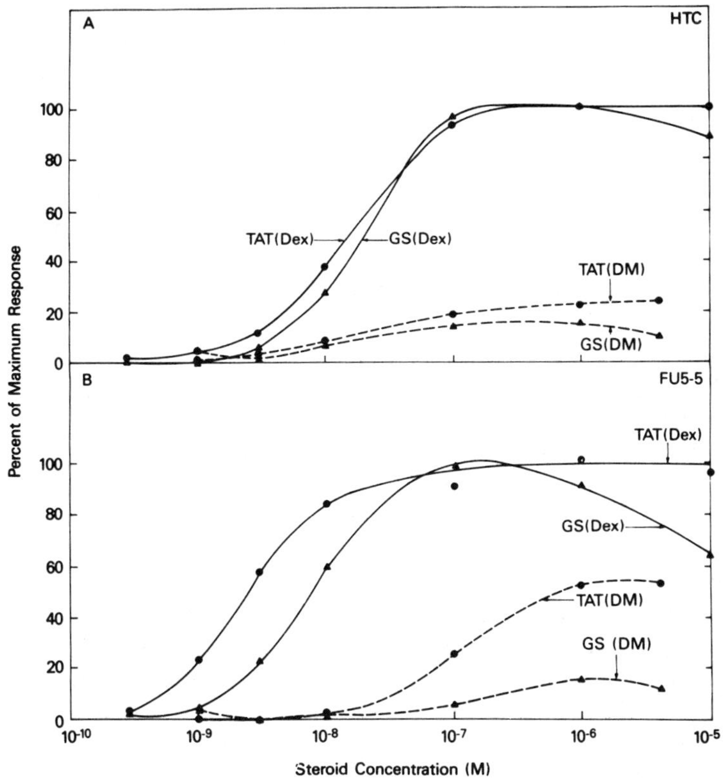

Fig. 5 — Induction of TAT and GS in HTC (A) and Fu5-5 (B) cells by dexamethasone and Dex-Mes (DM). Quadruplicate monolayer cultures of (A) HTC cells or (B) Fu5-5 cells were incubated at 37°C with various concentrations of dexamethasone (——) or Dex-Mes (- - -). After 16–20 h, the cells were harvested and the specific activities of TAT (●) and GS (▲) in cell lysates were determined, expressed as a percentage of full induction, and plotted as a function of the steroid concentration. (From Mercier *et al.*, 1986.)

tion of receptor–steroid complexes in HTC cells at 37°C, as expressed by the ratio of cytoplasmic to nuclear complexes, was quite different for dexamethasone and Dex-Mes, i.e. 0.93±0.16 ($n = 8$) for non-covalent [^{3}H]dexamethasone-bound receptors and 3.21±0.78 ($n = 6$) for covalent [^{3}H]Dex-Mes-labeled complexes (Simons *et al.*, 1983; Simons and Miller, 1986). In parallel experiments, the efficiency of nuclear translocation (the reciprocal of the ratio of cytoplasmic:nuclear complexes) for the non-covalent [^{3}H]dexamethasone complexes was 3.42±0.69 ($n = 6$) times greater than that of the covalent [^{3}H]Dex-Mes complexes (Simons and Miller, 1986). This decreased nuclear binding efficiency of covalent Dex-Mes labeled receptors has been observed in Fu5-5 cells (P. A. Miller and S. S. Simons Jr., in preparation) and mouse fibroblasts (F. D. Sistare, G. L. Hager, and S. S. Simons Jr., in preparation) and is independent of the amount of covalent labeling of whole cell receptors. There also appears to be a difference in the percentage of total cellular complexes that resist extraction from nuclei with 0.3 M NaSCN (5–10% of the non-covalent dexamethasone complexes vs. ~0% of the covalent Dex-Mes complexes; no Dex-Mes labeled complexes were found in the NaSCN extracted nuclei (see Fig. 4; Simons and Miller, 1986).

A similar decreased nuclear binding has been observed both for the non-covalent complexes of the antiglucocorticoid RU 486 in mouse thymus (W7TB) (Bourgeois *et al.*, 1984) and rat thymus cells (Moguilewsky and Philibert, 1984) and for the non-covalent complexes of the antiandrogen flutamide in mouse kidneys (Kontula *et al.*, 1985).

The existence of nuclear bound Dex-Mes labeled receptors indicated that Dex-Mes antiglucocorticoid activity was not due simply to the total inactivation of the receptor after covalent labeling. In an effort to explain this decreased nuclear binding of Dex-Mes labeled complexes and to determine its role in the whole cell irreversible antiglucocorticoid actions of Dex-Mes, some of the current models of antisteroid action have been examined. An early model proposed that the initially bound steroid was not stable and rapidly dissociated from the unactivated receptor–steroid complex (Raynaud *et al.*, 1980). However, the observed irreversible antiglucocorticoid activity of Dex-Mes in HTC cells, together with the labeling of about 57% of the whole-cell receptors (Fig. 4) and the arguments of Section I.B, indicate that it is the covalent complex that is responsible for Dex-Mes antagonist activity. Another popular hypothesis, that Dex-Mes covalently labels a form of the receptor that cannot be activated (Rousseau, 1984), or that the covalent complex itself cannot be activated, can be eliminated since Dex-Mes labeled receptors can be activated to a DNA-binding species with an efficiency that is essentially identical to that seen with non-covalent dexamethasone-bound receptors (Simons *et al.*, 1983). Under certain cell-free conditions (dilution of receptors at 0°C in pH 8.8 buffer containing 50 mM β-mercaptoethanol), the rate of activation of covalent complexes to a DNA-binding species was 2.6-fold slower than that of the non-covalent dexamethasone-bound complexes (Simons and Miller, 1984). However, such a small difference in the rate of activation at 0°C is unlikely to be of significance at 37°C in whole cells, especially since the final amounts of activated complexes at 0°C were similar. Less cell-free activation of Dex-Mes labeled vs. triamcinolone acetonide bound IM-9 receptors after 30 min at 23°C has been reported (Smith *et al.*, 1986).

An alteration in the affinity of the covalent complexes for DNA could result in antiglucocorticoid activity, similar to what is seen for the non-covalent [^{3}H]dexamethasone complexes of the glucocorticoid resistant nt$^-$ and nt^i receptor mutants of mouse lymphoma S49 cells (Yamamoto *et al.*, 1974). However, the amount of NaCl required to disrupt the binding to calf thymus DNA of non-covalent and covalent HTC cell complexes differed only by about 15 mM NaCl (Simons and Miller, 1984), which was much less than that observed for wild type vs. nt$^-$ S49 receptors (Δ = 50–60 mM NaCl (Yamamoto *et al.*, 1974)).

While the binding affinities of covalent and non-covalent complexes to 'non-specific' calf thymus DNA were very similar (Simons and Miller, 1984), the affinity differences for biologically active DNA sequences (such as the 300 bp immediately upstream of the CAP site in the 5′ long terminal repeat (LTR) of MMTV that are required for glucocorticoid regulation of MMTV transcription (Scheidereit *et al.*,

1983; Pfahl *et al.*, 1983; Payvar *et al.*, 1983; Scheidereit and Beato, 1984) could have been much greater. However, calf thymus DNA, intact MMTV LTR DNA, and six approximately equal-sized restriction fragments spanning the entire MMTV LTR were all found to bind covalent [^{3}H]Dex-Mes-Labeled HTC cell receptors with approximately the same relative affinity and order of preference as non-covalent dexamethasone-bound complexes (Miller *et al.*, 1984). Thus the differences in the whole-cell biological activity between Dex-Mes and dexamethasone and the decreased nuclear binding of Dex-Mes labeled receptors cannot be ascribed to (1) a rapid dissociation of steroid from receptors, (2) a decreased ability or efficiency in the cell-free activation of covalent complexes, (3) a significantly altered affinity of the covalent Dex-Mes labeled complexes for either non-specific or biologically relevant DNA sequences, or (4) a redistribution of the covalent complexes to a nuclear fraction resisting NaSCN extraction. Therefore, if the decreased nuclear binding of covalent complexes does reflect some alteration of the receptor that also gives rise to its irreversible antiglucocorticoid activity, this lesion must lie at some other point at or before the nuclear binding step.

C. Future directions of research

The simplest explanation for the decreased nuclear binding of Dex-Mes labeled receptors is that covalent attachment of Dex-Mes in the binding site of the receptor attenuates the ability of the covalent complexes to complete whatever steps are required for normal nuclear binding. Spelsberg *et al.* (1984) and Ross and Ruh (1984) have documented the importance of chromosomal proteins in the nuclear binding of progesterone and estrogen receptor–steroid complexes. Different nuclear proteins appear to be involved in the binding of estrogen vs. antiestrogen complexes (Singh *et al.*, 1986). Such studies in HTC and Fu5-5 cells may help to explain why covalent Dex-Mes and non-covalent dexamethasone complexes display nearly identical DNA binding properties but give very different amounts of nuclear binding in whole cells. Similar studies will be required to understand why those covalent Dex-Mes labeled complexes which are present in the nucleus are not active in producing the normal glucocorticoid response. Preliminary data indicate both that nuclear bound, Dex-Mes labeled complexes do not cause the appearance of the same nuclease hypersensitive site as do dexamethasone bound complexes and that treatment of the cells with excess Dex-Mes blocks the induction of the hypersensitive site by dexamethasone (H. Richard-Foy, F. D. Sistare, S. S. Simons Jr. and G. L. Hagar, in preparation). These results could be due to the Dex-Mes labeled complexes' binding either to different nuclear acceptor sites or to the correct sites but in a manner that does not cause the proposed conformational changes leading to nuclease hypersensitivity. Studies involving footprinting, cross-linking, and competition of complex binding and transcription initiation would be invaluable in investigating these issues.

The above questions involve the mechanism of action of the irreversible antiglucocorticoid Dex-Mes. It remains to be seen whether similar or different steps are followed by reversible antiglucocorticoids which are less toxic to whole animals (Chrousos *et al.*, 1982) and would have clinical utility (e.g. RU 486 (Gaillard *et al.*, 1984)).

IV. USE OF AFFINITY LABELS TO DETERMINE THE PHYSICAL PROPERTIES OF GLUCOCORTICOID RECEPTORS

A. Specificity of photo- and electrophilic affinity labels

Determination of the physical properties of photoaffinity labeled receptors may be difficult for several reasons. First, the yield of labeled receptors is usually ≤10% (see Chapter 2; Simons and Thompson, 1982). Second, the photoaffinity labels for glucocorticoid receptors that have been described, i.e. diazo ketones (Govindan and Manz, 1980) and dieneones (Westphal *et al.*, 1981; Nordeen *et al.*, 1981), can give numerous reactive intermediates (Coxan and Halton, 1974; Gronemeyer and Govindan, 1986) and thus numerous products in the reactions with each amino acid in the steroid binding cavity. Third, photoaffinity labels are quite non-specific in their reactions and will label numerous amino acids in a given protein (Brems *et al.*, 1981). Thus it is highly probable that photoaffinity labeled receptors will be chemically heterogeneous with each species being formed in very low yield.

The reaction of electrophilic affinity labels with proteins generally gives high yields of the covalent complex with the labeling occurring at specific functional groups of only 1–3 amino acids (Chin and Warren, 1972; Simons and Thompson, 1982; Simons *et al.*, 1987). The only electrophilic affinity labels of glucocorticoid receptors that have been described to date are CM and Dex-Mes (Simons *et al.*, 1980b; Simons, 1981; Simons and Thompson, 1981; Eisen *et al.*, 1981). Dex-Mes gives nearly quantitative yields (98%±11%) of cell-free labeled receptors (Simons and Miller, 1984) and is very specific in its reactions with nucleophiles. Dex-Mes in acetone solution reacts with ionized thiols at rates that are ≥500-fold greater than with un-ionized thiols or with basic solutions of alcohols, amines, acids, imidazole, phenol, thiocyanate anion, guanadinium group, dA, dG, or dC (Simons *et al.*, 1980a; Eisen *et al.*, 1981). In pH 8.8 aqueous solutions at 0°C, [^{3}H]Dex-Mes labeling of total protein is inhibited by ≥96% by the addition of 50 mM β-mercaptoethanol (Simons *et al.*, 1983). The labeling of bovine serum albumin (BSA) at room temperature in pH 8.5 is selective for the one free–SH group (Cys-34) and this labeling is blocked by the thiol-specific reagent methyl methanethiosulfonate (MMTS) (Simons, 1987). All the labeling of HTC cell cytosol proteins including the glucocorticoid receptor is also blocked by MMTS (Simons, 1987). Thus Dex-Mes selectively reacts with thiols under many conditions.

B. Identification of the amino acids of glucocorticoid receptors that react with Dex-Mes

Under conditions similar to those used with BSA, MMTS is equally effective in preventing [^{3}H]Dex-Mes labeling, and [^{3}H]dexamethasone binding, of cell-free glucocorticoid receptors (Simons, 1987). These results argue that Dex-Mes labeling of the glucocorticoid receptor occurs at the same free –SH groups(s) that are involved in dexamethasone binding. Harrison *et al.*, (1983) have shown that there is at least one other free –SH group in the receptor from AtT-20 cells since Dex-Mes labeled receptors are retained on sulfhydryl columns. Similar results have been obtained by Bodwell *et al.* (1984a,b), who found that one of the free thiols not labeled by Dex-Mes is also involved in DNA binding.

Limit proteolysis of [^{3}H]Dex-Mes labeled receptors with trypsin, chymotrypsin, of *S.taphylococcus aureus* V8 protease each produced a unique labeled fragment at an apparent molecular weight of 1700–1500 daltons (the covalently attached steroid artifactually decreases M_r by about 700 daltons; Simons, 1987). The V8 protease and chymotrypsin limit digest fragments each contained all the radioactivity that was initially present in the undigested 98K receptor. The trypsin limit digest fragment contained about 50% of the initial radioactivity, indicating the presence of an internal trypsin site that is inefficiently cleaved to give fragments that are too small to remain on the SDS–polyacrylamide gel. Thus all the [^{3}H]Dex-Mes labeling of receptor –SH groups appears to occur in one small segment of the receptor (Simons, 1987).

Purification of the V8 protease limit digest fragment was thwarted by its extreme hydrophobicity (Simons *et al.*, 1987). However, a comparison of the sequential Edman degradation data of each limit protease digest (see Chapter 2, Fig. 6) with the

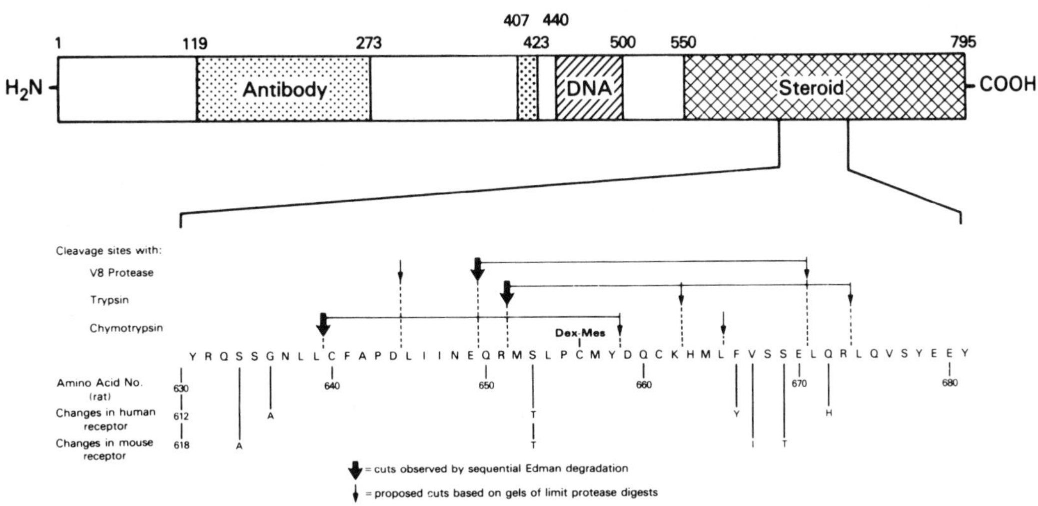

Fig. 6 — Structure of the rat glucocorticoid receptor. The various domains of the receptor (antibody binding sites (□), DNA binding domain (□), and steroid binding domain (□)) in the intact receptor (Miesfeld *et al.*, 1986) are shown on top. Antibodies 250 (Okret *et al.*, 1984) and BUGR2 (Gametchu and Harrison, 1984) recognize amino acids contained between 119–273 and 407–423 respectively (K. R. Yamamoto, personal communication). Below is given the single-letter amino acid sequence, and numbering, of the region surrounding Cys–656 of the HTC cell receptor (Miesfeld *et al.*, 1986). The heavy arrows (➡) indicate the cleavage sites of V8 protease, trypsin, and chymotrysin that were identified by sequential Edman degradation. The light arrows (↓) show the probable additional cleavage sites that would account for the observed sizes of the single protease limit digest fragments (indicated by ——) and the other small species seen in the single and double protease limit digests. Those amino acids which are different in the human (Hollenberg *et* al., 1985) and mouse (Danielson *et al.*, 1986) glucocorticoid receptor are identified. (From Simons *et al.*, 1987.)

known sequence of the rat glucocorticoid receptor (Miesfeld *et al.*, 1986) gave Cys-656 as the only possible [^{3}H]Dex-Mes labeled amino acid (Fig. 6) (Simons *et al.*, 1987). Considering the greater than 95% homology between the steroid binding domains of rat (Miesfeld *et al.*, 1986), mouse (Danielsen *et al.*, 1986), and human (Hollenberg *et al.*, 1985) receptors, it can be predicted that Dex-Mes labels Cys-644 of the mouse receptor and Cys-638 of the human receptor (Fig. 6). There is also considerable homology between the steroid binding domains of glucocorticoid, progesterone (Loosfelt *et al.*, 1986; W. T. Schrader *et al.*, personal communication), estrogen (Krust *et al.*, 1986), and mineralocorticoid (R. M. Evans *et al.*, personal communication) receptors. In each receptor, the amino acid that corresponds to Cys-656 of the rat glucocorticoid receptor has been changed to some other amino acid. Consistent with this is the observation that Dex-Mes has not yet been found to affinity label any other class of receptors (W. T. Schrader, personal communication; J. M. Harmon, personal communication).

Given the large size of the steroid binding domain of all the steroid receptors, it is probable that a highly ordered tertiary structure is formed which envelops much or most of the steroid molecule. Similar conclusions had been based on thermodynamic studies of steroid binding to the glucocorticoid receptor (Wolff *et al.*, 1978). Affinity columns containing steroids that are covalently attached to a matrix by the C-21 position of the steroid are very effective in binding glucocorticoid receptors (Lustenberger *et al.*, 1981; Grandics *et al.*, 1982; Govindan and Gronemeyer, 1984). Furthermore, charged groups at the C-21 position of glucocorticoid steroids do not seem to affect dramatically the binding affinity to the receptor if there is a reasonable distance between the C-21 carbon atom and the charged group (Pons *et al.*, 1985, and references therein). Thus it is likely that Cys-656, which reacts with the C-21 position of Dex-Mes, resides near the opening of a cavity which constitutes the steroid binding site (Fig. 7). Now that one amino acid has been definitively associated with the steroid binding function of glucocorticoid receptors, it should be easier to use other techniques (e.g. other affinity labels, point mutagenesis, and X-ray crystallography of crystallized receptors) to identify other amino acids in the steroid binding cavity, and in other functional domains, of the receptor.

C. Identification of glucocorticoid receptors by affinity labeling

Affinity labeling has greatly increased the number of methods which can be employed to identify glucocorticoid receptors and to separate them from non-receptor molecules. Labeling of the receptor with [^{3}H]Dex-Mes has been found not to alter the apparent M_r of the receptor on SDS–polyacrylamide gels (Mendel *et al.*, 1986a). In the absence of proteolysis (see below) and mutated receptors (see Chapter 9), only one high affinity, low capacity binding macromolecule with an apparent molecular weight of about 98K has been affinity labeled (cf. Simons and Thompson, 1981; Nordeen *et al.*, 1981, Eisen *et al.*, 1981; Dellweg *et al.*, 1982; Simons *et al.*, 1983; Foster *et al.*, 1983; Gehring and Holtz, 1983; Thompson *et al.*, 1983; Housley and Pratt, 1983; Gronemeyer, *et al.*, 1983). The data indicating that this 98K species is the receptor have been summarized in Section I.B. Even in the absence of demonstrable proteolysis, small amounts of specifically labeled, lower M_r species are consistently seen. These species are thought to result either from proteolysis of the receptor in the intact cell (or immediately after cell lysis) (Reichman *et al.*, 1984) or from alternative translational start sites in the intact receptor mRNA (Miesfeld *et al.*, 1986).

Proteolysis is a constant complication in identifying the intact, full-sized gluco-

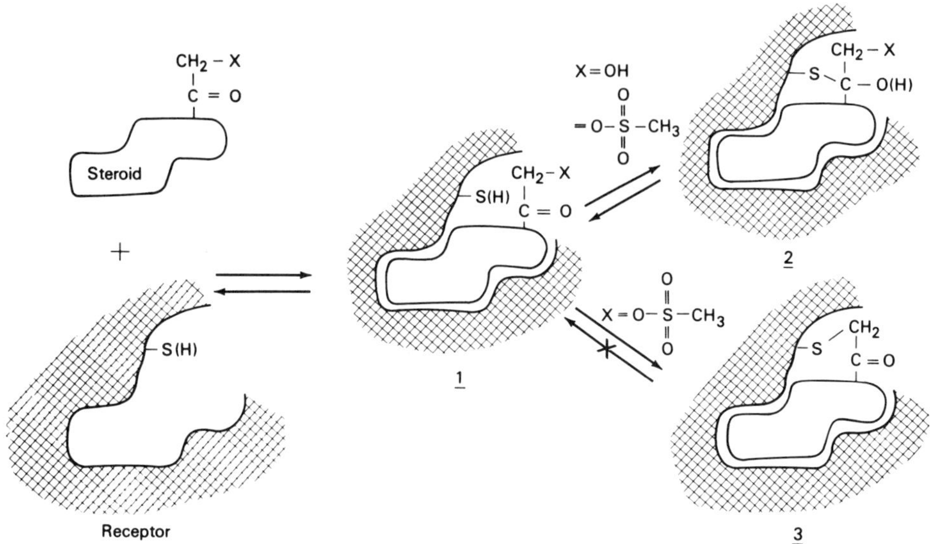

Fig. 7 — Proposed mechanism for the formulation of a covalent glucocorticoid receptor—steroid complex with Dex-Mes. Initially formed non-covalent complex **1** can give the covalent hemi-thiolketal **2** in a rapidly reversible reaction. With Dex-Mes, irreversible nucleophilic attack of the ionized sulphydryl group of Cys-656 at the C-21 of Dex-Mes gives the covalent adduct **3**. (From Simons and Thompson, 1981.)

corticoid receptor. SDS–polyacrylamide gel analysis of affinity labeled receptors appears to be the best method to assess proteolysis. Sucrose gradients and column chromatography studies have been used to assess receptor proteolysis from frozen, but not fresh, tissues (Sherman *et al.*, 1983). However, Mendel *et al.* (1985) have found that [^{3}H]Dex-Mes labeled receptors prepared from fresh rat thymus are fully degraded to a 50K complex during a 2 h incubation at 3°C, even though this 50K complex appears as an intact 8 nm complex when analyzed by gel filtration. Apparently the receptor fragments have sufficient affinity for each other that they do not dissociate during chromatography. Identical results were obtained even when the thymus receptors were prepared and incubated in the presence of 20 nM Na_2MoO_4, which is thought to block receptor proteolysis (Hazato and Murayama, 1981; Sherman *et al.*, 1983).

The proteolysis of receptors from frozen samples (Sherman *et al.*, 1983) may be a result of mechanical homogenization of the thawed sample; Dex-Mes labeling of receptors prepared from frozen (but not homogenized) HTC cells gave almost no proteolytic fragments (see Chapter 2). Furthermore, when similarly prepared crude [^{3}H]Dex-Mes labeled receptors were heated at 25°C in the absence of Na_2MoO_4 for up to 5.5 h, there was no proteolysis (Fig. 8A). Low concentations of Ca^{2+}, however, do cause extensive proteolysis of these labeled receptors (Fig. 8B; A. H. Cavanaugh and S. S. Simons Jr., unpublished data), perhaps through the calcium activated protease described by Sherman *et al.* (1978). While Na_2MoO_4 is not needed in these

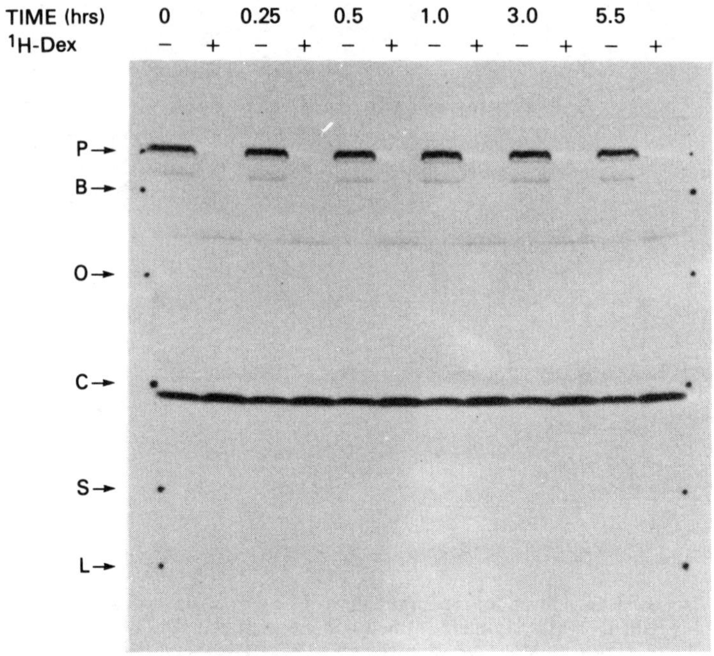

Fig. 8A — Conditions affecting endogenous digestion of Dex-Mes labeled HTC cell receptors. (A) Fluorograph of gel showing the stability of Dex-Mes labeled, 98K receptor in crude cytosol at 25°C without molybdate. HTC cell cytosol in $TAPS_0$ buffer (25 mM TAPS, 1 mM EDTA, 10% (v/v) glycerol) was labeled at pH 8.8, 0°C, with 5×10^{-8} M [^{3}H]Dex-Mes without (−) or with (+) 4×10^{-6} M [^{1}H]dexamethasone for 2½ h. All samples were then adjusted to contain 50 mM β-mercaptoethanol (to consume unreacted [^{3}H]Dex-Mes) and 4×10^{-6} M [^{1}H]dexamethasone and incubated to 0°C for 45 min before raising the temperature to 25°C. Aliquots were removed at the indicated times and chromatographed on 10.8% SDS–polyacrylamide gels.

Ca^{2+}-free receptor preparations to prevent proteolysis, it is of interest that the presence of molybdate causes the non-specific covalent labeling of a species at about 37K (Fig. 8C). The identity of this new labeled molecule is not known.

Affinity labeled receptors are uniquely suited to follow proteolysis because many proteases appear to attack preferentially the amino-terminal portion of the glucocorticoid receptor. Since this region contains all the antigenic domains (Miesfeld *et al.*, 1986) of those antibodies (Okret *et al.*, 1984; Gametchu and Harrison, 1984) which have been characterized (Fig. 6), it is not possible to use antibodies to follow the fate of the carboxyl-terminal portion of the receptor at M_r values lower than about 42K (Eisen *et al.*, 1985). In contrast, the steroid binding domain and the amino acids around the Dex-Mes labeled Cys-656 of the rat receptor are one of the last regions to be digested (Fig. 9). Thus, [^{3}H]Dex-mes labeled receptors were used to help localize the antigenic region for the BUGR2 monoclonal antibody to a 16K fragment that contains the DNA binding, but not the steroid binding, region of the receptor (Eisen *et al.*, 1985). Dex-Mes labeled human (Smith and Harmon, 1987) and mouse

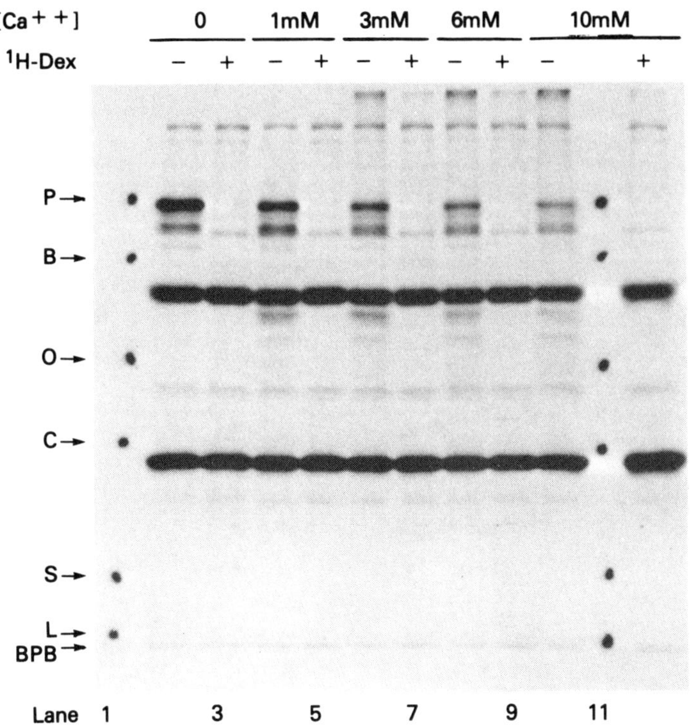

Fig. 8B — Conditions affecting endogenous digestion of Dex-Mes labeled HTC cell receptors. (B) Fluorograph of gel showing the effect of Ca^{2+} on stability of Dex-Mes labeled, 98K receptor. HTC cell cytosol was labeled as in (A) with 24×10^{-8} M [^{3}H]Dex-Mes without (−) with (+) 1.1×10^{-5} M [^{1}H]dexamethasone for 2½ h at 0°C. The covalent complexes were activated by diluting with an equal volume of pH 8.8 $TAPS_0$ buffer, adjusting all samples to contain 50 mM β-mercaptoethanol and 1.1×10^{-5} M [^{1}H]dexamethasone, and then heating at 20°C for 30 min. The activated complexes were cooled to 0°C, treated with enough 0.3 M $CaCl_2$ to give the indicated final concentration of Ca^{2+}, and incubated for 2 h at 0°C before being chromatographed on 10.8% SDS–polyacrylamide gels.

(Distelhorst *et al.*, 1987) receptors have been used to confirm the functional domain results first seen with rat receptor (Reichman *et al.*, 1984; Eisen *et al.*, 1985).

When proteolysis of the receptor is taken into consideration, affinity labeling has proved to be a powerful method of identifying glucocorticoid receptors. Affinity labeling has been used to establish the presence of receptor in tissues with unusual properties (Foster *et al.*, 1983; Gadson *et al.*, Schlechte *et al.*, 1985; Distelhorst *et al.*, 1985) and to argue that a smaller 72K protein that co-purifies with the receptor during affinity chromatography is not a proteolytic fragment of the 96K receptor (Wrange *et al.*, 1984). Epstein–Barr virus transformation of human lymphocytes caused a 5-fold increase in glucocorticoid receptor number but labeling with Dex-Mes showed that there had been no change in the M_r of the receptor (Tomita *et al.*, 1985). Co-migration of [^{3}H]Dex-Mes labeled receptor and a^{32}P-labeled protein on SDS–polyacrylamide gels has been taken as evidence that the receptor can be phosphorylated in whole and broken cell systems (Housley and Pratt, 1983; Singh and Moudgil, 1985; Sanchez and Pratt, 1986).

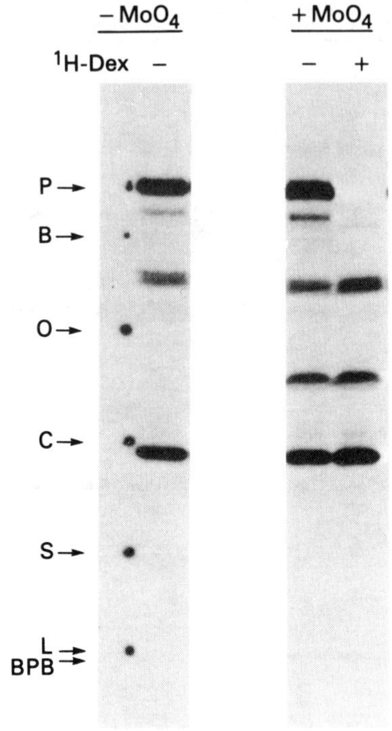

Fig. 8C — Conditions affecting endogenous digestion of Dex-Mes labeled HTC cell receptors. (C) Fluorograph of gel showing the effect of molybdate on [^{3}H]Dex-Mes labeling of HTC cell cytosol. Crude HTC cell receptors were labeled as in (A) with 11×10^{-8} M [^{3}H]Dex-Mes without (–) or with (+) 8×10^{-6} M [^{1}H]dexamethasone in the absence and presence of 20 mM Na_2MoO_4. The samples were then chromatographed on 10.8% SDS–polyacrylamide gels. The molecular weight markers (P, B, O, C, S, L) are those of Fig. 4 and were similarly located by overlaying with UltEmit (New England Nuclear). BPB indicates the bromphenol blue dye front.

Affinity labeling has greatly facilitated the purification of receptors by providing a means of following the receptor at various stages (Westphal *et al.*, 1981; Payvar *et al.*, 1983; Govindan and Gronmeyer, 1984; Singh and Moudgil, 1984; Wrange *et al.*, 1984; Housley *et al.*, 1985). Affinity labeling has also been used to determine whether the purified receptor is intact or has been degraded (Wetsphal *et al.*, 1981; Kovacic-Milivojevic *et al.*, 1985).

The final stages of cloning of the glucocorticoid receptor have benefited from the use of affinity labeling. Dex-Mes labeled, and dexamethasone photoaffinity labeled, receptors were found to be immunoprecipitated by those antibodies that were then used to select c-DNA clones producing immunoreactive material, thus confirming the specificity of the antibodies for the glucocorticoid receptor and strengthening the conclusion that the immunoselected clones coded for antigenic regions of the receptor (Govindan *et al.*, 1985). Even more conclusive was the elegant demonstration that [^{3}H]Dex-Mes labeling of cytosol extracts of receptor negative HTC cells that had been stably transfected with a suspected receptor cDNA clone affected a specifically labeled, 98K species that could be immunoprecipitated by BUGR2 monoclonal antireceptor antibody (Miesfeld *et al.*, 1986).

D. Examination of receptor heterogeneity with affinity labeling

Receptor heterogeneity has always been a formal possibility to explain the different actions of glucocorticoid steroids in different tissues (Webb *et al.*, 1985). When this question was addressed using affinity labeling, only one receptor molecule was observed and it was apparently of the same size for rat, mouse, and human tissues, as determined by SDS–polyacrylamide gel electrophoresis (Nordeen *et al.*, 1981; Foster *et al.*, 1983; Reichman *et al.*, 1984; Govindan *et al.*, 1985; Hoschutzky and Pongs, 1985; Northrop *et al.*, 1985). However, this method is only as accurate as the resolution of the gels. When the receptors from these species were recently cloned, it was evident that there were minor differences in amino acid composition (cf. Fig. 6) and size: rat, 795 amino acids (Miesfeld *et al.*, 1986); mouse, 783 amino acids (Danielsen *et al.*, 1986); human, 777 amino acids (Hollenberg *et al.*, 1985). One common feature for all of these glucocorticoid receptors (and also for the rabbit progesterone receptor (Logeat *et al.*, 1985; Loosfelt *et al.*, 1986)) is that the observed molecular weight of affinity labeled receptor is significantly larger than the calculated value (e.g. [^{3}H]Dex-Mes labeled rat receptor, 98 600 ± 1500 daltons (SD, $n = 28$; Reichman *et al.*, 1984); calculated $M_r = 87\,512$ (Miesfeld *et al.*, 1986)). This difference is not due to the covalently bound Dex-Mes (Mendel *et al.*, 1986a); it could be an artifact of the SDS–polyacrylamide gel system (Reichman *et al.*, 1984; Simons, 1985; Mendel *et al.*, 1986a). Alternatively, the M_r discrepancy could be indicative of post-translational modification of the receptor. There are four potential *N*-glycosylation sites in the rat glucocorticoid receptor (Miesfeld *et al.*, 1986) and lectins specific for mannose and fucose have been reported to decrease the steroid binding affinity to crude rat liver receptor preparations (Blanchardie *et al.*, 1986).

Protease digestion of affinity labeled receptors has been quite successful in further examining receptor homology. It has been used to demonstrate the difference between progesterone and glucocorticoid receptors (Gronemeyer *et al.*, 1983) and to show that a smaller, specifically labeled ~83K species is closely related to the full sized 98K receptor (Wrange *et al.*, 1984). No differences were found between rat and HTC cell receptors (Reichman *et al.*, 1984), but several differences were observed in the immunogenic and steroid binding domains of rat and human receptors (Hotschutzky and Pongs, 1985) which were subsequently confirmed when the receptors were cloned (Miesfeld *et al.*, 1986; Danielsen *et al.*, 1986).

There is considerable size heterogeneity among mutant receptors. These receptors are discussed in Chapter 9. Functional heterogeneity of receptors, in the form of unactivated and activated receptor–steroid complexes, is well documented. When affinity labeled receptors are analysed on SDS–polyacrylamide gels, there is no difference in the M_r of unactivated and activated complexes (Simons *et al.*, 1983; Reichman *et al.*, 1984; Distelhorst and Benutto, 1985). Thus, considering the above-noted limitations of gel analysis, one can conclude that activation does not cause any major changes in the size of the denatured unactivated 98K receptor protein.

Currently there are three general models of receptor–steroid complex activation, each of which could result in different proteolytic digestion patterns for activated and

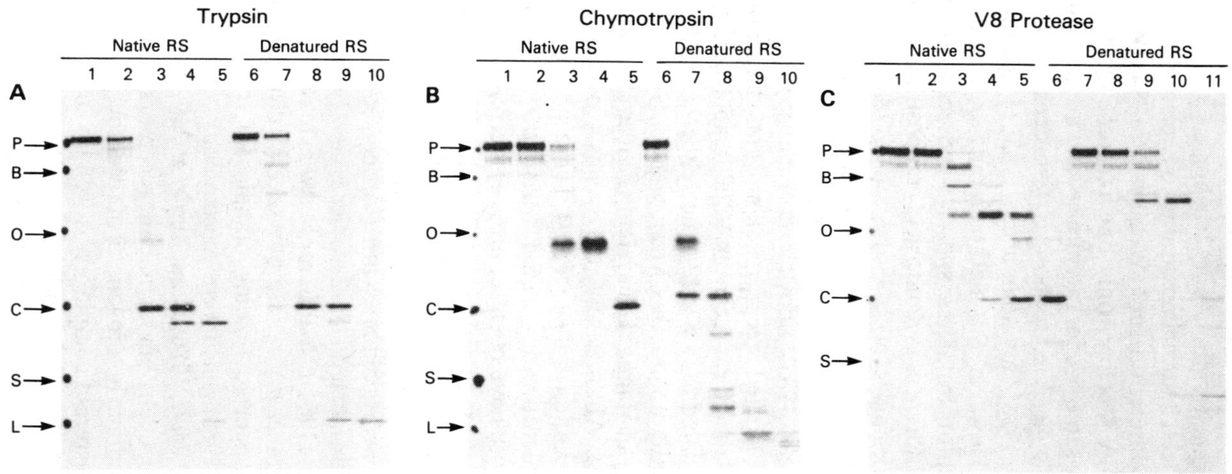

Fig. 9 — Fluorographs of digestion of activated, [^{3}H]Dex-Mes labeled, HTC cell receptor by (A) trypsin, (B) chymotrypsin, and (C) *S. aureus* V8 protease. Activated, affinity labeled receptors that were partially purified by DNA–cellulose chromatography (Simons *et al.*, 1983) were treated in the native (lanes 1–5 in (A) and (B), lanes 1–6 in (C) or denatured state (30 min at 22°C in SDS–dithiothreitol) (lanes 6–10 in (A) and (B), lanes 7–11 in (C)) with protease for 20 min at 10°C. All samples were then quick frozen in dry ice and thawed only long enough to permit mixing of an equal volume of 2× SDS sample buffer. These samples were heated at 100°C for 5 min before being analyzed on SDS–polyacrylamide gels. The gels were subsequently stained with Coomassie Blue and fluorographed. The concentrations of trypsin and chymotrypsin in (A) and (B) used to digest the native, activated complexes in lanes 1–5 were 0, 0.063, 0.5, 4, and 100 μg/ml; the concentrations used to digest the denatured, activated complexes in lanes 6–10 were 0, 0.5, 4, 20, and 100 μl/ml. In (C), the concentrations of V8 protease used in lanes 2–6, and lanes 7–11, were 0 μg/ml, 1.6 μg/ml, 0.8 μg/ml, 40 μg/ml, and 200 μg/ml respectively. The molecular weight markers (P, B, O, C, S, L) are those of Fig. 4 and were similarly located by overlaying with UltEmit (New England Nuclear). (From Reichman *et al.*, 1984.)

unactivated receptors; oligomer dissociation (Raaka and Samuels, 1983; Vedeckis, 1983; Distelhorst and Benutto, 1985), conformational change (Godowski *et al.*, 1987), or covalent modification of one or more amino acids (Sando *et al.*, 1979; Leach *et al.*, 1979; Schmidt and Litwack, 1982) (Fig. 10). Gehring and coworkers

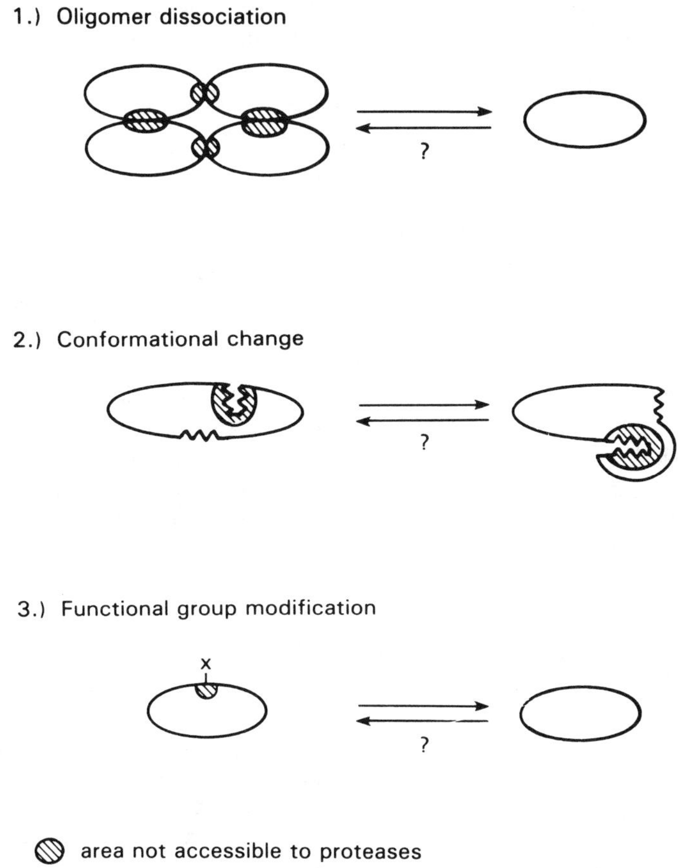

Fig. 10 — Models of activation of glucocorticoid receptor–steroid complex. For the model of oligomer dissociation, it is not necessary that all the subunits be the steroid receptor. (From Simons, 1985.)

(Gehring and Holtz, 1983; Gehring and Arndt, 1985) have used photoaffinity labeling to show that the oligomeric form of the unactivated S49 cell receptor contains only one steroid binding molecule. Thus, under conditions where chymotrypsin digestion of the 9–10S unactivated complex caused all of the denatured 98K receptors (on SDS–polyacrylamide gels) to be reduced by ~56K to a 42K species, the native oligomeric 325K receptors (on Sephacryl S-300 columns) were reduced by only 40K to a 285K species. However, detailed analysis of the proteolytic digestion patterns of native or denatured, activated and unactivated, [^{3}H]Dex-Mes labeled HTC cell receptors on SDS–polyacrylamide gel revealed no differences down to M_r= 14 000 (Reichman *et al.*, 1984). These protease data do not support any of the models of Fig. 10. It is possible, however, that the first few cuts cause both the activated and unactivated forms of the native receptor to assume a common structure to give identical digestion patterns. In this respect, it would be interesting to see whether chymotrypsin digestion of unactivated complexes gave a 42K fragment that had DNA binding properties that were similar to those of the 42K fragment from activated complexes.

Further digestion of the denatured, [^{3}H]Dex-Mes labeled, activated and unactivated complexes gave identical patterns in the region of M_r = 14 000–1400 (Simons, 1987). This region is intimately involved in activation since deletion of the carboxy-terminal ~200 amino acids gives mutant receptors that are constitutively activated, even in the absence of added steroid (Godowski *et al.*, 1987). Therefore, the protease digestion results suggest that activation does not involve any covalent modifications in the area of the receptor close to the affinity labeled Cys-656 (Simons, 1987). Similarly, Smith and Harmon (1987) found no difference in the p*I* of the trypsin 29K and 27K meroreceptor fragments of covalently labeled, activated and unactivated complexes and concluded that activation did not produce any charge modifications in this larger steroid binding region of the receptor.

Affinity labeled receptors from a given source may be homogeneous in size but they are heterogeneous in charge. Unactivated, [^{3}H]Dex-Mes labeled HeLa cell (human) receptors migrated as 4–8 species on 2D gels with identical molecular weights but with p*I* values for the native receptors that ranged from 6.5 to 7.2 (Cidlowski and Richon, 1984). These receptors were obtained from whole cells that were treated with [^{3}H]Dex-Mes but most of the covalent labeling undoubtedly occurred after cell lysis since the yield of labeled receptors was increased simply by raising the pH of the lysis buffer. In contrast, unactivated IM-9 cell (human) receptors that were labeled by [^{3}H]Dex-Mes under cell-free conditions at 0°C and then immunoprecipitated were resolved into two forms (p*I* = 5.7 and 6.0–6.5) of identical M_r on 2D gels but only when using non-equilibrium pH gel electrophoresis (NEPHGE) of denatured receptors (in 9.5 M urea) in the first dimension (Smith and Harmon, 1985; Smith *et al.*, 1986). If the initially labeled receptors were not immunoprecipitated before gel analysis, a third form at p*I* = 5.2 was also observed. No change in the p*I* of any of the species was observed after being exposed to activating conditions but only the most basic form was found to bind to DNA (Smith *et al.*, 1986). These results were interpreted as being inconsistent with activation involving a covalent modification of the receptor; however, small (M_r ≤1000 daltons) charge-neutral modifications or intramolecular rearrangements would not be detected by those techniques. Initial studies with molybdate suggested that activation involves dephosphorylation of the unactivated receptor (Sando *et al.*, 1979; Leach *et al.*, 1979), which is consistent with the most basic form of the Dex-Mes labeled receptor's being the DNA binding form. However, Smith *et al.* (1986) argue against this since the differences in p*I* values of the receptor–[^{3}H]Dex-Mes isoforms are larger than expected to be caused by a single phosphate group and since treatment of the labeled receptors with calf intestinal alkaline phosphatase did not alter the observed p*I* values. Protease digestion studies have localized these p*I* differences to the 27K meroreceptor fragment (Smith and Harmon, 1987). Pratt and coworkers have seen a similar charge heterogeneity in native mouse L-cell receptors in the range of p*I* = 5.3–6.2, which is thought to be due to phosphorylation of one or more serines (Housley and Pratt, 1983). Cell-free phosphorylation was seen for steroid-free or steroid-bound receptors in the unactivated or activated state (Sanchez and Pratt, 1986). Thus the functional significance of receptor phosphorylation and charge heterogeneity is not yet understood.

The p*I* values for native, Dex-Mes labeled receptors have recently been found to be much more acidic (i.e. 4.4–5.7) on agarose isofocusing gels. The unactivated complexes appear to have a p*I* of 4.4 (Danze *et al.*, 1987). Agarose gels are thought to circumvent the problems of hydrophobic interactions and protein aggregation that may artifactually yield higher p*I* values on polyacrylamide gels (Danze *et al.*, 1987). It will be interesting to see whether these more acidic p*I* values are also obtained with denatured complexes or are due to other proteins associated with the native receptor.

V. USE OF AFFINITY LABELS TO IDENTIFY SPECIES THAT PHYSICALLY INTERACT WITH THE GLUCOCORTICOID RECEPTOR

Antireceptor antibodies can be used to study those molecules that appear to be associated with steroid-free and steroid-bound receptors. With affinity labeling steroids, it is possible to study only those molecules that may be associated with steroid-bound receptors. Up to now, this distinction has not been of major interest. However, with the report of a form of the wild-type glucocorticoid receptor that does not bind steroid in cells that have been deprived of ATP (Mendel *et al.*, 1986b), this distinction may assume greater importance.

A. Heat shock protein (hsp90)

Recent studies have identified hsp90 as a component of most preparations of unactivated glucocorticoid receptors and steroid receptors in general (Catelli *et al.*, 1985). Considering the similar size of steroid receptors and hsp90, it is therefore not surprising that there have been mistaken identifications of the two species (Catelli *et al.*, 1985; Sullivan *et al.*, 1985; Sargan *et al.*, 1986). hsp90 and the glucocorticoid receptor are sometimes very difficult to separate (Mendel *et al.*, 1986a), especially if there has been any proteolysis of the receptor (Housley and Pratt, 1983; Housley *et al.*, 1985; Sanchez *et al.*, 1986). Just to confuse the issue further, anti-receptor and anti-hsp90 antibodies each remove both glucocorticoid receptor and hsp90 from solutions of unactivated receptors, and both the receptor and hsp90 are phosphorylated (Housley *et al.*, 1985; Sanchez *et al.*, 1985; Mendel *et al.*, 1986a; Sanchez *et al.*, 1986). However, only the receptor is affinity labeled by Dex-Mes (Housley *et al.*, 1985; Mendel *et al.*, 1986a) or photoaffinity labeled by dexamethasone (Gehring and Arndt, 1985).

A succinct study using proteolysis of photoaffinity labeled receptors showed that the 9–10S unactivated glucocorticoid complex contains just one steroid binding subunit (Gehring and Arndt, 1985). The rest of the unactivated complex appears to consist of a dimer of the hsp90 (Koyasu *et al.*, 1986) and perhaps a 59K protein that interacts with hsp90 and has been found associated with several steroid receptors

(Tai *et al.*, 1986). The binding of hsp90 is of low affinity and it dissociates from the receptor in the absence of molybdate (Housley *et al.*, 1985). hsp90 is no longer associated with the activated complexes (Mendel *et al.*, 1986a; Sanchez *et al.*, 1986). However, given the very high intracellular concentrations of hsp90, it has yet to be firmly established whether the association of hsp90 with unactivated receptors is biologically relevant or an artifact of the cell-free preparations. Functional studies of cell-free translation products of the cloned glucocorticoid receptor and further studies of receptor–antisteroid complexes should help to resolve this question.

B. RNA

Glucocorticoid receptor–steroid complexes can be activated by RNase (Chong and Lippman, 1982; Hutchens *et al.*, 1982; Tymoczko and Phillips, 1983) and purified, unactivated receptor preparations contain a ^{32}P-labeled 24K species that is thought to be a small polynucleotide (Grandics *et al.*, 1984). This 24K species has recently been shown to be a protein that is phosphorylated on serine (Kovacic-Milivojevic and Vedeckis, 1986). More direct evidence that cytoplasmic glucocorticoid receptors are associated with RNA is provided by the facts (1) that formaldehyde crosslinked, Dex-Mes labeled receptors display a sedimentation value that is greater than that of non-cross linked receptors and a density similar to that of hnRNA– and mRNA–protein complexes, (2) that RNase treatment of the crosslinked receptors eliminated the new, faster sedimenting species, (3) that the crosslinked receptors co-sediment with the ^{14}C label in cytosol preparations of cells grown in ^{14}C uridine to label the RNA, and (4) that, when UV-light crosslinked, [^{3}H]Dex-Mes labeled receptors from cells grown with ^{14}C uridine were digested with RNase and then analyzed on SDS–polyacrylamide gels, the ^{3}H-labeled 98K receptor band also contained some ^{14}C label (Economidis and Rousseau, 1985).

The biological significance of these observations has been further investigated using highly purified, unactivated, affinity labeled receptors (Kovacic-Milivojevic and Vedeckis, 1986). RNase treatment of [^{3}H]Dex-Mes labeled receptors, either before or after UV irradiation to crosslink bound RNA, produced no change in the molecular weight of the receptor protein on SDS–polyacrylamide gels (no hsp90 was seen in these preparations, however). These, plus other data, led Vedeckis and coworkers to conclude that the unactivated complexes do not contain RNA (see Webb *et al.*, 1986; Tymoczko *et al.*, 1986 for arguments to the contrary) and that the association with RNA occurs after the activation of receptor–glucocorticoid complexes (Kovacic-Milivojevic and Vedeckis, 1986). This interpretation is also consistent with the known ability of activated complexes to bind to RNA (Rousseau *et al.*, 1975; Miller *et al.*, 1984). Recently, activated complexes have been found to bind preferentially to t-RNA (Tymoczko *et al.*, 1986; Ali and Vedeckis, 1987). It will be of interest to see whether this binding of receptor–steroid complexes to RNA is sequence specific and thus could be involved in steroid-mediated stabilization of the mRNAs of regulated genes (Mayo and Palmiter, 1985; Chomczynski *et al.*, 1986).

C. DNA and chromatin

The association of receptor–steroid complexes with DNA and chromatin that occurs after activation is well established. The affinity of unactivated complexes for DNA or chromatin is much lower but has not been quantitated. [^{3}H]Dex-Mes labeled, activated glucocorticoid receptors bind to calf thymus DNA with an affinity that is about half of that for dexamethasone bound receptors (Simons and Miller, 1984). However, the relative affinities of covalent Dex-Mes labeled, and non-covalent dexamethasone bound, receptors for the biologically active glucocorticoid regulatory elements of the LTR of MMTV vs. biologically inactive DNA sequences are the same. Thus affinity labeling has not altered the ability of activated complexes to discriminate among different DNA binding sequences (Miller *et al.*, 1984). Affinity labeled receptors have not yet been reported in other types of DNA binding experiments such as methylation protection or protection of DNA digestion with DNase I or exonuclease III.

D. Future directions of research

There will always be new situations in which the current affinity labels can be used in established protocols, e.g. ascertaining the presence of the receptor in new target cells, identifying possible new steroid-binding forms of protease digestion fragments of the receptor, and assisting in development of new receptor purification procedures. The main new uses of affinity labels in probing receptor structure will come from new affinity labels which may react with amino acids other than Cys-656 of the rat receptor and from identification of the amino acids that are modified by photoaffinity labels. A new affinity label that possessed full glucocorticoid activity would be especially useful. So far, the only steroid receptor affinity label found to be a full agonist is a hexestrol derivative with an aziridine-containing side chain, which is an affinity label for estrogen receptors (Katzenellenbogen *et al.*, 1987).

Recent reports indicate that non-receptor molecules in addition to hsp90 may be associated with receptors in ways that modify or transmit the biological activity of the receptor. These non-receptor molecules can be of cytoplasmic origin or can be found in the nucleus and almost certainly include some chromatin-associated molecules that are intimately involved in steroid hormone regulation of gene transcription (cf. Cordingley *et al.*, 1987). Crosslinking studies with both agonist and antagonist affinity labels would be invaluable in investigating these associated non-receptor molecules. Protein–protein crosslinking studies with the affinity labeled receptors may identify nearby proteins in the chromatin binding sites for receptors. Such studies could be extremely useful in testing the looping model for how regulatory elements influence transcription initiation from remote DNA sequences (Ptashne, 1986; Echols, 1986). DNA–protein crosslinking studies (Elsner *et al.*, 1985; Solomon and Varshavsky, 1985; Hockensmith *et al.*, 1986) could directly identify those nucleotides involved in receptor–steroid complex binding to DNA. Indirect evidence could be gained from DNase footprinting and exonuclease III digestion experiments.

One area that has received very little study so far is the fate of nuclear bound complexes. Recycling and processing models have been proposed, but studies with glucocorticoids have been inhibited by two facts. First, activated complexes do not exchange bind steroid and Mendel *et al.* (1986b) have recently observed that ATP depleted cells contain nuclear receptors that cannot bind steroid. Thus, it is difficult to determine whether a decrease in nuclear [^{3}H]receptor–steroid complexes is due to a simple dissociation of the steroid, to a change in receptor affinity for the steroid, or to a destruction of receptors. Second, proteolytic degradation of complexes is associated with a loss of steroid binding activity and of antibody reactive sites so that the ability to follow degraded receptors by these methods may be rapidly lost. Affinity labeling steroids can help to answer these questions since they do not dissociate and since the steroid binding cavity is among the last portions of the glucocorticoid receptor to be digested by those proteases that have been examined. Whole-cell degradation experiments with affinity labeled estrogen receptors were surprisingly uninformative — no significant amounts of labeled species smaller than the intact 63K labeled receptor were observed (Monsma *et al.*, 1984). Similar results with glucocorticoid receptor would not be expected in view of the protease studies with cell-free labeled receptors (Reichman *et al.*, 1984; Simons, 1987).

A major area of future interest concerns the biological activity of affinity labeled receptors and how it differs from that of non-covalent complexes. Possible directions of research in this field have been discussed in Section III.C.

In summary, affinity labels for glucocorticoid receptors have proven to be invaluable in answering many questions, some of which could not be addressed by any other currently available technique. For this reason, affinity labeling has now become another standard tool for studying steroid hormone action. The future promises to reveal even more uses and benefits of this methodology.

REFERENCES

Ali, M. and Vedeckis, W. V. (1987) The glucocorticoid receptor protein binds to transfer RNA. *Science* **235**, 467–470.

Blanchardie, P., Lustenberger, P., Denis, M., Orsonneau, J. L. and Bernard, S. (1986) Interaction of rat liver glucocorticoid receptor with lectins: Is the glucocorticoid receptor a glycoprotein? *J. Steroid Biochem.* **24**, 263–267.

Bodwell, J. E., Holbrook, N. J. and Munck, A. (1984a) Sulfhydryl-modifying reagents reversibly inhibit binding of glucocorticoid-receptor complexes to DNA–cellulose. *Biochemistry* **23**, 1392–1398.

Bodwell, J. E., Holbrook, N. J. and Munck, A. (1984b) Evidence for distinct sulfhydryl groups associated with steroid- and DNA-binding domains of rat thymus glucocorticoid receptors. *Biochemistry* **23**, 4237–4242.

Bohn, M. C., Bloom, E., Goldstein, M. and Black, I. B. (1984) Glucocorticoid regulation of phenylethanolamine *N*-methyltransferase (PNMT) in organ culture of superior cervical ganglia. *Development Biol.* **105**, 130–136.

Bourgeois, S., Pfahl, M. and Baulieu, E.-E. (1984) DNA binding properties of glucocorticosteroid receptors bound to the steroid antagonist RU-486. *EMBO J.* **3**, 751–755.

Brems, D. N., Bruenger, E. and Rilling, H. C. (1981) Isolation and characterization of a photoaffinity-labeled peptide from the catalytic site of prenyltransferase. *Biochemistry* **20**, 3711–3718.

Carlstedt-Duke, J., Okret, S., Wrange, O. and Gustafsson, J.-A. (1982) Immunochemical analysis of the glucocorticoid receptor: identification of a third domain separate from the steroid-binding and DNA-binding domains. *Proc. Natl. Acad. Sci. USA* **79**, 4260–4264.

Catelli, M. G., Binart, N., Jung-Testas, I., Renoir, J. M., Baulieu, E. E., Feramisco, J. R. and Welch, W. J. (1985) The common 90-kd protein component of non-transformed '8S' steroid receptors is a heat-shock protein. *EMBO J.* **4**, 3131–3135.

Chin, C.-C. and Warren, J. C. (1972) Synthesis of 2α-, 6α-, and 6β-bromoprogesterone and study of the binding site of 20β-hydroxysteroid dehydrogenase. *Biochemistry* **11**, 2720–2726.

Chomczynski, P., Qasba, P., and Topper, Y. J. (1986) Transcriptional and post-transcriptional roles of glucocorticoid in the expression of the rat 25 000 molecular weight casein gene. *Biochem. Biophys. Res. Comm.* **134,** 812–818.

Chong, M. T. and Lippman, M. E. (1982) Effects of RNA and ribonuclease on the binding of estrogen and glucocorticoid receptors from MCF-7 cells to DNA–cellulose. *J. Biol. Chem.* **257**, 2996–3002.

Chowdhry, V. and Westheimer, F. H. (1979) Photoaffinity labeling of biological systems. *Ann. Rev. Biochem.* **48**, 293–325.

Chrousos, G. P., Cutler, G. B., Jr., Sauer, M., Simons, S. S. Jr. and Loriaux, D. S. (1982) Development of antiglucocorticoids with potential clinical usefulness. In: Lee, H. J. and Fitzgerald, T. J. (eds.) *Progress in Research and Clinical Applications of Corticosteroids.* Heyden and Son, PA, pp. 152–176.

Cidlowski, J. A. and Richon, V. (1984) Evidence for microheterogeneity in the structure of human glucocorticoid receptors. *Endocrinol.* **115**, 1588–1597.

Cordingley, M. G., Riegel, A. T. and Hager, G. L. (1987) Steroid-dependent interaction of transcription factors with the inducible promoter of mouse mammary tumor virus *in vivo. Cell* **48**, 261–270.

Coxon, J. M. and Halton, B. (1974) *Organic Photochemistry.* Cambridge University Press, London, pp. 86–107, 188–189.

Danielsen, M., Northrop, J. P. and Ringold, G. M. (1986) The mouse glucocorticoid receptor: mapping of functional domains by cloning, sequencing and expression of wild-type and mutant receptor proteins. *EMBO J.* **5**, 2513–2522.

Danze, P.-M., Formstecher, P., Richard, C. and Dautrevaux, M. (1987) Microheterogeneity of agonist and antagonist glucocorticoid receptor complexes detected by isoelectric focusing and modifications induced by receptor activation. *Biochim. Biophys. Acta* **927**, 129–138.

Dellweg, H.-G., Hotz, A., Mugele, K. and Gehring, U. (1982) Active domains in wild-type and mutant glucocorticoid receptors. *EMBO J.* **1**, 285–289.

Distelhorst, C. W. and Benutto, B. M. (1985) Activation of the rat liver cytosol glucocorticoid receptor by Sephacryl S-300 filtration in the presence and absence of molybdate. Physical properties of the receptor and evidence for an activation inhibitor. *J. Biol. Chem.* **260**, 2153–2159.

Distelhorst, C. W., Benutto, B. M. and Griffith, R. C. (1985) A single common electrophoretic abnormality of glucocorticoid receptors in human leukemia cells. *Blood* **66**, 679–685.

Distelhorst, C. W., Kullman, L. and Wasson, J. (1987) Characterization of glucocorticoid receptors in S49 mouse lymphoma cells by affinity labeling with [^{3}H]dexamethasone 21-mesylate. *J. Steroid Biochem.* **26**, 59–65.

Echols, H. (1986) Multiple DNA–protein interactions governing high-precision DNA transactions. *Science* **233**, 1050–1056.

Economidis, I. V. and Rousseau, G. G. (1985) Association of the glucocorticoid hormone receptor with ribonucleic acid. *FEBS Letters* **181**, 47–52.

Eisen, H. J., Schleenbaker, R. E. and Simons, S. S. Jr. (1981) Affinity labeling of the rat liver glucocorticoid receptor with dexamethasone 21-mesylate: identification of covalently labeled receptor by immunochemical methods. *J. Biol. Chem.* **256**, 12920–12925.

Eisen, L. P., Reichman, M. E., Thompson, E. B., Gametchu, B., Harrison, R. W. and Eisen, H. J. (1985) Monoclonal antibody to the rat glucocorticoid receptor. Relationship between the immunoreactive and DNA-binding domain. *J. Biol. Chem.* **260**, 11805–11810.

Elsner, H., Buchardt, O., Moller, J. and Nielsen, P. E. (1985) Photochemical crosslinking of protein and DNA in chromatin. *Anal. Biochem.* **149**, 575–581.

Foster, C. M., Eisen, H. J. and Bloomfield, C. D. (1983) Covalent labeling of rat thymocyte and human lymphoid glucocorticoid receptor. *Cancer Research* **43**, 5273–5277.

Gadson, P. F., Russell, J. D. and Russell, S. B. (1984) Glucocorticoid receptors in human fibroblasts derived from normal dermis and keloid tissue. *J. Biol. Chem.* **259**, 11236–11241.

Gaillard, R. C., Riondel, A. Muller, A. F., Herrmann, W. and Baulieu, E.-E. (1984) RU 486: A steroid with antiglucocorticosteroid activity that only disinhibits the human pituitary–adrenal system at a specific time of day. *Proc. Natl. Acad. Sci. USA* **81**, 3879–3882.

Gametchu, B. and Harrison, R. W. (1984) Characterization of a monoclonal antibody to the rat liver glucocorticoid receptor. *Endocrinol.* **114**, 274–279.

Gehring, U. and Holtz, A. (1983) Photoaffinity labeling and partial proteolysis of wild-type and variant glucocorticoid receptors. *Biochemistry* **22**, 4013–4018.

Gehring, U. and Arndt, H. (1985) Heteromeric nature of glucocorticoid receptors. *FEBS Letters* **179**, 138–142.

Godowski, P. J., Rusconi, S., Miesfeld, R. and Yamamoto, K. R. (1987) Glucococorticoid receptor mutants that are constitutive activators of transcriptional enhancement. *Nature* **325**, 365–368.

Govindan, M. V. and Gronemeyer, H. (1984) Characterization of the rat liver glucocorticoid receptor purified by DNA–cellulose and ligand affinity chromatography. *J. Biol. Chem.* **259**, 12915–12924.

Govindan, M. V. and Manz, B. (1980) Biochemistry of glucocorticoid action, involvement of specific receptors. *Allergologie* **3**, 204–213.

Govindan, M. V., Devic, M., Green, S., Gronemeyer, H. and Chambon, P. (1985) Cloning of the human glucocorticoid receptor cDNA. *Nucleic Acids Research* **13**, 8293–8304.

Grandics, P., Gasser, D. L. and Litwack, G. (1982) Monoclonal antibodies to the glucocorticoid receptor. *Endocrinol.* **111**, 1731–1733.

Grandics, P., Miller, A., Schmidt, T. J. and Litwack, G. (1984) Phosphorylation *in vivo* of rat hepatic glucocorticoid receptor. *Biochem. Biophys. Res. Comm.* **120**, 59–65.

Groman, E. V., Schultz, R. M. and Engel, L. L. (1975) Catalytic competence, a new criterion for affinity labeling. Demonstration of the reversible enzymatic interconversion of estrone and estradiol-17 β covalently bound to human placental estradiol-17 β dehydrogenase. *J. Biol. Chem.* **250**, 5450–5454.

Gronemeyer, H. and Govindan, M. V. (1986) Affinity labeling of steroid hormone receptors. *Mol. Cell. Endo.* **46**, 1–19.

Gronemeyer, H., Harry, P. and Chambon, P. (1983) Evidence for two structurally related progesterone receptors in chick oviduct cytosol. *FEBS Letters* **156**, 287–292.

Harrison, R. W., Woodward, C. and Thompson, E. (1983) Titratable effects of *p*-chloromercuriphenyl sulfonate, a thiol-attacking reagent on glucocorticoid receptor binding. *Biochim. Biophys. Acta* **759**, 1–6.

Hazato, T. and Murayama, (1981) Protection of steroid hormone receptors by protease inhibitors. *Biochem. Biophys. Res. Comm.* **98**, 488–493.

Hockensmith, J. W., Kubasek, W. L., Vorachek, W. R. and von Hippel, P. H. (1986) Laser cross-linking of nucleic acids to proteins. Methodology and first applications to the phage T4 DNA replication system. *J. Biol. Chem.* **261**, 3512–3518.

Hollenberg, S. M., Weinberger, C., Ong, E. S., Cerelli, G., Oro, A., Lebo, R., Thompson, E. B., Rosenfeld, M. G., and Evans, R. M. (1985) Primary structure and expression of a functional human glucocorticoid receptor cDNA. *Nature* **318**, 635–641.

Horwitz, K. B. (1985) The antiprogestin in RU 38,486: Receptor-mediated progestin versus antiprogestin actions screened in estrogen-insensitive $T47D_{co}$ human breast cancer cells. *Endocrinology* **116**, 2236–2245.

Horwitz, K. B., Koseki, Y. and McGuire, W. L. (1978) Estrogen control of progesterone receptor in human breast cancer: role of estradiol and antiestrogen. *Endocrinol.* **103**, 1742–1751.

Hoschutzky, H. and Pongs, O. (1985) Characterization of glucocorticoid receptor in HeLa-S_3 cells. *Biochemistry* **24**, 7348–7356.

Housley, P. R. and Pratt, W. B. (1983) Direct demonstration of glucocorticoid receptor phosphorylation by intact L-cells. *J. Biol. Chem.* **258**, 4630–4635.

Housley, P. R., Sanchez, E. R., Westphal, H. M., Beato, M. and Pratt, W. B. (1985) The molybdate-stabilized L-cell glucocorticoid receptor isolated by affinity chromatography or with a monoclonal antibody is associated with a 90–92-kDa nonsteroid-binding phosphoprotein. *J. Biol. Chem.* **260**, 13810–13817.

Hutchens, T. W., Markland, F. S. and Hawkins, E. F. (1982) RNA induced reversal of glucocorticoid receptor activation. *Biochem Biophys. Res. Comm.* **105**, 20–27.

Jakoby, W. B. and Wilchek, M. (1977) Affinity labeling. *Methods in Enzymol.* **46** .

Katzenellenbogen, J. A. (1977) Affinity labeling as a technique in determining hormone mechanisms. In: G. Litwack, (ed.) *Biochemical Actions of Hormones.* Vol. 4. Academic Press, New York, pp. 1–84.

Katzenellenbogen, J. A., Carlson, K. E., Elliston, J. F., Salituro, F. G., Simpson, D. M. and Zablocki, J. A. (1987) Covalent labeling agents for the estrogen receptor. *Cell Biol.* **11A**, B117.

Kontula, K. K., Seppanen, P. J., van Duyne, P., Bardin, C. W., and Janne, O. A. (1985) Effect of a nonsteroidal antiandrogen, flutamide, on androgen receptor dynamics and ornithine decarboxylase gene expression in mouse kidney. *Endocrinol.* **116**, 226–233.

Kovacic-Milivojevic, B. and Vedeckis, W. V. (1986) Absence of detectable ribonucleic acid in the purified, untransformed mouse glucocorticoid receptor. *Biochemistry* **25**, 8266–8273.

Kovacic–Milivojevic, B., LaPointe, M. C., Reker, C. E. and Vedeckis, W. V. (1985) Ribonucleic acid is a component of the oligomeric, transformed mouse AtT-20 cell glucocorticoid receptor. *Biochemistry* **24**, 7357–7366.

Koyasu, S., Nishida, E., Kadowaki, T., Matsuzaki, F., Iida, K., Harada, F., Kasuga, M., Sakai, H. and Yahara, I. (1986) Two mammalian heat shock proteins, HSP90 and HSP100, are actin-binding proteins. *Proc. Natl. Acad. Sci. USA* **83**, 8054–8058.

Krust, A., Green, S., Argos, P., Kumar, V., Walter, P., Bornert, J.-M. and Chambon, P. (1986) The chicken oestrogen receptor sequence: homology with v-erbA and the human oestrogen and glucocorticoid receptors. *EMBO J.* **5**, 891–897.

Lamontagne, N., Mercier, L., Pons, M., Thompson, E. B. and Simons, S. S. Jr. (1985) Glucocorticoid *vs* antiglucocorticoid activity: can a single functional group modification of glucocorticoid steroids always convey antiglucocorticoid activity? *Endocrinology* **114**, 2252–2263.

Lanoux, M. J., Cleland, W. H., Mendelson, C. R., Carr, B. R. and Simpson, E. R. (1985) Factors affecting the conversion of androstenedione to estrogens by human fetal hepatocytyes in monolayer culture. *Endocrinology* **117**, 361–368.

Leach, K. L., Dahmer, M. K., Hammond, N. D., Sando, J. J. and Pratt, W. B. (1979) Molybdate inhibition of glucocorticoid receptor inactivation and transformation. *J. Biol. Chem.* **254**, 11884–11890.

Lewis, G. D., Campbell, W. B. and Johnson, A. R. (1986) Inhibition of prostaglandin synthesis by glucocorticoids in human endothelial cells. *Endocrinology* **119**, 62–69.

Logeat, F., Pamphile, R., Loosfelt, H., Jolivet, A., Fournier, A. and Milgrom, E. (1985) One-step immunoaffinity purification of active progesterone receptor. Further evidence in favor of the existence of a single steroid binding subunit. *Biochemistry* **24**, 1029–1035.

Loosfelt, H., Atger, M., Misrahi, M., Guiochon-Mantel, A., Meriel, C., Logeat, F., Benarous, R. and Milgrom, E. (1986) Cloning and sequence analysis of rabbit progesterone-receptor complementary DNA. *Proc. Natl. Acad. Sci. USA* **83**, 9045–9049.

Lustenberger, P., Formstecher, P. and Dautrevaux, M. (1981) Purification of rat liver glucocorticoid receptor by affinity chromatography: design of a suitable adsorbent. *J. Steroid Biochem.* **14**, 697–703.

Mathis, J. M., Prough, R. A. and Simpson, E. R. (1986) Synergistic induction of monooxygenase activity by glucocorticoids and polycyclic aromatic hydrocarbons in human fetal hepatocytes in primary monolayer culture. *Arch. Biochem. Biophys.* **244**, 650–661.

Mayo, K. E. and Palmiter, R. D. (1985) Glucocorticoid regulation of metallothionein gene expression. *Biochemical Actions Hormones* **12**, 69–88.

Mendel, D. B., Holbrook, N. J. and Bodwell, J. E. (1985) Degradation without apparent change in size of molybdate-stabilized nonactivated glucocorticoid–receptor complexes in rat thymus cytosol. *J. Biol. Chem.* **260**, 8736–8740.

Mendel, D. B., Bodwell, J. E., Gametchu, B., Harrison, R. W. and Munck, A.

(1986a) Molybdate-stabilized nonactivated glucocorticoid–receptor complexes contain a 90-kDa non-steroid-binding phosphoprotein that is lost on activation. *J. Biol. Chem.* **261**, 3758–3763.

Mendel, D. B., Bodwell, J. E. and Munck, A. (1986b) Glucocorticoid receptors lacking hormone-binding activity are bound in nuclei of ATP-depleted cells. *Nature* **324**, 478–480.

Mercier, L., Thompson, E. B. and Simons, S. S. Jr. (1983) Dissociation of steroid binding to receptors and steroid induction of biological activity in a glucocorticoid responsive cell. *Endocrinology* **112**, 601–609.

Mercier, L., Miller, P. A. and Simons, S. S. Jr. (1986) Antiglucocorticoid steroids have increased agonist activity in those hepatoma cell lines that are more sensitive to glucocorticoids. *J. Steroid Biochem.* **25**, 11–20.

Miesfeld, R., Rusconi, S., Godowski, P. J., Maler, B. A., Okret, S., Wikstrom, A.-C. Gustafsson, J.-A. and Yamamoto, K. R. (1986) Genetic complementation of a glucocorticoid receptor deficiency by expression of cloned receptor cDNA. *Cell* **46**, 389–399.

Miller, P. A., Ostrowski, M. C. , Hager, G. L. and Simons, S. S. Jr. (1984) Covalent and noncovalent receptor–glucocorticoid complexes preferentially bind to the same regions of the long terminal repeat of murine mammary tumor virus proviral DNA. *Biochemistry* **23**, 6883–6889.

Moguilewsky, M. and Philibert, D. (1984) RU 38486: potent antiglucocorticoid activity correlated with strong binding to the cytosolic glucocorticoid receptor followed by an impaired activation. *J. Steroid Biochem.* **20**, 271–276.

Monsma, F. J. Jr., Katzenellenbogen, B. S., Miller, M. A., Ziegler, Y. S. and Katzenellenbogen, J. A. (1984) Characterization of the estrogen receptor and its dynamics in MCF-7 human breast cancer cells using a covalently attaching antiestrogen. *Endocrinol.* **115**, 143–153.

Nordeen, S. K., Lan, N. C., Showers, M. O. and Baxter, J. D. (1981) Photoaffinity labeling of glucocorticoid receptors. *J. Biol. Chem.* **256**, 10503–10508.

Northrop, J. P., Gametchu, B., Harrison, R. W. and Ringold, G. M. (1985) Characterization of wild type and mutant glucocorticoid receptors from rat hepatoma and mouse lymphoma cells. *J. Biol. Chem.* **260**, 6398–6403.

Okret, S., Wikstrom, A.-C., Wrange, O., Andersson, B. and Gustafsson, J.-A. (1984) Monoclonal antibodies against the rat liver glucocorticoid receptor. *Proc. Natl. Acad. Sci. USA* **81**, 1609–1613.

Payvar, F., DeFranco, D., Firestone, G. L., Edgar, B., Wrange, O., Okret, S., Gustafsson, J.-A. and Yamamoto, K. R. (1983) Sequence-specific binding of glucocorticoid receptor to MTV DNA to sites within and upstream of the transcribed region. *Cell* **35**, 381–392.

Pfahl, M., McGinnis, D., Hendricks, M., Groner, B. and Hynes, N. E. (1983) Correlation of glucocorticoid receptor binding sites on MMTV proviral DNA with hormone inducible transcription. *Science* **222**, 1341–1343.

Pons, M., Robinson, T. E. J., Mercier, L., Thompson, E. B. and Simons, S. S. Jr. (1985) Formation of a fluorescent glucocorticoid receptor–steroid complex in HTC cell cytosol. *J. Steroid Biochem.* **23**, 267–273.

Ptashne, M. (1986) Gene regulation by proteins acting nearby and at a distance. *Nature* **322**, 697–701.

Raaka, B. M. and Samuels, H. H. (1983) The glucocorticoid receptor in GH1 cells. Evidence from dense amino acid labeling and whole cell studies for an equilibrium model explaining the influence of hormone on the intracellular distribution of receptor. *J. Biol. Chem.* **258,** 417–425.

Raynaud, J. P., Bouton, M. M., Moguilewsky, M., Ojasoo, T., Philibert, D., Beck, G., Labrie, F. and Mornon, J. P. (1980) Steroid hormone receptors and pharmacology. *J. Steroid Biochem.* **12**, 143–157.

Reichman, M. E., Foster, C. M., Eisen, L. P., Eisen, H. J., Torain, B. F. and Simons, S. S. Jr. (1984) Limited proteolysis of covalently labeled glucocorticoid receptors as a probe of receptor structure. *Biochem.* **23**, 5376–5384.

Richardson, M. R., Mitchell, M. D., MacDonald, P. C. and Casey, M. L. (1986) Glucocorticosteroid regulation of prostaglandin biosynthesis in human myometrial smooth muscle cells in monolayer culture. *J. Steroid Biochemistry* **25**, 521–526.

Ross, P. Jr. and Ruh, T. S. (1984) Binding of the estradiol–receptor complex to reconstituted nucleoacidic protein from calf uterus. *Biochim. Biophys. Acta* **782**, 18–25.

Rousseau, G. G. (1984) Control of gene expression by glucocorticoid hormones. *Biochem. J.* **224**, 1–12.

Rousseau, G. G., Higgins, S. J., Baxter, J. D., Gelfand, D. and Tomkins, G. M. (1975) Binding of glucocorticoid receptors to DNA. *J. Biol. Chem.* **250**, 6015–6021.

Sanchez, E. R. and Pratt, W. B. (1986) Phosphorylation of L-cell glucocorticoid receptors in immune complexes: Evidence that the receptor is not a protein kinase. *Biochemistry* **25**, 1378–1382.

Sanchez, E. R., Toft, D. O., Schlesinger, M. J. and Pratt, W. B. (1985) Evidence that the 90-kDa phosphoprotein associated with the untransformed L-cell glucocorticoid receptor is a murine heat shock protein. *J. Biol. Chem.* **260**, 12398–12401.

Sanchez, E. R., Housley, P. R. and Pratt, W. B. (1986) The molybdate-stabilized glucocorticoid binding complex of L-cells contains a 98–100 kdalton nonsteroid-binding phosphoprotein that is part of the murine heat-shock complex. *J. Steroid Biochem.* **24**, 9–18.

Sando, J. J., Hammond, N. D., Stratford, C. A. and Pratt, W. B. (1979) Activation of thymocyte glucocorticoid receptors to the steroid binding form. The roles of reducing agents, ATP, and heat-stable factors. *J. Biol. Chem.* **254**, 4779–4789.

Sargan, D. R., Tsai, M.-J. and O'Malley, B. W. (1986) Hsp 108, a novel heat shock inducible protein of chicken. *Biochemistry* **25**, 6252–6258.

Scheidereit, C. and Beato, M. (1984) Contents between hormone receptor and DNA double helix within a glucocorticoid regulatory element of mouse mammary tumor virus. *Proc. Natl. Acad. Sci. USA* **81**, 3029–3033.

Scheidereit, C., Geisse, S., Westphal, H. M. and Beato, M. (1983) The glucocorticoid receptor binds to define nucleotide sequences near the promoter of mouse mammary tumor virus. *Nature* **304**, 749–752.

Schlechte, J. A., Simons, S. S. Jr., Lewis, D. A. and Thompson, E. B. (1985) [^{3}H]Cortivazol: A unique high affinity ligand for the glucocorticoid receptor. *Endocrinology* **117**, 1355–1362.

Schmidt, T. J. and Litwack, G. (1982) Activation of the glucocorticoid–receptor complex. *Physiological Review* **62**, 1131–1192.

Sherman, M. R., Pickering, L. A., Rollwagen, F. M. and Miller, L. K. (1978) Mero-receptors: proteolytic fragments of receptors containing the steroid-binding site. *Fed. Proc.* **37**, 167–173.

Sherman, M. R., Moran, M. C., Tuazon, F. B. and Stevens, Y.-W. (1983) Structure, dissociation, and proteolysis of mammalian steroid receptors. Multiplicity of glucocorticoid receptor forms and proteolytic enzymes in rat liver and kidney cytosols. *J. Biol. Chem.* **258**, 10366–10377.

Simons, S. S. Jr. (1981) The first irreversible antiglucocorticoid. US Patent No. 4,296,206.

Simons, S. S. Jr. (1985) Affinity labeling steroids as biologically active probes of glucocorticoid receptor structure and function. In: (Moudgil, V. K. (ed.). *Molecular Mechanism of Steroid Hormone Action.* de Gruyter, Berlin, pp. 111–140.

Simons, S. S. Jr. (1987) Selective covalent labeling of cysteines in bovine serum albumin and in HTC cell glucocorticoid receptors by dexamethasone 21-meslate. *J. Biol. Chem.*, in press.

Simons, S. S. Jr. and Miller, P. A. (1984) Comparison of DNA binding properties of activated, covalent and noncovalent glucocorticoid receptor–steroid complexes from HTC cells. *Biochemistry* **23**, 6876–6882.

Simons, S. S. Jr. and Miller, P. A. (1986) Affinity-labeling steroids as biologically active probes of antiglucocorticoid hormone action. *J. Steroid Biochem.* **24**, 25–32.

Simons, S. S. Jr. and Thompson, E. B. (1981) Dexamethasone 21-mesylate: an affinity label of glucocorticoid receptors from rat hepatoma tissue culture cells. *Proc. Natl. Acad. Sci. USA*, **78**, 3541–3545.

Simons, S. S. Jr. and Thompson, E. B. (1982) Affinity labeling of glucocorticoid receptors: New methods in affinity labeling. In: Litwack, G. (ed.) *Biochemical Actions of Hormones*, Vol. 9. Academic Press, New York, pp. 221–254.

Simons, S. S. Jr. and Yen, P. M. (1987) Variations in agonist activity among antiglucocorticoid steroids and its relation to glucocorticoid regulated genes. In: Spelsberg, T. and Kumar, R. (eds.) *Sterol/Steroid Hormone Action.* Martinus Nijhoff, Norwell, MA.

Simons, S. S. Jr., Pons, M. and Johnson, D. F. (1980a) α-Keto mesylate: a reactive thiol-specific functional group. *J. Org. Chem.* **45**, 3084–3088.

Simons, S. S. Jr., Thompson, E. B. and Johnson, D. F. (1980b) Unique long acting antiglucocorticoid in whole and broken cell systems. *Proc. Natl. Acad. Sci. USA* **77**, 5167–5171.

Simons, S. S. Jr., Schleenbaker, R. E. and Eisen, H. J. (1983) Activation of covalent, affinity labeled, glucocorticoid receptor–steroid complexes. *J. Biol. Chem.* **258**, 2229–2238.

Simons, S. S. Jr., Pumphrey, J. G., Rudikoff, S. and Eisen, H. J. (1987) Identification of cysteine-656 as the amino acid of HTC cell glucocorticoid receptors that is covalently labeled by dexamethasone 21-mesylate. *J. Biol. Chem.*, in press.

Singh, R. K., Ruh, M. F., Butler, W. B. and Ruh, T. S. (1986) Acceptor sites on chromatin for receptor bound by estrogen versus antiestrogen in antiestrogen-sensitive and -resistant MCF-7 cells. *Endocrinology* **118**, 1087–1095.

Singh, V. B. and Moudgil, V. K. (1984) Protein kinase activity of purified rat liver glucocorticoid receptor. *Biochem. Biophys. Res. Comm.* **125**, 1067–1073.

Singh, V. B. and Moudgil, V. K. (1985) Phosphorylation of rat liver glucocorticoid receptor. *J. Biol. Chem.* **260**, 3684–3690.

Smith, A. C. and Harmon, J. M. (1985) Multiple forms of the glucocorticoid receptor steroid binding protein identified by affinity labeling and high-resolution two-dimensional electrophoresis. *Biochemistry* **24**, 4946–4951.

Smith, A. C. and Harmon, J. M. (1987) Structural organization of the human glucocorticoid receptor determined by one- and two-dimensional gel electrophoresis of proteolytic receptor fragments. *Biochemistry* **26**, 646–652.

Smith, A. C., Elsasser, M. S. and Harmon, J. M. (1986) Analysis of glucocorticoid receptor activation by high resolution two-dimensional electrophoresis of affinity-labeled receptor. *J. Biol. Chem.* **261**, 13285–13292.

Solomon, M. J. and Varshavsky, A. (1985) Formaldehyde-mediated DNA–protein crosslinking: A probe for *in vivo* chromatin structures. *Proc. Natl. Acad. Sci. USA* **82**, 6470–6474.

Spelsberg, T. C., Gosse, B. J., Littlefield, B. A., Toyoda, H. and Seelke, R. (1984) Reconstibtion of nativelike nuclear acceptor sites of the avian oviduct progesterone receptor: evidence for involvement of specific chromatin proteins and specific DNA sequences. *Biochemistry* **23**, 5103–5113.

Sullivan, W. P., Vroman, B. T., Bauer, V. J., Puri, R. K., Riehl, R. M., Pearson, G. R. and Toft, D. O. (1985) Isolation of steroid receptor binding protein from chicken oviduct and production of monoclonal antibodies. *Biochemistry* **24**, 4214–4222.

Tai, P.-K. K., Maeda, Y., Nakao, K., Wakim, N. G., Duhring, J. L. and Faber, L. E. (1986) A 59-kilodalton protein associated with progestin, estrogen, androgen, and glucocorticoid receptors. *Biochemistry* **25**, 5269–5275.

Terada, N. and Oka, T. (1983) Cortisol 21-mesylate exerts glucocorticoid action on the induction of milk protein synthesis in cultured mammary gland. *Horm. Metabol. Res* **15**, 508–512.

Thompson, E. B., Zawydiwski, R., Brower, S. T., Eisen, H. J., Simons, S. S. Jr., Schmidt, T. J., Schlechte, J. A., Moore, D. E., Norman, M. R. and Harmon, J. M. (1983) Properties and function of human glucocorticoid receptors in steroid-sensitive and resistant leukemic cells. In: Ericksson, H. and Gustafsson, J.-A. (eds.) *Steroid Hormone Receptors: Structure and Function.* Elsevier, Amsterdam, pp. 171–194.

Titani, K., Koide, A., Hermann, J., Ericsson, L. H., Kumar, E. S., Wade, R. D., Walsh, K. A., Neurath, H. and Fischer, E. H. (1977) Complete amino acid sequence of rabbit muscle glycogen phosphorylase. *Proc. Natl. Acad. Sci. USA* **74**, 4762–4766.

Tomita, M., Chrousos, G. P., Brandon, D. D., Ben-Or, S., Foster, C. M., de Vough, L., Taylor, S., Loriaux, D. L., and Lipsett, M. B. (1985) Glucocorticoid receptors in Epstein–Barr virus-transformed human lymphocytes. *Horm. Metabol. Res.* **17**, 674–678.

Tymoczko, J. L. and Phillips, M. M. (1983) The effects of ribonuclease on rat liver dexamethasone receptor: increased affinity for deoxyribonucleic acid and altered sedimentation profile. *Endocrinology* **112**, 142–149.

Tymoczko, J. L., Anderson, E. E., Lee, J. H. and Unger, A. L. (1986) Studies with chymotrypsin and RNAase showing a heterooligomeric structure of the glucocorticoid receptor complex from rat liver which is stabilized by low molecular weight factor. *Biochem. Biophys. Acta* **888**, 296–305.

Vedeckis, W. V. (1983) Subunit dissociation as a possible mechanism of glucocorticoid receptor activation. *Biochemistry* **22**, 1983–1989.

Wasner, G. and Simons, S. S. Jr. (1987) Differential sensitivity of HTC and Fu5-5 cells for induction of tyrosine aminotransferase by 3′,5′-cyclic adenosine monophosphate. *Mol. Endo.* **1**, 109–120.

Webb, M. L., Miller-Diener, A. S. and Litwack, G. (1985) Purification, characterization, and activation of the glucocorticoid–receptor complex from rat kidney cortex. *Biochemistry* **24**, 1946–1952.

Webb, M. L., Schmidt, T. J., Robertson, N. M. and Litwack, G. (1986) Evidence for an association of a ribonucleic acid with the purified, unactivated glucocorticoid receptor. *Biochem. Biophys. Res. Comm.* **140**, 204–211.

Weisz, A., Buzard, R. L., Horn, D., Lis, M. P., Dunkerton, L. V. and Markland, F. S. Jr. (1983) Steroid derivatives for electrophilic affinity labeling of glucocorticoid binding sites: interaction with the glucocorticoid receptor and biological activity. *J. Steroid Biochem.* **18**, 375–382.

Westley, B., May, F. E. B., Brown, A. M. C., Krust, A., Chambon, P., Lippman, M. E. and Rochefort, H. (1984) Effects of antiestrogen on the estrogen-regulated pS2 RNA and the 52- and 160-kilodalton proteins in MCF7 cells and two tamoxifen-resistant sublines. *J. Biol. Chem.* **259**, 10030–10035.

Westphal, H. M., Fleischmann, G. and Beato, M. (1981) Photoaffinity labeling of steroid binding proteins with unmodified ligands. *Eur. J. Biochem.* **119**, 101–106.

Wolff, M. E., Baxter, J. D., Kollman, P. A., Lee, D. L., Kuntz, I. D., Bloom, E., Matulich, D. T., and Morris, J. (1978) Nature of steroid–glucocorticoid receptor interactions: thermodynamic analysis of the binding reaction. *Biochemistry* **17**, 3201–3208.

Wrange, O., Okret, S., Radojcic, M., Carlstedt-Duke, J. and Gustafsson, J.-A. (1984) Characterization of the purified activated glucocorticoid receptor from rat liver cytosol. *J. Biol. Chem.* **259**, 4534–4541.

Yamamoto, K. R., Stampfer, M. R. and Tomkins, G. M. (1974) Receptors from glucocorticoid-sensitive lymphoma cells and two classes of insensitive clones: physical and DNA-binding properties. *Proc. Natl. Acad. Sci. USA* **71**, 3901–3905.

9

Glucocorticoid receptor mutants

U. Gehring
Institut für Biologische Chemie, Universität Heidelberg, Heidelberg, FRG

I. INTRODUCTION

Mutant receptors have played an important role in the elucidation of the mechanisms involved in the induction of cellular responses to steroid hormones. In particular, mutant glucocorticoid receptors have been discovered in cell variants which were selected for hormone unresponsiveness. While most cell types of the mammalian organism respond to glucocorticoids in ways which are not very conspicuous in terms of morphology and can only be detected by biochemical or immunological technqiues, various cells of the lymphoid system react quite dramatically by growth inhibition and cell lysis. In some cells the response is limited to the inhibition of cell proliferation, in others an initial phase of diminished growth rate is follwed by cell death. Hormone induced lysis, of course, can quite easily be used for selecting cell variants which are resistant to the cytolytic glucocorticoid effect. In many instances such resistant cells turned out either to be defective in receptor production or to harbour mutant receptors. The method of photoaffinity labeling of hormone binding polypeptides has been of great importance for the biochemical characterization of some of these receptor mutants.

II. CELL VARIANTS WITH ALTERED GLUCOCORTICOID RESPONSIVENESS

Even though the phenomenon of glucocorticoid inducible lymphocytolysis was recognized several decades ago not much is known yet about the biochemistry involved beyond the receptor system (Gehring, 1986). For the selection of resistant variants several lymphoid cell lines in continuous culture have been used, mostly of mouse and human origin. Fig. 1A shows a growth experiment with the murine thymic

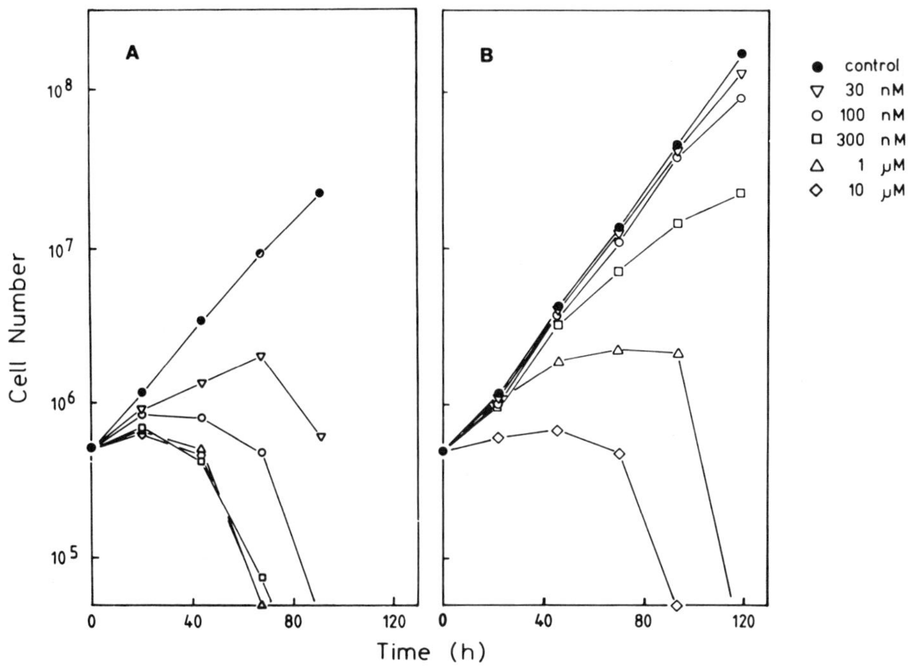

Fig. 1 — Glucocorticoid effect on the growth of S49.1 cells. Wild-type cells (A) and cells of a low-sensitivity clone (B) were cultured in the presence of dexamethasone at the concentrations indicated. Cells which excluded trypan blue were scored. (Data from Gehring *et al.*, 1982).

lymphoma line S49.1 in which increasing concentrations of the glucocorticoid dexamethasone were added to replica cultures. After a lag of several hours the cells slowed down or stopped multiplying and after about another day — depending on the dose of the hormone — cell killing set in. Inhibition of cell proliferation has been shown to lead to accumulation of cells in the G_1 phase of the cell cycle (Gehring *et al.*, 1976). Cytolysis, therefore, appears to occur while the cells are arrested in G_1.

Under the conditions of Fig. 1A the majority of cells die at dexamethasone concentrations of 30 nM and above; the resistant cells contained in the original population can be grown up in suspension culture if appropriate care is being taken. If selection is carried out in this way a mixture of resistant cells is obtained. It may therefore be of advantage to combine the selection for hormone resistance with cell cloning. This method gives rise to many different sublines which may differ in their phenotypes. Fig. 2 shows such a serial cloning experiment with increasing steroid

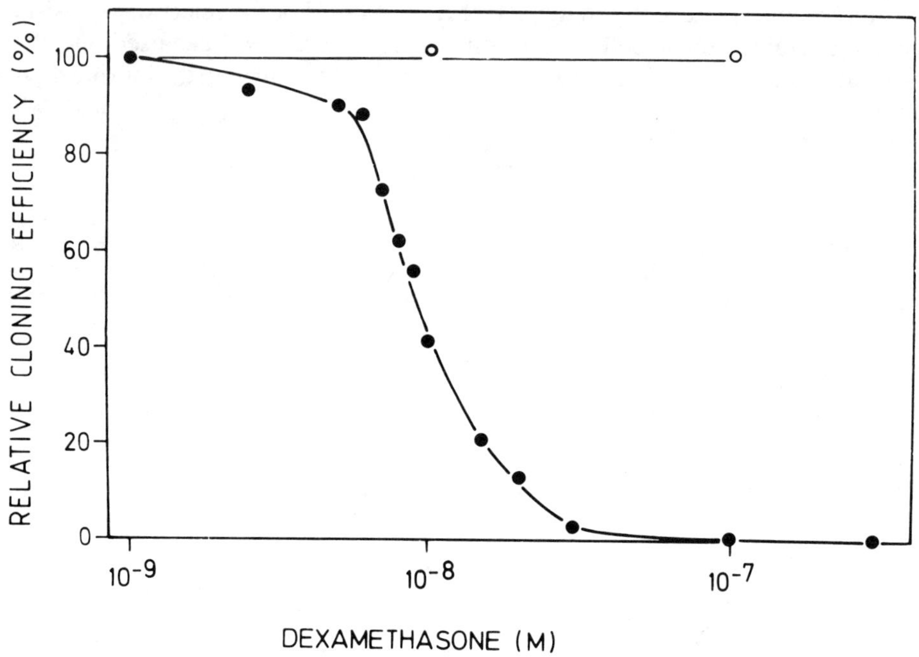

Fig. 2 — Glucocorticoid effect on the cloning of S49.1 cells. Wild-type cells (●) were cloned in soft agar over a feeder layer of mouse embryo fibroblasts with dexamethasone added to the agar. One clone which came up in the presence of 0.1 μM dexamethasone (○) was recloned.

concentrations. At high doses of dexamethasone, i.e. 0.1–1 μM, only very few colonies came up. These were completely resistant to the cytolytic hormone effect as is seen in Fig. 2 for one such clone. A large number of resistant subclones of S49.1 was isolated in this way (Yamamoto *et al.*, 1976; Gehring, 1980a) and each one was investigated for its receptors as described below.

The same cloning method has also been used for selecting variants of decreased sensitivity (Gehring *et al.*, 1982). In this case S49.1 subclones were picked which grew up in the presence of 10 nM dexamethasone. The growth behavior of one such cell clone in the presence of dexamethasone at various concentrations is shown in Fig. 1B. Clearly, much higher steroid concentrations are required for the same response to occur than in wild-type cells. These variants turned out to contain significantly decreased levels of glucocorticoid receptors with unchanged affinity for the steroid and normal subcellular distribution. This observation points to a quantitative relationship between cellular receptor content and hormonal sensitivity.

In a subsequent study (Gehring *et al.*, 1984) a series of independently isolated mouse lymphomas of widely differing glucocorticoid sensitivities were shown to display a gradient in their cellular receptor levels. For the majority of the cell lines a direct correlation was obtained between hormone sensitivity on the one hand and the number of cellular receptor sites and the receptors' affinities for the steroid on the other. Thus the receptor is a mjaor quantitative determinant for responsiveness. Some cell lines used in this study, however, behaved differently in that they were of either higher or lower sensitivity than expected from their receptors. In these lines the mechanism of cell inhibition and lysis appears to be quantitatively different from that of the other cell lines. It is interesting to note in the context of this discussion that variants of mouse lymphoma cells have been obtained (Mermod *et al.*, 1981) which are supersensitive to glucocorticoids in that they are killed faster than the parental cells but nevertheless contain the same levels of otherwise normal receptors. Other factors in addition to receptors must therefore play a crucial role in hormone responsiveness.

Another line of observation is concerned with the cellular regulation of receptor levels and hence hormonal sensitivity. Some years ago it was found that certain lymphoid cells in culture may first respond to glucocorticoids with cytolysis but the surviving cells may resume cell proliferation even in the presence of high concentrations of hormone (Harris and Baxter, 1979). In these cells treatment with steroid induces resistance. A similar transient cytolytic effect was also seen in some WEHI-7 lymphoma cell variants selected for glucocorticoid resistance (Danielsen and Stallcup, 1984) and it was shown by whole-cell binding assays that the hormone caused a reversible down-regulation of receptors. This decrease in receptor content thus allows the cells to escape from the lethal hormone effect. With receptor-specific cDNA probes which recently became available it is now possible to search for changes in receptor mRNA levels in cells under different physiological conditions. In fact, a recent study documents a 50–95% decrease in glucocorticoid receptor mRNA content in rat hepatoma cells and rat liver following treatment with dexamethasone (Okret *et al.*, 1986).

In addition to the S49.1 line described above, several other lymphoid cells in culture have also been used for selecting unresponsive variants, most notably the glucocorticoid sensitive murine thymic lymphoma line WEHI-7 (Bourgeois and Newby, 1977; Pfahl *et al.*, 1978; Bourgeois and Gasson, 1985) and the human lymphoblastic leukemia cell line CEM-C7 (Norman and Thompson, 1977; Harmon and Thompson, 1981; Thompson *et al.*, 1986). By and large the variants obtained from these lines correspond to those obtained in the S49.1 cell system, which will be described in more detail below.

A. Types of receptor defects

For the classification of glucocorticoid resistant lymphoma cell variants a straighforward method has been developed (Sibley and Tomkins, 1974; Yamamoto *et al.*, 1976). This consists of a whole-cell binding assay followed by crude cell fractionation. The vast majority of resistant variants of S49.1 turned out to have low or negligible hormone binding activity in this assay as well as in experiments with cell extracts or isolated nuclei. This phenotype was collectively called r^-, 'receptorless' (Sibley and Tomkins, 1974) or 'receptor deficient' (Yamamoto *et al.*, 1976). It may be due to a defect in the hormone binding site of the receptor protein such that the affinity for the ligand is greatly reduced or virtually absent. Alternatively, the receptor molecule as such might be missing. By use of receptor-specific monoclonal antibodies it was possible to distinguish between these alternatives (Westphal *et al.*, 1984; Northrop *et al.*, 1985). A full-size receptor polypeptide devoid of hormone binding ability was detected in those r^- variants of S49.1 which were investigated; this defective receptor will be called R^{s^-}.

In two types of receptor abnormalities discovered in S49.1 variants the binding of hormone is normal or almost normal but the subcellular distribution of receptors differs from that of the wild type (Table 1). In the type called nt^- the receptors (R^{nt^-})

Table 1 — Classification of resistance phenotypes

Cell type	Nuclear binding
Wild type	50–60%
r^- resistance	—
nt^- resistance	8–20%
nt^i resistance	80–95%

Binding of [^{3}H]dexamethasone to intact cells and crude cell fractionation were carried out as described in Part I of this book. Percentage nuclear binding is expressed relative to total cellular binding of hormone.

are 'nuclear transfer deficient' in the sense of defective nuclear binding and decreased affinity for general DNA relative to wild-type receptors (R^{wt}). By contrast, in the phenotype referred to as nt^i the receptors (R^{nt^i}) exhibit 'increased nuclear transfer' in the sense of increased binding to nuclei or chromatin and abnormally high affinity for DNA (Yamamoto *et al.*, 1976; Gehring, 1980b).

The distinction between these receptor types became very clear by the use of chromatography on DNA–cellulose (Yamamoto *et al.*, 1974, 1976), a technique described in more detail in Part I of this book. As shown in Fig. 3 and summarized in Table 2, nt^- receptor complexes require lower and nt^i receptors higher salt concentrations respectively to be eluted from DNA–cellulose as compared with the wild type. It should be pointed out that DNA–cellulose chromatography as used here detects the interaction of wild-type and mutant receptors with general and mostly unspecific DNA only. Binding of receptors to specific DNA sequences called 'glucocorticoid response elements' has of course attracted a lot of interest during recent years (Ringold, 1985; Yamamoto, 1985; Beato, 1986; Pfahl, 1986). Because of the ease with which receptors come off the DNA the interaction with these specific sequences can so far only be studied *in vitro* if purified receptor preparations are being used. The bulk of DNA within a eukaryotic cell is certainly of the unspecific type and consequently most receptor binding is to unspecific sequences owing to general affinity for DNA and chromatin.

Interestingly, one of the nt^- variants of S49.1 cells turned out to contain a receptor which in addition to defective DNA binding has decreased affinity for the

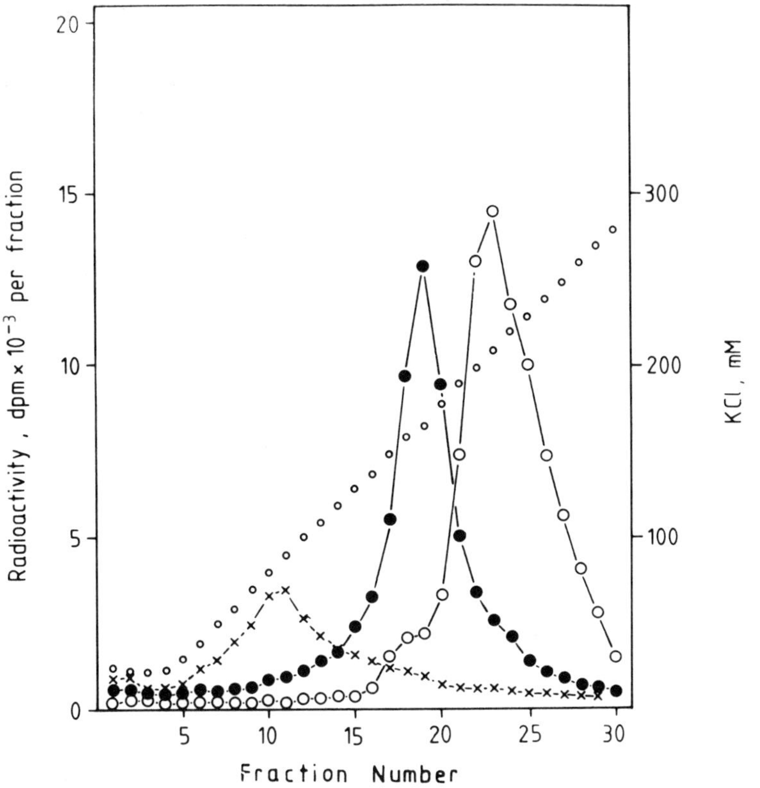

Fig. 3 — DNA–cellulose chromatography of wild-type and mutant receptors. S49.1 wild-type (●), nt^- (x), and nt^i (○) receptor complexes with tritiated triamcinolone acetonide were activated by a 30 min incubation at 20°C and chromatographed on DNA–cellulose (Gehring, 1980b). ○, KC1 concentrations.

hormonal ligand due to an increased rate of dissociation of receptor–hormone complexes (Spindler-Barth and Gehring, 1982). This receptor was indeed recently shown to harbour two amino acid exchanges in separate domains of the molecule, each of them causing different lesions (Danielson *et al.*, 1986).

Among glucocorticoid resistant variants of the human line CEM-C7 an interesting type of receptor defect has turned up which has so far not been found in mouse lymphoma cell systems (Schmidt *et al.*, 1980; Harmon *et al.*, 1985). It has been designated 'activation labile' because these receptors are extremely unstable under conditions which normally activate receptor steroid complexes to forms which are able to bind to DNA. Physiologically activation occurs at body temperature (Munck and Foley, 1979; Marković and Litwack, 1980) but with cell extracts other conditions, for example exposure to high ionic strength, may also cause activation. Under such conditions conformational changes occur which in the case of activation labile receptors lead to rapid loss of the hormonal ligand. A recent study shows that these receptors behave normally if the ligand is covalently attached to the receptor polypeptide by affinity labeling (Eisen *et al.*, 1987).

Table 2 — DNA binding properties and polypeptide molecular weights of receptor types

Receptor	DNA–cellulose chromatography: KCL required for elution (mM)		Molecular weight of hormone-labeled polypeptide	
	Native	After chymotrypsin	Native	After chymotrypsin
S49.1 wild type	175	236	94 700	38 000
S49.1 nt$^-$ (clone 22R)	75	129	94 000	37 400
S49.1 nt$^-$ (clone 83R)	86	87	94 000	37 400
S49.1 nt^i (clone 55R)	225	229	40 300	39 000
S49.1 nt^i (clone 143R)	210	209	40 800	41 000
WEHI-7 wild type	186	233	94 000	38 000
P1798 wild type	193	—	93 000	—
P1798 nt^i	212	—	38 500	—

Data from Dellweg *et al.*, 1982; Gehring and Hotz, 1983.

B. Photoaffinity labeling of receptors

An exciting observation was made when wild-type and mutant receptors of the nt$^-$ and nt^i types were subjected to photoaffinity labeling with radiolabeled hormonal ligands (Nordeen *et al.*, 1981; Dellweg *et al.*, 1982; Gehring and Hotz, 1983). Electrophoresis in SDI containing polyacrylamide gels revealed a labeled polypeptide band of $M_r = 40\ 000$ for nt^i receptors while the wild-type and nt$^-$ receptors had $M_r = 94\ 000$ (Fig. 4, Table 2). The same was also found for R$^{nt^i}$ contained in a variant of the mouse lymphoma P1798 (Table 2).

Photoaffinity labeling of glucocorticoid receptors was first carried out with the gestagen promegestone as ligand (Nordeen *et al.*, 1981). This steroid, like dexamethasone, triamcinolone acetonide, and the naturally occurring glucocorticoids, contains the α,β-unsaturated ketone structure which can be excited by long wavelength UV light as described in detail in Part I of this book. The high affinity glucocorticoids dexamethasone and triamcinolone acetonide offer the great advantage for photoaffinity labeling of glucocorticoid receptors that they can be used at lower concentrations than the weak antagonist promegestone and excess unbound steroid can be removed prior to illumination of the samples (Dellweg *et al.*, 1982; Gehring and Hotz, 1983). Consequently, covalent labeling is much cleaner and more specific; only the receptor is being labeled as it has the highest affinity for the ligand. Indeed, Nordeen *et al.* (1981) obtained considerable labeling of other cellular components by promegestone in extacts of S49.1 and hepatoma cells. A recovery of about 10% crosslinked receptors on photolabeling with triamcinolone acetonide has been reported (Dellweg *et al.*, 1982); however, this may be an overestimate due to some unbound hormone trapped in the protein precipitated with trichloroacetic acid.

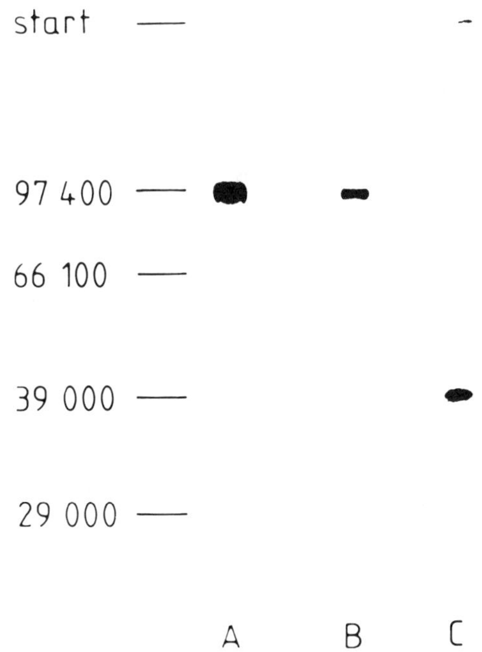

Fig. 4 — SDS gel electrophoresis of photoaffinity labeled receptors. Complexes of S49.1 wild-type (A), nt$^-$ (B), and nt^i (C) receptors with tritiated triamcinolone acetonide were subjected to photolabeling and subsequent SDS–polyacrylamide gel electrophoresis. Radiolabeled bands were detected by fluorography. (Data from Dellweg *et al.*, 1982.)

Photolabeling of receptors with high affinity glucocorticoids needs to be compared with chemical affinity labeling described in Chapter 8. The use of dexamethasone 21-mesylate has the advantage that the yield of covalent cross-linking is much greater than that obtainable by photolabeling. However, glucocorticoid receptors have a lower affinity for dexamethasone mesylate than for dexamethasone or triamcinolone acetonide; some purification of receptors may therefore be required because other components may also react with the mesylate derivative.

Glucocorticoid receptors are known to be rather susceptible to degradation by endogenous or exogenous proteases (Wrange and Gustafsson, 1978; Dellweg *et al.*, 1982; Andreasen and Gehring, 1981; Gehring and Hotz, 1983; Sherman *et al.*, 1983; Reichman *et al.*, 1984). Interestingly, mild treatment of wild-type receptors of S49.1, WEHI-7, and CEM-C7 cells with chymotrypsin produced receptor fragments with properties similar to those of nt^i mutant receptors. Photoaffinity labeling and SDS gel electrophoresis revealed labeled polypeptides of $M_r \sim$ 38 000 (Table 2). These receptor fragments bound more tightly to DNA–cellulose than native wild-type receptors and eluted with about 235 mM KCl (Table 2, Fig. 5). Similar receptor fragments were also obtained from nt$^-$ receptors but these did not show dramatic increases in their affinities for DNA (Table 2). When trypsin was used instead of chymotrypsin it was found to produce hormone binding fragments of $M_r \sim$ 29 000 and 27 000 from wild-type, nt$^-$, and nt^i receptors (Gehring and Hotz, 1983). These

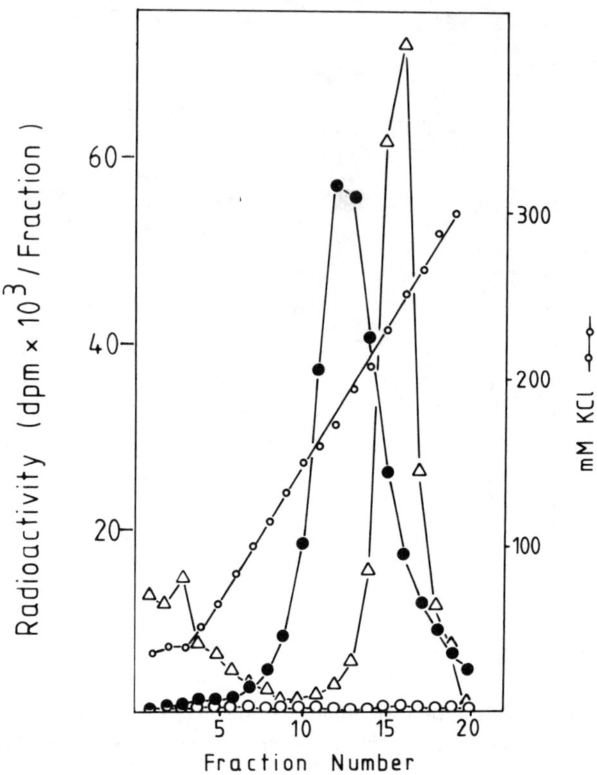

Fig. 5 — DNA-cellulose chromatography of protease treated receptors. S49.1 wild-type receptor complexes with tritiated triamcinolone acetonide were activated and chromatographed as in Fig. 3 either in the native state (●) or following a mild treatment with α-chymotrypsin (△) or trypsin (○). (Data from Dellweg *et al.*, 1982; Gehring and Hotz, 1983).

fragments were devoid of DNA binding ability as is seen in Fig. 5 for the wild type. Tryptic digestion, therefore, is able to separate the domains for hormone binding and for DNA binding while chymotrypsin leaves these domains joint but cleaves off another domain from wild-type and nt^- receptor molecules.

III. A DOMAIN MODEL FOR GLUCOCORTICOID RECEPTORS

Comparison of wild-type and nt^i receptors as well as the above-mentioned proteolysis studies have led to a domain model for the wild-type glucocorticoid receptor polypeptide which is shown schematically in Fig. 6 (Dellweg *et al.*, 1982; Gehring and Hotz, 1983; Gehring, 1984, 1986). Quite obviously, the receptor molecule comprises functional domains for hormone binding and for interaction with DNA or chromatin. In addition, the wild-type receptor contains a third domain which in Fig. 6 is designated M. This part of the molecule was thought to function by modulating the receptor's interaction with DNA or chromatin. In the absence of the M domain, as pointed out above, the truncated nt^i receptor and the partially proteolyzed wild-type

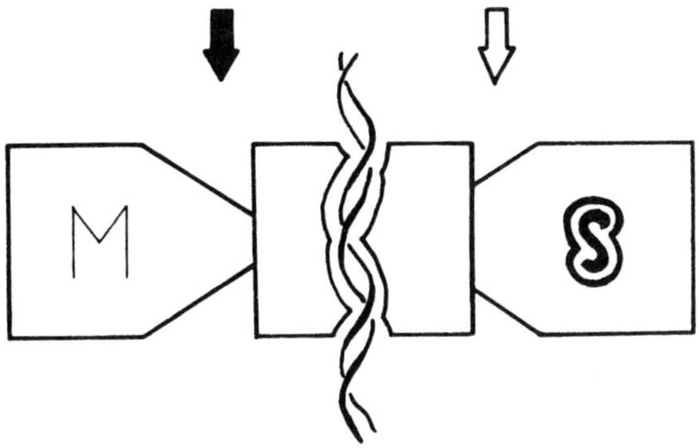

Fig. 6 — Domain model of the wild-type receptor. The DNA binding domain is located in the center and the steroid hormone binding domain (S) to the right. The domains are not drawn in proportion to their lengths within the polypeptide chain. The arrows indicate linker regions which are particularly susceptible to proteases such as chymotrypsin (➡) and trypsin (⇨).

receptor bind to general DNA with abnormally high affinities. Recent sequence studies revealed within this domain a significant preponderance of acidic amino acids (Danielsen *et al.*, 1986) while the rest of the molecule rather appears to be dominated by a cluster of basic amino acids within the DNA binding domain. Thus it is quite obvious that the electrostatic interaction of receptors with DNA is changed significantly on elimination of the M domain. It appeared reasonable to assume that the M domain might aid in discriminating between specific and non-specific DNA binding sites by exerting some kind of allosteric effect on the DNA binding domain, thus limiting the receptor's affinity for general, non-specific DNA (Ringold, 1985; Gehring, 1986). The non-functional state of nt^i receptors in variant lymphoma cells may be explained by too tight binding to the vast excess of general DNA with practically no receptor molecules left for interaction with specific sites. This leaves open the possibility that in a less complex system than glucocorticoid-induced lymphocytolysis and with excess receptor some effect might be seen even if the truncated receptor molecule were used (see below).

The biochemical analysis of wild-type and mutant glucocorticoid receptors showed that the domains for hormone and for DNA binding are adjacent and that the M domain is distally located on the receptor polypeptide. The spacial order of these three domains became known when receptor cDNAs were sequenced recently (Hollenberg *et al.*, 1985; Weinberger *et al.*, 1985; Miesfeld *et al.*, 1985, 1986; Danielsen *et al.*, 1986). It is clear now that the M domain occupies the amino-terminal half of the molecule. The carboxy-terminal portion is quite hydrophobic in nature and is involved in hormone binding. The DNA binding domain occupies a stretch of about 80 amino acids near the center of the molecule with the other two domains on either side. Thus far the amino acid sequences of the glucocorticoid receptors of man, mouse, and rat with 777, 783, and 795 amino acid residues respectively have been described (Hollenberg *et al.*, 1985; Danielsen *et al.*, 1986; Miesfeld *et al.*, 1986). These sequences are extensively homologous. Amino acid differences are more frequent within the M domains than within the domains for hormone and DNA binding. The most striking difference is the presence of a stretch of 8 and 19 glutamine residues within the M domains of the mouse and rat receptors respectively which is absent from the human receptor. The biological significance of this sequence, if any, is not known.

Perhaps the most interesting part of the glucocorticoid receptor polypeptide is the DNA binding domain. This domain shows striking homology in sequence with other steroid hormone receptors (Green and Chambon, 1986), with thyroid hormone receptors (Sap *et al.*, 1986; Weinberger *et al.*, 1986), and with the product of the v-erb-A oncogene of avian erythroblastosis virus (Weinberger *et al.*, 1985; Bishop, 1986). This homology will be dealt with in more detail in Part III of this book. The DNA binding domain is rich in lysine and arginine residues and, moreover, contains an interesting clustering of cysteines. It has therefore been proposed (Berg, 1986) that the contact with DNA is brought about by folding up of the polypeptide chain into finger-like structures similar to those detected as tandemly repeated units in transcription factor IIIA from *Xenopus laevis* (Miller *et al.*, 1985; Brown *et al.*, 1985). The DNA binding domain of steroid hormone receptors may form two such structures involving cysteine and histidine residues. While in the transcription factor Zn^{2+} ions are involved in stabilizing these finger structures there is at present no information available about whether glucocorticoid receptors require Zn^{2+} or any other divalent cation for biological activity.

Interestingly, one of the nt^- mutants of S49.1 cells has recently been shown to contain an amino acid replacement of arginine by histidine in position 484 (Danielsen *et al.*, 1986), i.e. within the DNA binding domain (Fig. 7). This single amino acid

Fig. 7 — Amino acid positions in wild-type and mutant receptors. In the S49.1 wild-type receptor, amino acids Arg 484, Glu 546, Cys 644, and Tyr 770 are indicated. R^{nt^-} refers to the double mutant described in the text and R^{s^-} to the receptor polypeptide devoid of hormone binding. The amino terminus is on the left and the carboxy terminus on the right. (Data from Danielsen *et al.*, 1986).

exchange renders the receptor protein unfunctional. Either arginine 484 in the wild-type receptor is directly and specifically involved in interacting with DNA or a histidine residue placed into this position disturbs the correct formation of the second finger-like structure of the DNA binding domain (Danielsen *et al.*, 1986). Targeted amino acid exchanges in this part of the molecule as can be introduced by site-directed mutagenesis will shed light on this problem.

The hormone binding domains of steroid and thyroid hormone receptors are much longer than the corresponding DNA binding domains. They show some homology which, however, is less pronounced than that found with the DNA binding domains (Green and Chambon, 1986). The steroid ligands probably establish contact with several amino acid side groups within the hormone binding domains. It will be interesting to find out about the individual amino acid residues involved in different hormone receptors and to see which ones are conserved, for example, between glucocorticoid and progesterone receptors or glucocorticoid and mineralocorticoid receptors whose ligands are quite related in their chemical structures. So far the best studied example is the glucocorticoid receptor.

A recent report shows that cysteine residue 656 of the rat receptor is covalently modified by the chemical affinity label dexamethasone mesylate (Simons *et al.*, 1987) and is therefore located within or very close to the hormone binding site. This residue corresponds to cysteine 644 of the mouse glucocorticoid receptor. More information about the hormone binding site came from the analysis of receptor mutants of S49.1 lymphoma cells (Danielson *et al.*, 1986). In the receptor polypeptide R^{s^-} described above the inability to bind the hormone was found to be due to a single amino acid substitution of glycine for glutamic acid in position 546 (Fig. 7). Tyrosine at position 770 in the wild type also contributes to hormone binding (Fig. 7). This amino acid was found to be exchanged against asparagine in the double mutant already described which is defective in DNA binding and has an increased rate of dissociation of the steroid ligand from the receptor–hormone complex. These observations clearly show that in the hormone binding domain a fairly long part of the polypeptide chain consisting of a minimum of 225 amino acid residues is intricately folded in order to form the steroid binding site. The data also show that the domain model of Fig. 6, crude as it may be, is able to explain some of the receptor mutants obtained in lymphoid cells selected for glucocorticoid resistance.

IV. DELETION MUTATIONS INTRODUCED INTO RECEPTOR cDNAs

Following the isolation and elucidation of cDNAs which specify glucocorticoid receptors of several mammalian species it is now possible to produce at will various new mutations by *in vitro* DNA manipulations. These derivatives can be tested for biological activity on transfection into cultured cells of the r^- phenotype. Unfortunately, the lymphoid cells described in this chapter turned out to be particularly difficult to transfect. Rat hepatoma cell variants and the monkey kidney cell line CV-1 and its derivative COS-7, all of which grow as attached cells, have therefore been used in transfection studies.

One series of deletions concerns the M domain of the glucocorticoid receptor. Using the rat receptor cDNA several deletions were produced and each one was stably transfected into r^- rat hepatoma cells which carry several integrated copies of the glucocorticoid inducible mouse mammary tumor virus (Miesfeld *et al.*, 1987). It was found that elimination of part of the M domain or its whole, i.e. amino acids 1–406 from the total of 795 residues, resulted in synthesis of mutilated receptor molecules which nevertheless were able to mediate hormone induction of an endogenous liver-specific gene and of mammary tumor virus RNA. Depending on the deletion, however, the level of induction was significantly reduced. The truncated receptor with the largest deletion produced a 5- to 6-fold lower induction ratio than the full-length polypeptide even though it was expressed to a much higher level as a result of the specific plasmid construction. Similar deletions in the sequence coding for the M domain of the mouse receptor were also constructed and were cotransfected into COS-7 cells together with the mouse mammary tumor virus LTR region coupled to a bacterial indicator gene (Danielsen *et al.*, 1986, 1987). These deletions were again found to be qualitatively functional for gene activation. Even though these data clearly argue against an absolute requirement of the M domain for receptor function they show that this domain modulates the extent of hormonal responsiveness. The contribution of the M domain to more complex receptor mediated responses such as the cytolysis effect in lymphoid cells remains to be elucidated. It is also possible that the M domain plays a more decisive role in glucocorticoid controlled repression of genes like those for α_1-fetoprotein, pro-opiomelanocortin and others.

The hormone binding domain has also been studied by deletion analysis. In one approach cDNA fragments were transcribed *in vitro* using SP6 RNA polymerase, the mRNAs produced were translated in a reticulocyte lysate system, and the products were assayed for dexamethasone binding (Rusconi *et al.*, 1986; Godowski *et al.*, 1987). It was found that deletion of 5 or 29 amino acids from the carboxy terminus reduced hormone binding by 50- to 100-fold. In another approach a series of rat receptor cDNAs bearing nested deletions were transiently transfected into CV-1 cells together with a glucocorticoid inducible indicator gene construct (Rusconi *et al.*, 1986; Godowski *et al.*, 1987). Deletions of as few as 27 carboxy-terminal amino acids and up to 180 residues failed to activate the enhancer function of the glucocorticoid response element in the presence or the absence of hormone. A very exciting result was obtained when even larger deletions were analyzed. Receptor derivatives from which 200–270 carboxy-terminal amino acids were missing showed constitutive activation of the glucocorticoid response element with elevated transcription of the indicator gene even in the absence of hormone. This observation suggests that in the wild type the unliganded hormone binding domain serves to suppress directly or indirectly the DNA binding domain. The steroid acts by relieving this inhibition. Following hormone binding a structural change in the receptor molecule may occur which leads to unmasking of the DNA binding domain which in itself is already preformed in its functional configuration.

An an extension of the deletion studies described above double mutations have been constructed on the cDNA level which on transfection into CV-1 or COS-7 cells produced receptor derivatives truncated on both amino-terminal and carboxy-terminal ends of the molecule (Miesfeld *et al.*, 1987). A polypeptide of only 150 amino acids was shown to be sufficient for constitutive activation. This receptor fragment encompasses a core of 86 amino acids which has been delineated by maximum deletions on either end of the molecule and which is the DNA binding domain.

V. IMMUNOCHEMICAL STUDIES WITH GLUCOCORTICOID RECEPTORS

Purified wild-type receptors of rat liver and human IM-9 cells were used by several groups to produce polyclonal and monoclonal antibodies. Interestingly, most of these antibodies recognize determinats within the M domain; this part of the molecule has therefore also been called the immunoactive domain (Carlstedt-Duke *et al.*, 1982; Weinberger *et al.*, 1985). When mutant receptors R^{nt^i} of S49.1 and P1798 mouse lymphoma cells were tested with some of these antibodies they were found not to react (Okret *et al.*, 1983; Westphal *et al.*, 1984; Northrop *et al.*, 1985). By contrast, R^{nt^-} and R^{s^-} mutant products were recognized by the antibodies (Westphal *et al.*, 1984; Northrop *et al.*, 1985). Perhaps the most widely used antibodies at present are the IgG monoclonals mab 49 (Westphal *et al.*, 1982) and BUGR-1 (Gametchu and Harrison, 1984). Both were obtained by immunizing mice with purified receptor from rat liver; however, they recognize different epitopes. Table 3

Table 3 — Antibody reactivity of rodent glucocorticoid receptors

Monoclonal antibody	R^{wt}	R^{wt} after chymotrypsin	R^{nt^i}	R^{nt^-}	R^{s^-}
mab 49	+	−	−	+	+
BUGR-1	+	+	−	+	+

The wild-type receptor R^{wt} was from rat liver or S49.1 cells; mutant receptors were of S49.1 origin. Data from Westphal *et al*, 1984; Northrop *et al.*, 1985; Eisen *et al.*, 1985; M. Rexin and U. Gehring, unpublished.

shows that BUGR-1 still reacts with the wild-type receptor of mouse and rat origin after elmination of the M domain by chymotryptic cleavage while mab 49 does not. This suggests that the BUGR-1 epitope is either within the DNA binding domain of the receptor or very close to it. Furthermore, R^{nt^i} is different from R^{wt} at this very location, i.e. the nt^i mutation leads to destruction of this epitope. By contrast, the mab 49 epitope is located within the M domain.

As described above, removal of the M domain from wild-type receptors by proteolytic cleavage results in increased affinity for general DNA. It was therefore of interest to see whether tagging of the M domain with antibody might influence the DNA binding properties of the wild-type receptor. The immune complex with mab 49 was indeed found to chromatograph on DNA–cellulose similarly to R^{nt^i} and chymotrypsin treated R^{wt} (Gehring and Segnitz, 1987). Covering up of the M domain by an IgG molecule thus mimics its removal. This observation raises the possiblity that biochemical modifications of the M domain might affect hormone responsiveness within living cells by altering the receptor's general DNA binding properties.

VI. RECEPTOR GENOTYPES OF MOUSE LYMPHOMA CELLS

Wild-type S49.1 and WEHI-7 cells differ in that line WEHI-7 has two receptor alleles which produce wild-type receptors while S49.1 is functionally hemizygous (Bourgeois and Newby, 1977; Francke and Gehring, 1980). Hemizygosity for the receptor gene was first demonstrated by hybridization of S49.1 cells with another mouse lymphoma (EL4) that contains abnormal receptors of the nt^- type. While the original cell hybrids were glucocorticoid sensitive and had a full complement of chromosomes from both parental cell types, resistant segregants turned out to be devoid of wild-type receptors and had lost one chromosome that was of S49.1 origin (Gehring, 1980b; Francke and Gehring, 1980). Immunochemical studies showed that the other receptor allele in S49.1 cells carries a mutation which gives rise to the abnormal receptor R^{s^-} (Westphal *et al.*, 1984; Gehring, 1986). With only one functional receptor allele expressed in S49.1 wild-type cells it is now obvious why it was so easy to obtain glucocorticoid resistant variants carrying receptor defects.

The wild-type allele in S49.1, designated r^{wt} in Fig. 8, may undergo one of three

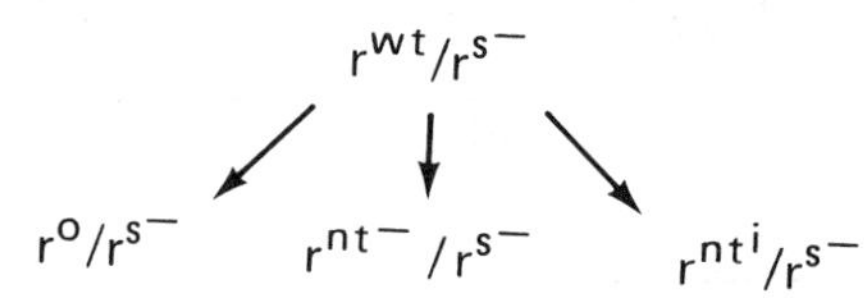

Fig. 8 — Receptor genotypes of wild-type and mutant S49.1 cells.

types of changes which have so far been recognized: it may mutate to r^{nt^-} or r^{nt^i} from which the mutant receptors R^{nt^-} and R^{nt^i} originate, or it may switch to the silent state r^o. Apparently, the r^o allele is shut off, since it does not lead to production of a detectable receptor polypeptide (Westphal *et al.*, 1984) or receptor mRNA (Miesfeld *et al.*, 1984, 1985). DNA analyses did not provide any evidence for deletions or DNA rearrangements within the r^o allele (Northrop *et al.*, 1986). The reason that this receptor allele is not expressed is unknown at present. So far only very few S49.1 variants of the r^- phenotype have been analyzed in detail; it is therefore possible that in addition to the kinds of changes of r^{wt} depicted in Fig. 8 this allele may also mutate to r^{s^-} in cell culture. Since the hormone binding domain of the receptor is quite large and complex, as pointed out above, it appears very likely that various spontaneous mutations may occur in this part of the molecule, all leading to R^{s^-} polypeptides.

The origin of the r^{s^-} mutation already present in wild-type S49.1 cells (Fig. 8) is not clear. It probably occurred as a spontaneous event very early in the history of the cell line or, perhaps more likely, in the S49 tumour from which the S49.1 cell line was established. Interestingly, the different receptor alleles in S49.1 may be expressed to different levels in different cell clones (Gehring *et al.*, 1984; Westphal *et al.*, 1984; Northrop *et al.*, 1986).

The origin of the truncated nt^i receptor remained open for speculation for quite some time and will only be fully understood when the sequence of the corresponding cDNA is available. The fact that wild-type and nt^i receptors can be present simultaneously in hybrid lymphoma cells excluded the possibility of overproduction of a protease in nt^i variant cells (Gehring, 1980a) but left open the possibility that, because of a mutated region, the putative nt^i receptor precursor might be particularly vulnerable to endogenous proteolysis. This has in fact been postulated and it was assumed that the nt^i receptor originally had the same size as the wild type (Gruol *et al.*, 1984; Bourgeois and Gasson, 1985). It became clear from recent studies with receptor cDNA probes that R^{nt^i} is being synthesized as a truncated polypeptide rather than being processed from a full-length molecule. In nt^i variant cells a major receptor mRNA was detected which is shorter by about 1.5 kb than the wild-type message and from which 5′ sequences are missing (Miesfeld *et al.*, 1984, 1985; Northrop *et al.*, 1986). Southern blot analyses did not reveal any gross DNA rearrangements or genomic deletions (Miesfeld *et al.*, 1984; Northrop *et al.*, 1986). Transcription of r^{nt^i} may therefore be initiated correctly and the truncated nt^i mRNA may be the result of aberrant splicing due to a mutational change. Such a splicing defect with translation starting from an internal methionine would certainly explain the destruction of the BUGR-1 epitope described above. It will be interesting to find out whether nt^i receptors are simply missing a long stretch of amino-terminal amino acids or whether they have new sequences at their amino termini which are not present in the wild type. Since a deletion of more than 400 residues from the amino terminus of the wild-type receptor was still functional (see above) it was important to ask whether nt^i receptors would also be able to mediate some hormonal response. In preliminary experiments nt^i variant cells were transfected with a DNA construct consisting of the glucocorticoid inducible promoter of mouse mammary tumor virus coupled to the prokaryotic gene for chloramphenicol acetyl transferase (U. Gehring and H. Losert, unpublished experiments). In these cells a hormone effect on the activity of the acetyle transferase was consistently observed; however, the ratio of induction was low, as one might expect.

In contrast to the S49.1 cell system, wild-type cells of line WEHI-7 are of the receptor geneotype r^{wt}/r^{wt} (Bourgeois and Newby, 1977; Bourgeois and Gasson, 1985). This is reflected in higher levels of receptor mRNA (Miesfeld *et al.*, 1984, 1985; Northrop *et al.*, 1986), higher numbers of cellular receptors (Bourgeois and Newby, 1977; Gehring *et al.*, 1984) and consequently lower hormone concentrations required to produce the cytolytic response (Bourgeois and Newby, 1977; Gehring *et al.*, 1984). The homozygous receptor state of wild-type WEHI-7 cells also explains why it was impossible to obtain glucocorticoid resistant variants in a single step without prior mutagenesis. Such resistant sublines, however, were obtained by using a two-step procedure in which cells of intermediate sensitivity similar to the S49.1 wild type were preselected (Bourgeois and Newby, 1977). Interestingly, a resistant subclone of WEHI-7 cells has been described in which the levels of hormone binding, immunoreactive material, and receptor mRNA are decreased in a parallel fashion (Northrop *et al.*, 1985, 1986). These cells probably contain a silent receptor allele r^o. Another WEHI-7 subclone was devoid of hormone binding and cross reacting material but still expressed a low level of receptor mRNA (Northrop *et al.*, 1985, 1986), indicating that multiple controls of receptor expression may exist in different cells.

VII. HIGH MOLECULAR WEIGHT RECEPTOR FORMS

If steroid receptor containing cell extracts is analyzed under very mild and non-denaturing conditions, receptor forms with molecular weights as large as 300 000–350 000 may be detected (for reviews, see Sherman and Stevens, 1984;

Vedeckis, 1985). This raises the problem as to how the steroid binding polypeptides which in the case of wild-type glucocorticoid receptors have $M_r \sim 94\,000$ are organized in such large structures. To this end nt^i mutant receptors with hormone binding polypeptides of $M_r \sim 40\,000$ have become very useful (Gehring and Arndt, 1985; Gehring *et al.*, 1987). It was reasoned that high molecular weight forms of these should be different in their hydrodynamic properties from wild-type receptors. If, for example, the large complexes of wild type and nt^i contained only one hormone binding polypeptide they should differ by a molecular mass of about 50 000; if, however, they contained two or more steroid carrying subunits the differences should be much greater, provided, of course, that the other components of the large complexes are the same in nt^i and wild-type receptors. Table 4 summarizes the

Table 4 — High molecular weight glucocorticoid receptors

Receptor type (treatment)	R_S(Å)	$s_{20,w}$(S)	M_r
S49.1 wild type (native)	81	9.5	325 000
S49.1 wild type (after chymotrypsin)	69	10.0	291 000
S49.1 nt^i (native)	71	9.5	285 000

Molecular weights were calculated from R_S and $s_{20,w}$. Data from Gehring and Arndt, 1985.

results obtained from gel filtration and sedimentation analyses and lists the Stokes radii (R_S), sedimentation coefficients ($s_{20,w}$) and molecular weights. The data indeed show a difference in molecular weights of about 40 000 between wild-type and nt^i receptors and thus suggest a heteromeric structure of the high molecular weight forms with only one hormone binding polypeptide per complex. Mild treatment of the wild-type receptor with chymotrypsin again produced a receptor form which was very similar to the nt^i mutant (Table 4). This supports the view that the associated macromolecular components are the same in wild-type and mutant receptors. It is interesting to note that a heteromeric structure has also been suggested for the rat liver receptor from experiments using a completely different approach (Okret *et al.*, 1985).

What then are the other subunits in the high molecular weight receptor forms? According to the above-mentioned differences in molecular weights they should amount to a molecular mass of about 230 000–250 000. Recent experiments point to two types of macromolecules. A non-hormone binding polypeptide of $M_r \sim 90\,000$ has been detected which is common to the large forms of several steroid receptors and which has been identified as the heat-shock protein hsp 90 (Joab *et al.*, 1984; Catelli *et al.*, 1985; Housley *et al.*, 1985; Sanchez *et al.*, 1985; Mendel *et al.*, 1986). In addition there is accumulating evidence that RNA is associated with large-sized receptor complexes (Economidis and Rousseau, 1985; Kovačič-Milivojević *et al.*, 1985; Webb and Litwack, 1986). Judged from the molecular weights of these associated molecules the large receptor complexes may consist of one hormone binding polypeptide, two molecules of hsp 90 and one molecule of RNA of the size of tRNA (Gehring *et al.*, 1987). However, such a structure would only be proven unequivocally by appropriate reconstitution experiments which so far have not been successful. Also, the functions of the additional receptor components in relation to receptor activity are not yet understood.

The high molecular weight receptor complexes are known to be forms which do not interact with DNA. Increasing the ionic strength in buffers results in dissociation of the large forms of wild-type and nt^{i} receptors and concomitant generation of DNA binding ability (Gehring *et al.*, 1987). This suggests that subunit dissociation leads to exposure of the DNA binding domain which — as pointed out above — appears to be a preformed structure. A relatively small region between positions 583 and 633 in the murine glucocorticoid receptor which shows homology to other receptors was proposed to be involved in forming the high molecular weight complexes (Danielsen *et al.*, 1986). The above-mentioned deletions resulting in hormone independent receptor activity (Godowski *et al.*, 1987; Miesfeld *et al.*, 1987) in fact include this region of homology.

ACKNOWLEDGEMENT

Studies carried out in the author's laboratory were supported by grants from the Deutsche Forschungsgemeinschaft.

REFERENCES

Andreasen, P. A. and Gehring, U. (1981) Activation and partial proteolysis of variant glucocorticoid receptors, studied by two-phase partitioning. *Eur. J. Biochem.* **120**, 443–449.

Beato, M. (1986) Steroid hormone receptors as gene regulators. In: Puett, D., Ahmad, F., Black, S., Lopez, D. M., Melner, M. H., Scott, W. A. and Whelan, W. J. (eds.) *Advances in Gene Technology: Molecular Biology of the Endocrine System. Proceedings of the 18th Miami Winter Symposium.* Cambridge University Press, Cambridge, pp. 250–253.

Berg, J. M. (1986) More metal-binding fingers. *Nature* **319,** 264–265.

Bishop, J. M. (1986) Oncogenes as hormone receptors. *Nature* **321,** 112–113.

Bourgeois, S. and Gasson, J. C. (1985) Genetic and epigenetic bases of glucocorticoid resistance in lymphoid cell lines. In: Litwack G. (ed.) *Biochemical Actions of Hormones*, Vol. 12. Academic Press, New York, pp. 311–351.

Bourgeois, S. and Newby, R. F. (1977) Diploid and haploid states of the glucocorticoid receptor gene of mouse lymphoid cell lines. *Cell* **11,** 423–430.

Brown, R. S., Sander, C. and Argos, P. (1985) The primary structure of transcription factor TFIIIA has 12 consecutive repeats. *FEBS Lett.* **186,** 217–274.

Carlstedt-Duke, J., Okret, S., Wrange, Ö. and Gustafsson, J.-Å. (1982) Immunochemical analysis of the glucocorticoid receptor: Identification of a third domain separate from the steroid-binding and DNA-binding domains. *Proc. Natl. Acac. Sci. USA* **79**, 4260–4264.

Catelli, M. G., Binart, N., Jung-Testas, I., Renoir, J. M., Baulieu, E. E., Fermamisco, J. R. and Welch, W. J. (1985) The common 90-kd protein component of non-transformed '8S' steroid receptors is a heat-stock protein. *EMBO J.* **4,** 3131–3135.

Danielsen, M. and Stallcup, M. R. (1984) Down-regulation of glucocorticoid receptors in mouse lymphoma cell variants. *Mol. Cell. Biol.* **4,** 449–453.

Danielsen, M., Northrop, J. P. and Ringold, G. M. (1986) The mouse glucocorticoid receptor: mapping of functional domains by cloning, sequencing and expression of wild-type and mutant proteins. *EMBO J.* **5**, 2513–2522.

Danielsen, M., Northrop, J. P. and Ringold, G. M. (1987) Mutational analysis of the glucocorticoid receptor. *J. Cell. Biochem. Suppl.* **11A**, 104.

Dellweg, H.-G., Hotz, A., Mugele, K. and Gehring, U. (1982) Active domains in wild-type and mutant glucocorticoid receptors. *EMBO J.* **1,** 285–289.

Economidis, I. V. and Rousseau, G. G. (1985) Association of the glucocorticoid hormone receptor with ribonucleic acid. *FEBS Lett.* **181,** 47–52.

Eisen, L. P., Reichman, M. E., Thompson, E. B., Gametchu, B., Harrison, R. W. and Eisen, H. J. (1985) Monoclonal antibody to the rat glucocorticoid receptor. *J. Biol. Chem.* **260,** 11805–11810.

Eisen, L. P., Elsasser, M. S. and Harmon, J. M. (1987) Activation labile glucocorticoid receptor mutants. *J. Cell. Biochem. Suppl.* **11A,** 105.

Francke, U. and Gehring, U. (1980) Chromosome assignment of a murine glucocorticoid receptor gene using intraspecies somatic cell hybrids. *Cell* **22,** 657–664.

Gametchu, B. and Harrison, R. W. (1984) Characterization of a monoclonal antibody to the rat liver glucocorticoid receptor. *Endocrinology* **114,** 274–279.

Gehring, U. (1980a) Cell genetics of glucocorticoid responsiveness. In: Litwack, G. (ed.) *Biochemical Actions of Hormones*, Vol. 7. Academic Press, New York, pp. 205–232.

Gehring, U. (1980b) Specific receptors control steroid sensitivity in lymphoma cell hybrids. *Mol. Cell. Endocrinol.* **20,** 261–274.

Gehring, U. (1984) Functional domains in wild-type and variant glucocorticoid receptors of lymphoid cells. In: Bresciani, F., King, R. J. B., Lippman, M. E., Namer, M. and Raynaud, J.-P. (eds.) *Progress in Cancer Research and Therapy*, Vol. 31, *Hormones and Cancer* 2. Raven Press, New York, pp. 245–254.

Gehring, U. (1986) Genetics of glucocorticoid receptors. *Mol. Cell. Endocrinol.* **48,** 89–96.

Gehring, U. and Arndt, H. (1985) Heteromeric nature of glucocorticoid receptors. *FEBS Lett.* **179,** 138–142.

Gehring, U. and Hotz, A. (1983) Photoaffinity labelling and partial proteolysis of wild-type and variant glucocorticoid receptors. *Biochemistry* **22,** 4013–4018.

Gehring, U. and Segnitz, B. (1987) Modulation of DNA binding of glucocorticoid receptors, submitted.

Gehring, U., Gray, J. W. and Tomkins, G. M. (1976) Corticosteroid effect on the cell cycle. In: Büchner, T., Göhde, W. and Schumann, J. (eds.) *Pulse-Cytophotometry*. European Press, Gent, pp. 284–289.

Gehring, U., Ulrich, J. and Segnitz, B. (1982) Lymphoma cell variants of decreased glucocorticoid sensitivity. *Mol. Cell. Endocrinol.* **28,** 605–611.

Gehring, U., Mugele, K. and Ulrich, J. (1984 Cellular receptor levels and glucocorticoid responsiveness of lymphoma cells. *Mol. Cell. Endocrinol.* **36,** 107–113.

Gehring, U., Mugele, K., Arndt, H. and Busch, W. (1987) Subunit dissociation and activation of wild-type and mutant glucocorticoid receptors, submitted.

Godowski, P. J., Rusconi, S., Miesfeld, R. and Yamamoto, K. R. (1987) Glucocorticoid receptor mutants that are constitutive activators of transcriptional enhancement. *Nature* **325,** 365–368.

Green, S. and Chambon, P. (1986) A superfamily of potentially oncogenic hormone receptors. *Nature* **324,** 615–617.

Gruol, D. J., Kempner, E. S. and Bourgeois, S. (1984) Characterization of the glucocorticoid receptor. *J. Biol. Chem.* **259,** 4833–4839.

Harmon, J. M. and Thompson, E. B. (1981) Isolation and characterization of dexamethasone-resistant mutants from human lymphoid cell line CEM-C7. *Mol. Cell. Biol.* **1,** 512–521.

Harmon, J. M., Thompson, E. B. and Baione, K. A. (1985) Analysis of glucocorticoid-resistant leukemic cells by somatic cell hybridization. *Cancer Res.* **45,** 1587–1593.

Harris, A. W. and Baxter, J. D. (1979) Variations in cellular sensitivity to glucocorticoid: Observations and Mechanisms. In: Baxter, J. D. and Rousseau, G. G. (eds.) *Glucocorticoid Hormone Action*. Springer-Verlag, Berlin, pp. 423–448.

Hollenberg, S. M., Weinberger, C., Ong, E. S., Cerelli, G., Oro, A., Lebo, R., Thompson, E. B., Rosenfeld, M. G. and Evans, R. M. (1985) Primary structure and expression of a functional human glucocorticoid receptor cDNA. *Nature* **318**, 635–641.

Housley, P. R., Sanchez, E. R., Westphal, H. M., Beato, M. and Pratt, W. B. (1985) The molybdate-stabilized L-cell glucocorticoid receptor isolated by affinity chromatography or with a monoclonal antibody is associated with a 90–92-kDa nonsteroid-binding phosphoprotein. *J. Biol. Chem.* **260,** 13810–13817.

Joab, I., Radanyi, C., Renoir, M., Buchou, T., Catelli, M.-G., Binart, N., Mester, J. and Baulieu, E. E. (1984) Common non-hormone binding component in non-transformed chick oviduct receptors of four steroid hormones. *Nature* **308,** 850–853.

Kovačič-Milivojević, B., LaPointe, M. C., Reker, C. E. and Vedeckis, W. V. (1985) Ribonucleic acid is a component of the oligomeric, transformed mouse AtT-20 cell glucocorticoid receptor. *Biochemistry* **24,** 7357–7366.

Marković, R. D. and Litwack, G. (1980) Activation of liver and kidney glucocorticoid-receptor complexes occurs in vivo. *Arch. Biochem. Biophys.* **202,** 374–379.

Mendel, D. B., Bodwell, J. E., Gametchu, B., Harrison, R. W. and Munck, A. (1986) Molybdate-stabilized nonactivated glucocorticoid-receptor complexes contain a 90-kDa non-steroid-binding phosphoprotein that is lost on activation. *J. Biol. Chem.* **261,** 3758–3763.

Mermod, J.-J., Intière, L., MacLeod, C. L. and Bourgeois, S. (1981) Characterization of a new type of thymoma variants supersensitive to dexamethasone. *J. Steroid Biochem.* **15,** 25–34.

Miesfeld, R., Okret, S., Wikström, A.-C., Wrange, Ö., Gustafsson, J.-Å. and Yamamoto, K. R. (1984) Characterization of a steroid hormone receptor gene and mRNA in wild-type and mutant cells. *Nature* **312,** 779–781.

Miesfeld, R., Rusconi, S., Okret, S., Wikström, A.-C., Gustafsson, J.-Å. and Yamamoto, K. R. (1985) preliminary analyses of the glucocorticoid receptor gene and its expression: a DNA-binding protein essential for hormone-dependent transcriptional enhancement. In: Calendar, R. and Gold, L. (eds.) *Sequence Specificity in Transcription and Translation*, *UCLA Symposia on Molecular and Cellular Biology*, Vol. 30. Alan R. Liss, New York, pp. 535–545.

Miesfeld, R., Rusconi, S., Godowski, P. J., Maler, B. A., Okret, S., Wikström, A.-C., Gustafsson, J.-Å. and Yamamoto, K. R. (1986) Genetic complementation of a glucocorticoid receptor deficiency by expression of cloned receptor cDNA. *Cell* **46,** 389–399.

Miesfeld, R., Godowski, P. J., Maler, B. A. and Yamamoto, K. R. (1987) Glucocorticoid receptor mutants that define a small region sufficient for enhancer activation. *Science*, in press.

Miller, J., McLachlan, A. D. and Klug, A. (1985) Repetitive zinc-binding domains in the protein transcription factor IIIA from *Xenopus* oocytes. *EMBO J.* **4,** 1609–1614.

Munck, A. and Foley, R. (1979) Activation of steroid hormone–receptor complexes in intact target cells in physiological conditions. *Nature* **278,** 752–754.

Nordeen, S. K., Lan, N. C., Showers, M. O. and Baxter, J. D. (1981) Photoaffinity labeling of glucocorticoid receptors. *J. Biol. Chem.* **256,** 10503–10508.

Norman, M. R. and Thompson, E. B. (1977) Characterization of a glucocorticoid-sensitive human lymphoid cell line. *Cancer Res.* **37,** 3785–3791.

Northrop, J. P., Gametchu, B., Harrison, R. W. and Ringold, G. M. (1985) Characterization of wild type and mutant glucocorticoid receptors from rat hepatoma and mouse lymphoma cells. *J. Biol. Chem.* **260,** 6398–6403.

Northrop, J. P., Danielsen, M. and Ringold, G. M. (1986) Analysis of glucocorticoid unresponsive cell variants using a mouse glucocorticoid receptor complementary DNA clone. *J. Biol. Chem.* **261,** 11064–11070.

Okret, S., Stevens, Y.-W., Carlstedt-Duke, J., Wrange, Ö., Gustafsson, J.-Å. and Stevens, J. (1983) Absence in glucocorticoid-resistant mouse lymphoma P1798 of a glucocorticoid receptor domain responsible for biological effects. *Cancer Res.* **43,** 3127–3131.

Okret, S., Wikström, A.-C. and Gustafsson, J.-Å. (1985) Molybdate stabilized glucocorticoid receptor: evidence for a receptor heteromer. *Biochemistry* **24,** 6581–6586.

Okret, S., Poellinger, L., Dong, Y. and Gustafsson, J.-Å. (1986) Down-regulation of glucocorticoid receptor mRNA by glucocorticoid hormones and recognition by the receptor of a specific binding sequence within a receptor cDNA clone. *Proc. Natl. Acad. Sci. USA* **83,** 5899–5903.

Pfahl, M. (1986) Interaction of the glucocorticoid receptor with specific DNA sequences. In: Litwack, G. (ed.) *Biochemical Actions of Hormones*, Vol. 13. Academic Press, New York, pp. 325–357.

Pfahl, M., Kelleher, R. J. and Bourgeois, S. (1978) General features of steroid resistance in lymphoid cell lines. *Mol. Cell. Endocrinol.* **10,** 193–207.

Reichman, M. E., Foster, C. M., Eisen, L. P., Eisen, H. J., Torain, B. F. and Simons, S. S. (1984) Limited proteolysis of covalently labeled glucocorticoid

receptors as a probe of receptor structure. *Biochemistry* **23,** 5376–5384.
Ringold, G. M. (1985) Steroid hormone regulation of gene expression. *Ann. Rev. Pharmacol. Toxicol.* **25,** 529–566.
Rusconi, S., Miesfeld, R., Godowski, P. J., Vanderbilt, J. N., Maler, B. A. and Yamamoto, K. R. (1986) Functional analysis of cloned glucocorticoid receptor sequences. In: Reznikoff, W. R., Burgess, R. R., Dahlberg, J. E., Gross, C. A., Record, T. M. and Wickens, M. P. (eds.) *RNA Polymerase and the Regulation of Transcription, Sixteenth Steenbock Symposium.* Elsevier, New York.
Sanchez, E. R., Toft, D. O., Schlesinger, M. J. and Pratt, W. B. (1985) Evidence that the 90-kDa phosphoprotein associated with the untransformed L-cell glucocorticoid receptor is a murine heat shock protein. *J. Biol. Chem.* **260,** 12398–12401.
Sap, J., Muñoz, A., Damm, K., Goldberg, Y., Ghysdael, J., Leutz, A., Beug, H. and Vennström, B. (1986) The c-erb-A protein is a high-affinity receptor for thyroid hormone. *Nature* **324,** 635–640.
Schmidt, T. J., Harmon, J. M. and Thompson, E. B. (1980) 'Activation-labile' glucocorticoid-receptor complexes of a steroid-resistant variant of CEM-C7 human lymphoid cells. *Nature* **286,** 507–510.
Sherman, M. R. and Stevens, J. (1984) Structure of mammalian steroid receptors: evolving concepts and methodological developments. *Ann. Rev. Physiol.* **46,** 83–105.
Sherman, M. R., Moran, M. C., Tuazon, F. B. and Stevens, Y.-W. (1983) Structure, dissociation and proteolysis of mammalian steroid receptors. *J. Biol. Chem.* **258,** 10366–10377.
Sibley, C. H. and Tomkins, G. M. (1974) Mechanisms of steroid resistance. *Cell* **2,** 221–227.
Spindler-Barth, M. and Gehring, U. (1982) Kinetics of glucocorticoid interaction with wild-type and variant receptors. *FEBS Lett.* **138,** 91–94.
Thompson, E. B., Sakaue, Y. and Harmon, J. M. (1986) In: Sakamoto, Y. and Isohashi, F. (eds.) *Glucocorticoid Hormone: Mechanisms of Action.* Japanese Science Society Press, Tokyo and Springer-Verlag, Berlin, pp. 119–135.
Vedeckis, W. V. (1985) Steroid hormone receptor structure in normal and neoplastic cells. In: Hollander, V. P. (ed.) *Hormonally Responsive Tumors.* Academic Press, New York, pp. 3–61.
Webb, M. L. and Litwack, G. (1986) Association of RNA with the glucocorticoid receptor and possible role in activation. In: Litwack, G. (ed.) *Biochemical Actions of Hormones*, Vol. 13. Academic Press, New York, pp. 379–402.
Weinberger, C., Hollenberg, S. M., Rosenfeld, M. G. and Evans, R. M. (1985) Domain structure of human glucocorticoid receptor and its relationship to the v-erb-A oncogene product. *Nature* **318,** 670–672.
Weinberger, C., Thompson, C. C., Ong, E. S., Lebo, R., Gruol, D. J. and Evans, R. M. (1986) The c-erb-A gene encodes a thyroid hormone receptor. *Nature* **324,** 641–646.
Westphal, H. M., Moldenhauer, G. and Beato, M. (1982) Monoclonal antibodies to the rat liver glucocorticoid receptor. *EMBO J.* **1,** 1467–1471.
Westphal, H. M., Mugele, K., Beato, M. and Gehring, U. (1984) Immunochemical characterization of wild-type and variant glucocorticoid receptors by monoclo-

nal antibodies. *EMBO J.* **3,** 1493–1498.

Wrange, Ö. and Gustafsson, J.-Å. (1978) Separation of the hormone- and DNA-binding sites of the hepatic glucocorticoid receptor by means of proteolysis. *J. Biol. Chem.* **253,** 856–865.

Yamamoto, K. R. (1985) Steroid receptor regulated transcription of specific genes and gene networks. *Ann. Rev. Genet.* **19,** 209–252.

Yamamoto, K. R., Stampfer, M. R. and Tomkins, G. M. (1974) Receptors and glucocorticoid-sensitive lymphoma cells and two classes of insensitive clones: physical and DNA-binding properties. *Proc. Natl. Acad. Sci. USA* **71,** 3901–3905.

Yamamoto, K. R., Gehring, U., Stampfer, M. and Sibley, C. H. (1976) Genetic approaches to steroid hormone action. *Recent Progr. Horm. Res.* **32,** 3–32.

10

Photoaffinity labeling of the chicken progesterone receptor

H. Gronemeyer
Institut de Chimie Biologique, Faculté de Médecine, Strasbourg, France

I. INTRODUCTION

For nearly 18 years the chicken progesterone receptor has been the subject of considerable controversy. Since the initial report by Sherman *et al.* (1970) concerning the existence of more than one protein which bound progestins in steroid receptor fashion in the chick oviduct cytosol, a variety of studies have been conducted to prove or disprove the co-existence of two (or more) progesterone receptor polypeptides. Although up to five different 'forms', termed I, II or B, III or

A, IV, and V or 'mero' receptor, have been described (Sherman *et al.*, 1976), it has been demonstrated that most of these forms are either aggregates (form I) or proteolytic fragments. For example, the 'mero' receptor with a molecular weight of approximately 20 kDa is the result of a specific proteolytic event which is induced by certain divalent cations such as Ca^{2+} (Sherman *et al.*, 1974, 1976, 1977, 1978). Excluding those 'forms' which correspond to proteolyzed and aggregated receptor proteins, two forms, generally termed A and B (or III and II), have been the subject of intensive analysis. One of the major contributors in this field was the group of O'Malley and a number of reviews have been presented summarizing their work on the A–B progesterone receptor system (Schrader, 1975; Schrader and O'Malley, 1977; Vedeckis *et al.*, 1978; Schrader *et al.*, 1981; Grody *et al.*, 1982; Weigel *et al.*, 1983; O'Malley, 1984). Briefly, they developed the concept that the chicken progesterone receptor is a 'metastable dimer composed of two non-identical subunits, termed A and B' (O'Malley, 1984). This dimer was thought to bind to chromatin via the interaction of the B subunit with non-histone proteins. The function of B was therefore that of a 'specifier' targeting the A–B dimer to certain 'acceptor sites' in the chromatin. Following this binding to chromatin the dimer dissociates and subsequently it is exclusively the A subunit which triggers the initiation of transcription (for this concept see O'Malley, 1984, and references therein).

The co-existence of two progesterone receptor forms was apparently supported by studies from Horwitz's group on the human T47D receptor which also detected two receptor polypeptides even when using *in situ* photoaffinity labeling (see Chapter 11 for details and references). This technique should minimize putative proteolytic events during preparation of the cytosol. However, results which are at variance with the idea of two structurally and functionally different progesterone receptor forms came from a series of experiments in which affinity labeling was the crucial technique used in the analysis of the structural and immunological similarities of the two chick oviduct progesterone receptor forms. These experiments will be described in detail below. Further criticism of the A–B concept came from studies of Milgrom's group who clearly demonstrated that the progesterone receptor A form, which can be observed equally in cytosolic preparations of rabbit uterus, is a proteolytic fragment of form B (Logeat *et al.*, 1985). Finally, cloning and expression of the chicken progesterone receptor cDNA and gene (see Part III of this book) fully support the conclusion derived from the crosslinking experiments described in this chapter, thus changing the classical concept of action for the A–B chicken progesterone receptor.

II. IDENTIFICATION

A. The ligand

Most of the crosslinking studies with the progesterone receptor have been carried out by photoaffinity labeling using the synthetic progestin R5020 (17α,21-dimethyl-19-nor-pregna-4,9-diene-3,20-dione, Fig. 1) as first described by Dure *et al.*, 1980). Irradiation with high power UV light at 300 nm activates the dienone chromophore and generates a highly reactive diradical-like species (see also Part I of this book). The chemical specificity of such excited ketones is very broad and both CH groups

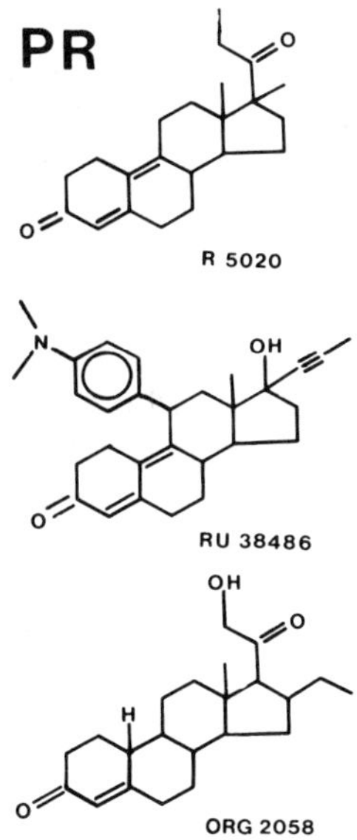

Fig. 1 — Structure of three steroids which have been used for UV crosslinking of the progesterone receptor (PR). Only R5020 can be used for the chicken homolog 1.

and nucleophiles are potential reactants. A major advantage for the photolabeling of biological systems is the fact that excited ketones are not able to react with the O–H bonds of water. Compared with the labeling efficiencies of electrophilic crosslinkers such as tamoxifen–aziridine for the estrogen receptor and dexamethasone–mesylate for the glucocorticoid receptor, which approach 90–95% under optimal conditions, R5020 photocrosslinking results in 5–10% of specifically bound hormones being covalently attached to the receptor protein. A significant advantage of this photolabeling reaction is, however, that it displays a high specificity such that the labeling can be performed even in crude cytosolic preparations (see below). Tissue culture cells such as T47D human breast tumor cells with high amounts of progesterone receptor may be labeled *in situ*, i.e. by exposing the intact cells after incubation with radioactively labeled R5020 to UV light (see Chapter 11).

Whereas R5020 has been used for the labeling of chicken, rabbit and human (T47D) progesterone receptors, the synthetic antiprogestin RU 38.486 (Fig. 1) does not bind to the chicken receptor and can be used only for the crosslinking of the mammalian homologues (Horwitz and Alexander, 1983; see also Chapter 11). As RU 38.486 contains the same dienone moiety as RU 5020, crosslinking proceeds probably via a similar photochemical route. Electrophilic affinity labeling of the progesterone receptor with bromoacetoxy derivatives has been reported but these compounds exhibited only a low selectivity. In partially purified receptor preparations, however, they efficiently labeled the receptor and it has been possible to determine the amino acid(s) which is (are) in close proximity to the electrophilic group (Holmes *et al.*, 1981; Holmes and Smith, 1983).

B. Two receptors are crosslinked in the cytosol

The preparation of chick oviduct cytosol is described in detail in Part I of this book. Crude cytosol is incubated with 5 nM [^{3}H]R5020 for approximately 2 h in an ice–water slurry. Thereafter the non-bound hormone is removed, for example by the dextran-coated charcoal method, and the preparation is directly UV irradiated. It is important to use buffers containing scavengers (e.g. Tris, DTT, β-mercaptoethanol) in order to quench the majority of radicals which are generated from the steroid but which do not covalently attach to the receptor. If they are not quenched, these radicals may react with other proteins which, if major constituents of a receptor preparation, may become (non-specifically) labeled. After irradiation, the cytosol is usually ammoniun sulfate precipitated and dialyzed against SDS–sample buffer. Following gel electrophoresis on 7.5% SDS–polyacrylamide gels, the gel is fixed and incubated in EN^{3}HANCE (NEN). The dried gel is exposed at −80°C. The exposure time should be long enough to reach a value of $>10^5$ cpm crosslinked × days. Under these conditions a signal similar to that shown in Fig. 2, lane 1, will be obtained and

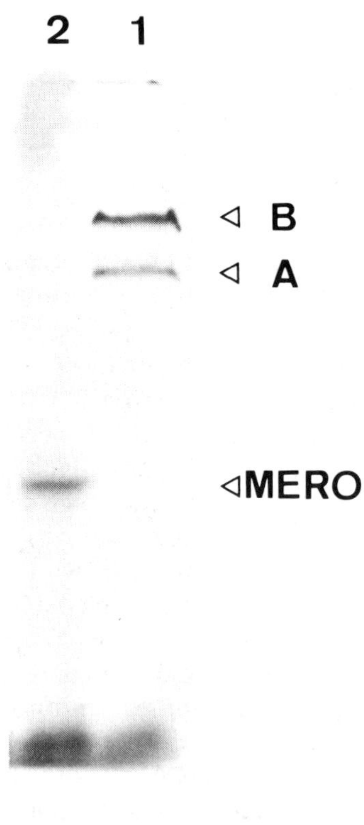

Fig. 2 — [^{3}H]R5020 crosslinked chick oviduct progesterone receptor. Crude cytosol was incubated and free hormone removed by dextran-coated charcoal and irradiated as described in Part I of this book. Fluorography was carried out before (lane 1) and after (lane 2) Ca^{2+} treatment of the cytosolic receptor as described in the text.

potential degradation products, due to improper handling, storage or cytosol preparations, will be seen as well. The high molecular weight signal in Fig. 2 corresponds to the receptor form B with a molecular weight of 109 kDa, and the lower signal is the 79 kDa form A. They are the two forms which have been extensively described and characterized by O'Malley's group (see above). In crude cytosol both forms are usually present in equal amounts and no significant variations have been noticed in preparations at different times thoughout the year. However, during conventional purification (see below), the A form is much less stable than form B, probably as a result of co-purified proteolytic activities in the form A preparation.

If a preparation such as that shown in lane 1 is treated on ice for a short time with millimolar concentrations of Ca^{2+}, both forms of the receptor are concomitantly degraded to yield the relatively stable intermediate 'mero' receptor of approximately 23 kDa as shown in Fig. 2, lane 2. In view of the results of Milgrom's group (Logeat *et al.*, 1985) showing that the rabbit progesterone receptor form A is a proteolytic fragment, we have tried, for both the chick oviduct and the T47D system, to minimize proteolysis and to shorten the time for purification. Despite this, two bands were observed routinely for the affinity labeled progesterone receptor with approximately similar intensities.

C. Peptide maps of A and B

The possibility of unequivocally identifying the progesterone receptor even in crude extracts permits experiments where a structural relationship between forms A and B can be investigated. The basis for such structural analyses are the so-called 'Cleveland maps', which are partial proteolytic digests of a pure protein (Cleveland *et al.*, 1977). These Cleveland maps correspond to the intermediates of proteolysis by a given protease and are highly specific because not only the sizes of the intermediates are characteristic of the protein under investigation but also their relative intensities. A chymotryptic attack, for example, is influenced by the nature of the amino acid residue next to the susceptible bond (Smyth, 1967) and differences in the relative intensities of the peptide bands are due to different rate constants of hydrolysis. Clearly, there is an influence of the ternary structure on hydrolysis; therefore digestion should be performed under identical conditions if different proteins are to be compared. It is, of course, difficult to ensure the native conformation of a protein, and it is experimentally easier to perform the digestion with denatured proteins. The power of Cleveland digests in analyzing the degree of structure relationships between proteins can be increased further by using several proteases, e.g. trypsin, chymotrypsin and *Staphylococcus* V8 protease.

In the first study of crosslinked receptor peptide maps (Gronemeyer *et al.*, 1983), TLCK–α-chymotrypsin was used to compare cytosolic chick oviduct progesterone receptor (forms A + B) with partially purified form A or form B. The maps, generated from partial proteolysis in the presence of 2 M urea with increasing amounts of enzyme, are depicted in Fig. 3. Several peptides are apparent between the size of the 'mero' receptor and the intact forms A and B. Interestingly, during partial proteolysis, no intensive bands are obvious below the molecular weight of the 'mero' receptor. Apparently the hormone binding domain, which constitutes

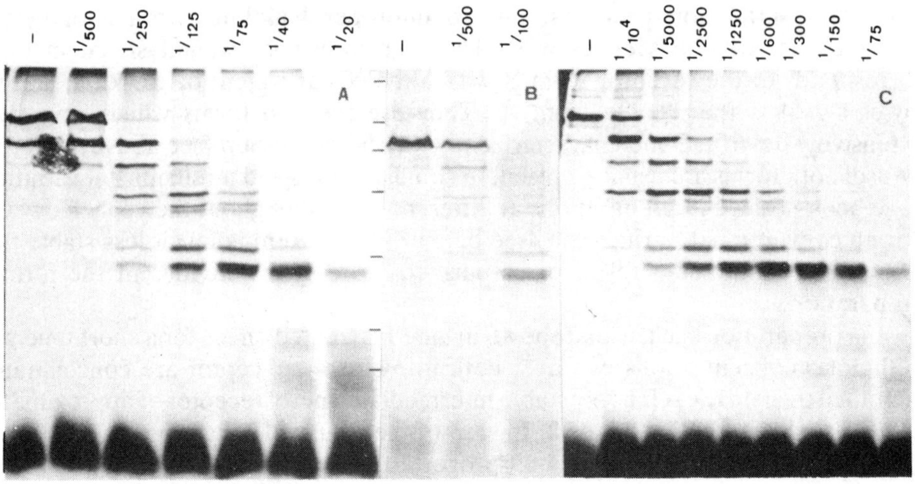

Fig. 3 — Peptide map of cytosolic (A), partially purified form A (B) and partially purified form B (C) chick oviduct progesterone receptor generated by limited proteolysis of [^{3}H]R5020 crosslinked receptor. (From Gronemeyer *et al.*, 1983.)

approximately one-third of the entire receptor (see Section III), is relatively resistant to protease attack.

Several interesting features are obvious from the Cleveland maps of the chicken progesterone receptor in Fig. 3.

(i) The map of cytosolic receptor in Fig. 2A is clearly the sum of the maps for form A (Fig. 3B) and form B (Fig. 3C). Thus the different purity of the receptors does not have any effect on the digests — except in changing the enzyme:protein ratio to obtain comparable maps.

(ii) The pattern seen for form A in Fig. 3B is a subset of that seen for form B in Fig. 3C.

(iii) No additional bands are seen in the form B map below a molecular weight corresponding to form A.

(iv) The relative intensities of the bands seen for both forms are very similar and characteristic for this receptor.

(v) The first step of proteolysis of form B appears to be the generation of a band very close to or identical with the band obtained for the non-digested form A.

From these data it was concluded that both forms share a high degree of structural similarity and it was proposed that form A might be partially identical to form B (Gronemeyer *et al.*, 1983).

In fact, as described in Chapter 15, it is clear now that form A is a part of form B starting at or close to the methionine at position 128. Taking the additional information that the steroid hormone binding domain is located in the carboxy-teminal region and that any deletions in this region abolish hormone binding, it is possible to determine the sites of proteolysis from the peptide maps and sequence data. They are shown schematically in Fig. 4. Interestingly, those sites which are most easily cleaved are in the flanking region or close to two functional units of the chicken progesterone receptor which have been determined by expression of

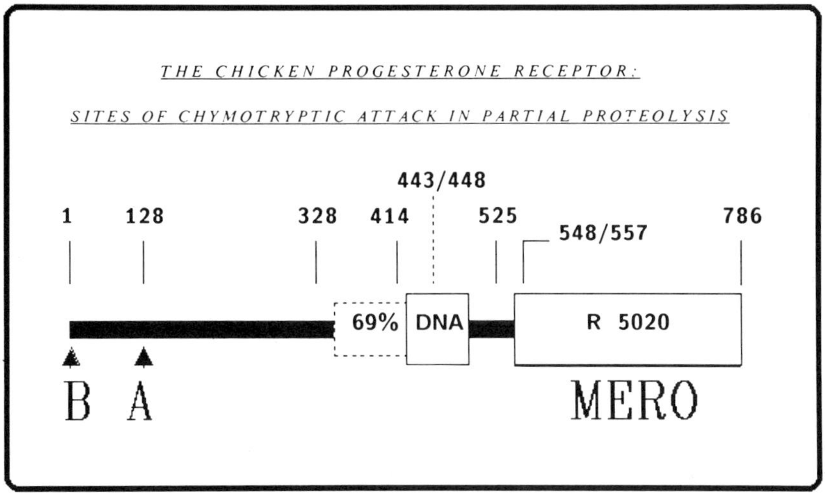

Fig. 4 — Localization of sites preferentially attached by proteolysis in the α-chymotrypsin, as deduced from peptide mapping and functional analysis of receptor mutants (see Section III).

mutated receptor molecules (Gronemeyer *et al.*, 1987). This is the hormone binding domain (amino acids 450–786) and a dominant peptide doublet corresponds to cuts at the approximate positions 548 and 557. No cuts appear under these conditions within this domain. The next cut in the N-terminal direction is located in the linker region (see Part III of this book) at position 525 and flanks, together with the cut at position 414 (see Fig. 4), the DNA binding domain. It is of interest to note that the two minor cuts at 443 and 448 occur between the two potential 'finger' structures of the DNA binding domain, indicating that there may be an exposed region between the fingers. Further towards the N terminal, there is a dominant chymotryptic site at position 328. Until now, the function of the N-terminal half of the receptor has been unclear but it is worth mentioning that a progesterone receptor mutant lacking this region is very inefficient in transactivating transcription (for details see Chapter 15). A sequence comparison between the rabbit (Loosfelt *et al.*, 1987), human (Misrahi *et al.*, 1987) and chicken progesterone receptors reveals a 69% homology block N terminal of the DNA binding region. Again the sites at 414 and 328 appear to flank this region which might correspond to an as-yet undefined functional block.

Reports, based on total tryptic maps of the two receptor forms, which concluded that the two forms are the products of different genes (Birnbaumer *et al.*, 1983a,b) are at variance with our current knowledge about the chicken progesterone receptor and they probably arise from a misinterpretation of experimental artefacts. If the [^{3}H]R5020 photoaffinity labeled progesterone receptor is digested with an excess of protease a final peptide of approximately 6 kDa which contains the radioactive label can be detected on dematuring gels (Fig. 5). This size corresponds to approximately 55 amino acids and is most likely the smallest digestable unit containing the covalently bound steroid.

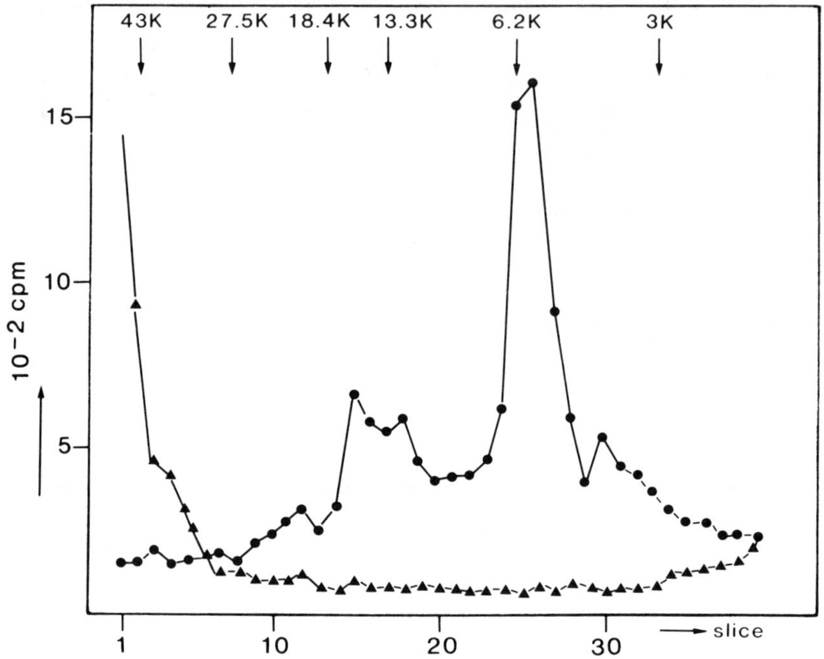

Fig. 5 — A peptide of approximately 6 kb is the final digestion product of crosslinked progesterone receptor. (From Gronemeyer *et al.*, 1983.)

III. ISOLATION

A. Classical techniques

Several purification protocols for the isolation of the chick oviduct progesterone receptor have been published and the reader is referred to reviews from O'Malley's group, cited in Section I. These methods involve steps in which both forms are purified together, such as ammonium sulfate precipitation and gel filtration, or methods designed to separate forms A and B. The latter are based on the different chromatographic behaviors of forms A and B; for example, on DEAE, phosphocellulose or DNA cellulose columns. We know now that the different bindings on charged matrices are due to a peculiar polyglutamic acid stretch present in form B and absent in form A. Interestingly, although it was already well accepted that form A but not form B binds to DNA cellulose, we were able to show that form B also exhibits significant DNA binding properties (Gronemeyer *et al.*, 1985) and DNAase footprinting studies could be performed (von der Ahe *et al.*, 1986).

B. Ligand affinity chromatography

Although potentially a most powerful technique, until recently no matrices that could be used for the isolation of steroid receptors were available. Indeed, the success with purification schemes using phosphocellulose and DNA–cellulose chromatography was a major reason for the diminished interest in ligand affinity chromatography. Additionally, for the chicken progesterone receptor, initial purification attempts gave ambiguous results: apparently only one progesterone receptor form was isolated which was neither form A nor form B. This purification was performed in the presence of molybdate which had been shown to prevent the 'activation', as well as degradation, of receptors (Nishigori and Toft, 1980; Shyamala and Leonard, 1980) and hence this protein was called the 'molybdate-stabilized' form of the progesterone receptor (Renoir *et al.*, 1982; Puri *et al.*, 1982). As it turned out, however, these results were due to the fact that a ternary binder of the receptor, later identified as the 90 kDa heat shock protein, was co-purified (Joab *et al.*, 1984; Dougherty *et al.*, 1984). Despite these initial misleading results, affinity chromatography of the progesterone receptor is an established method. For an extensive discussion and further information on this subject the reader is referred to a recent review by Formstecher and Lustenberger (1987).

C. Using photoaffinity labeling

In order to derive a receptor purification protocol in which at any stage the presence and integrity of the receptor can be checked, classical purification methods were combined with affinity labeling and SDS–polyacrylamide gel electrophoresis. As a base the most powerful classical purification method, differential DNA–cellulose and phosphocellulose chromatography, was chosen. The entire purification scheme is depicted in Fig. 6 and is described in detail in Chapter 3. The classical part of the purification resulted in a 1000-fold purified receptor with a 30% yield. This corresponds to a receptor which is approximately 5% pure. It is important to mention at this point that the purification was such that significant losses of receptor due to dissociated ligand were avoided as seen from the increasing values for the specific activity. This contrasts, for example, with the early purification scheme of Schrader *et al.* (1977) where the specific activity starts decreasing at the hydroxylapatite step.

At each step of purification the receptor was analyzed by photoaffinity labeling. For this the initial incubation of cytosol was carried out using radioactive R5020. The last step of the purification was initially designed to separate forms A and B on a tandem arrangement of DNA–cellulose and phosphocellulose columns. Unexpectedly, even in the presence of 100 mM salt, using this purification protocol, form B of the chicken progesterone receptor also bound to DNA–cellulose. This was convincingly demonstrated by affinity labeling as shown in Fig. 7. Therefore, in addition to the structural similarities, both forms of the receptor also displayed similar, although not identical, DNA binding properties.

The final purification step took advantage of the possibility of identifying the receptor even under denaturing conditions by the presence of the crosslinked radioactive hormone. Using SDS–polyacrylamide electrophoresis with continuous flow elution, form B could be directly isolated in an electrophoretically homogenous form (Gronemeyer *et al.*, 1985) as demonstrated in Fig. 8. By changing the parameters of electrophoresis, form A, which was initially isolated with two contaminating proteins, could also be purified to apparent homogeneity. One advantage of this purification system — in addition to the obvious advantage of being able to check the receptor at any stage, especially during the last steps — is the complete separation of form B and form A. It is apparent from Fig. 8 that there are at least ten fractions between both elution peaks where neither form A nor form B is present. This clear separation led directly to the immunological studies described below.

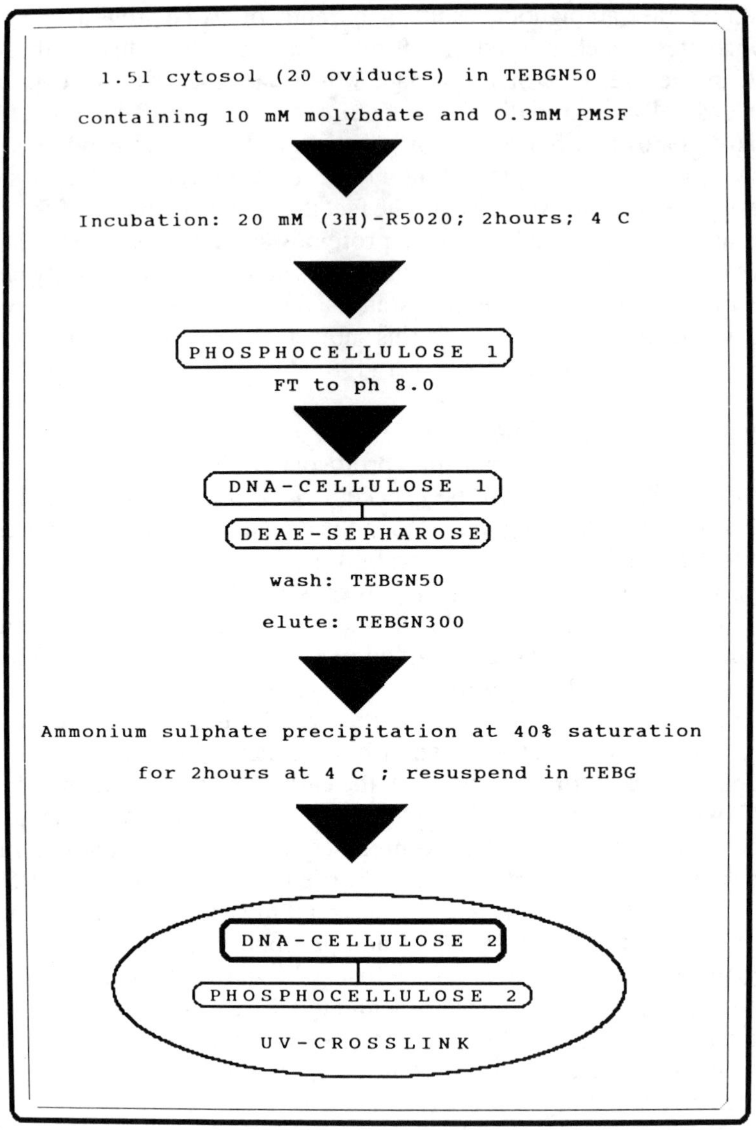

Fig. 6 — Flow chart of large-scale chick oviduct progesterone receptor purification yielding approximately 5% pure receptor.

IV. IMMUNOANALYSIS

Having established both structural and functional similarities between the two forms of the chicken progesterone receptor, the next question was whether they were also immunologically related. For this, polyclonal antisera were raised in rabbits, using the purified forms A and B as described above. It could be excluded that a serum

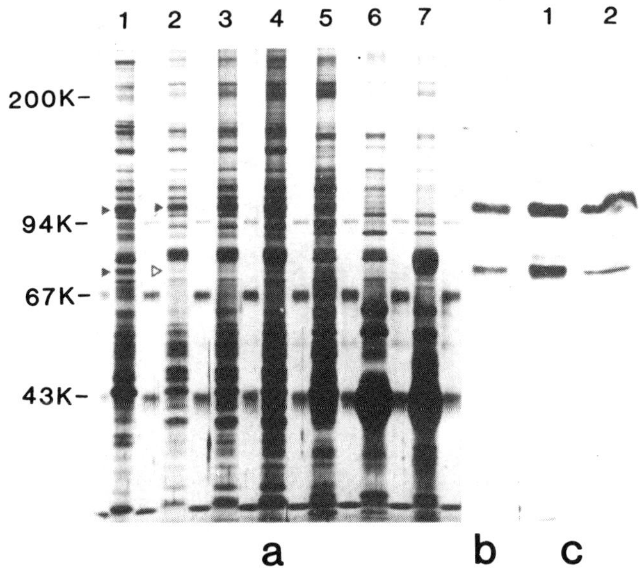

Fig. 7 — SDS-gel electrophoresis and photoaffinity labeling analysis of the progesterone receptor preparations obtained by partial purification. a, silver-stained 7.5% SDS-polyacrylamide gel of the following steps in purification 1: lane 1, DNA–cellulose column II eluate (total amount of protein, 7.9 μg; the filled arrowheads point to the bands of receptor form A (lower) and form B (upper)); Lane 2, phosphocellulose column II eluate (the filled arrowhead indicates form B; the open arrowhead marks the position where form A should appear; 5.8 μg protein in total); lanes 3 and 4, two and five times more total protein applied than in lane 2; lane 5, ammonium sulfate precipitate, 9.0 μg; lane 6, DEAE–Sepharose eluate, 10.2 μg; lane 7, 14 μg of cytosolic proteins. Between the lanes the protein standard mixture (Pharmacia) is applied; the molecular weights of the standards are indicated on the left. b, fluorograph of 7.5% SDS–gel of crosslinked DNA–cellulose II eluate; c, overexposure of fluorograph: lane 1, same as b; lane 2, 7.5% SDS–gel of photoaffinity labeled phosphocellulose II eluate. (From Gronemeyer *et al.*, 1985.)

obtained after immunization of form B contained antibodies against form A as a result of the presence of form A as minor contaminant in the immunogen (see Part I of this book). The results of a Western blot analysis of the progesterone receptor with the different antisera obtained are shown in Fig. 9. Surprisingly, an antiserum directed against the pure form B recognized also form A and an antiserum against form A detected with similar intensity form B. These results proved that, in addition to structural and functional similarities, the two forms shared common epitopes.

V. PEPTIDE ISOLATION AND SEQUENCING

The knowledge of peptide sequences is an important advantage if the cloning of the corresponding cDNA is envisaged. As explained in more detail in Part III of this book, not only can this information be used to screen libraries with synthetic oligonucleotide probes but also it gives additional proof if a cDNA clone has been obtained using, for example, antibodies and λgt11 expression libraries. Having

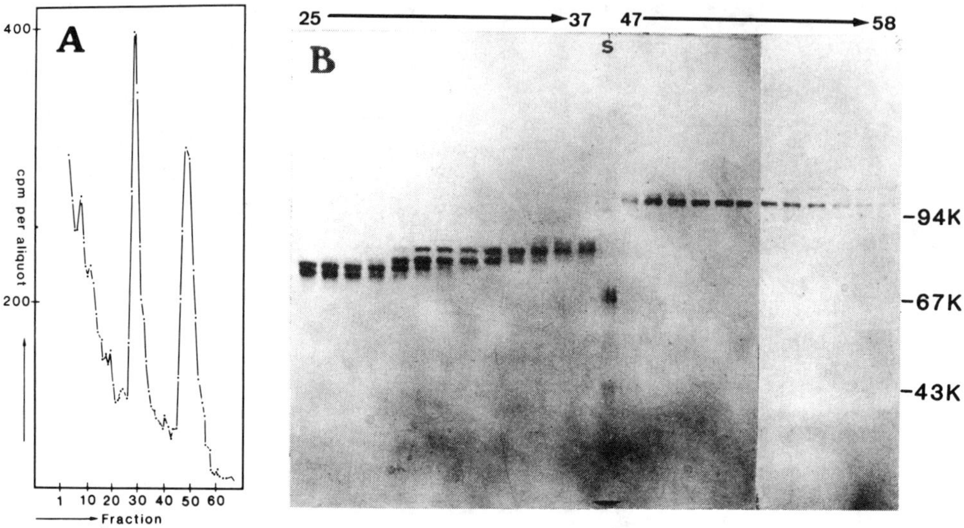

Fig. 8 — Preparative gel electrophoresis of 5% pure and photoaffinity labeled progesterone receptor. A, radioactivity profile obtained by liquid counting 5 μl/fraction. B, silver-stained SDS–7.5% polyacrylamide gel in which 5 μl of the radioactive peak fractions (indicatd at the top) were applied. The gel was run in such a manner that the bromphenol blue had left the gel in order to gain optimal separation. The radioactive peaks were in fractions 30 and 49. For further details see Chapter 3. s, standards. (From Gronemeyer *et al.*, 1985.)

established a purification protocol which could be used to isolate several hundred micrograms of chick oviduct progesterone receptor in apparently homogenous form, we decided to attempt the sequencing of this material. Form B was chosen for this project in preference to form A because the preparative yield is higher and a homogenous preparation (see Fig. 8) is obtained after only one cycle of preparative SDS–polyacrylamide electrophoresis.

The strategy chosen for sequencing was (i) to digest the receptor after *S*-carboxymethylation with trypsin, (ii) to separate the mixture of peptides on conventional HPLC columns and finally (iii) to purify individual peak areas of the first HPLC dimension by a recently developed multi-dimensional microbore HPLC peptide purification protocol (Simpson *et al.*, 1987, and references therein). The rationale for the individual steps was the following.

(A) The N terminus of electrophoretically purified proteins is usually blocked. In addition, the information obtained from N-terminal sequencing is often not valuable for cloning purposes because these sequences correspond to the 5′ coding region of a given cDNA which may be difficult to obtain by reverse transcription of poly(A^+) RNA by oligo(dT) priming (see also Part III of this book).

(B) Trypsin was found to be the enzyme of choice for the digestion of the receptor preparation, which incidentally was difficult to solubilize. Chymotrypsin and *Staphylococcus* V8 protease did not sufficiently digest the receptor to enable us to monitor the peaks during the first HPLC dimension.

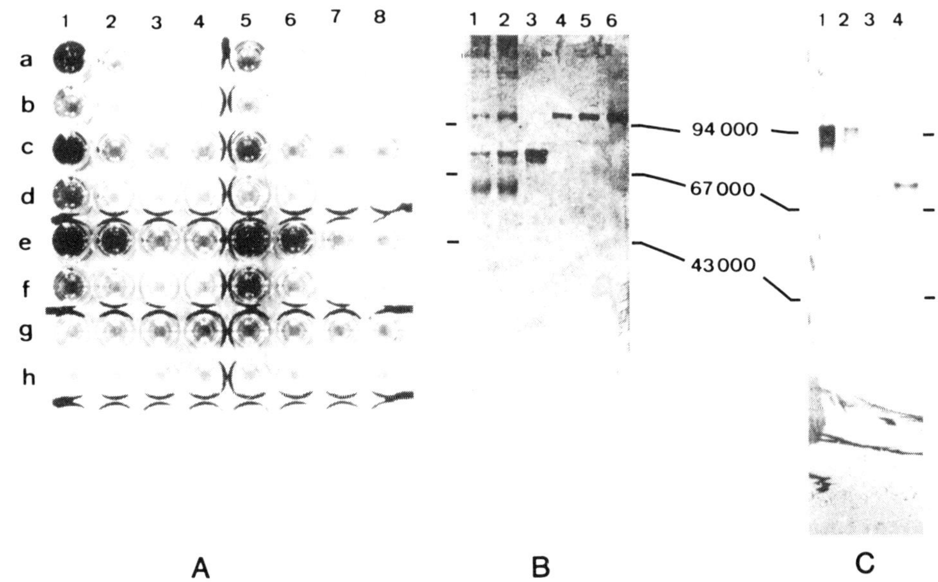

Fig. 9 — Immunoanalysis of progesterone receptor forms A and B. A, ELISA. 1.1 μg (1), 110 ng (2), 11 ng (3), and 1.1 ng (4) of preparative gel purified form A receptor and 1.4 μg (5), 140 ng (6), 14 ng (7), and 1.4 ng (8) of preparative gel-purified form B were absorbed on microtiter plates as described in the text. The plate was then probed with 10^{-2} diluted (a, c, e, g, h) or 10^{-3} diluted (b, d, f) antiserum from two rabbits immunized with form A receptor (a, b and c, d) or from one rabbit immunized with form B receptor (e, f). g and h are preimmune control bleeds. The three bottom rows of the microtiter plate are empty. Immune complexes are visualized using peroxidase-coupled sheep–anti-rabbit antibodies and *o*-phenylenediamine. B, immunoblot probed with anti-progesterone B receptor antiserum. A 10% SDS–gel was blotted onto nitrocellulose and treated as described in the text. Lane 1, 20 μg;, lane 2, 50 μg proteins of partially purified progesterone receptor (DNA–cellulose II eluate); lane 3, 2 μg of purified form A obtained from the first preparative gel electrophoresis. C, immunoblot probed with anti-progesterone A receptor antiserum. Lane 1, 2 μg; lane 2, 0.4 μg of preparative gel-purified form B; lane 3, 0.2 μg; lane 4, 1 μg of preparative gel-purified form A. (From Gronemeyer *et al.*, 1985.)

(C) Microbore techniques has to be used since only minimal amounts of the peptides were expected and the recoveries in the microbore technique are usually above 90%.

The result of the first-dimension reversed-phase HPLC of tryptic peptides of S–Cm–chicken progesterone receptor form B is shown in Fig. 10 (left). From the cDNA deduced amino acid sequence (see below) we know now that the number of tryptic fragments expected (longer than 1 residue) is around 62. This is in good agreement with the 68 peaks counted from the 206 nm elution profile (Fig. 10A, left) and collected manually. Rechromatography on microbore columns of peak fractions 31, 38, 49 and 66–67 using different supports resulted in the isolation of the five peptides T1–T5 in apparently homogeneous form whereas the second HPLC dimension gave multiple peaks for fractions 51–52 and 64–65. However, rechromatography of selected fractions yielded the pure peptides T6 and T7 (see Fig. 10). Amino acid sequence analysis revealed that T1 and T2 corresponded to autoproteolytic fragments of trypsin. The sequences T3–T7 and the yields of phenylthiohydantoin amino acids in each cycle are given in Fig. 11. The longest sequence obtained was 34

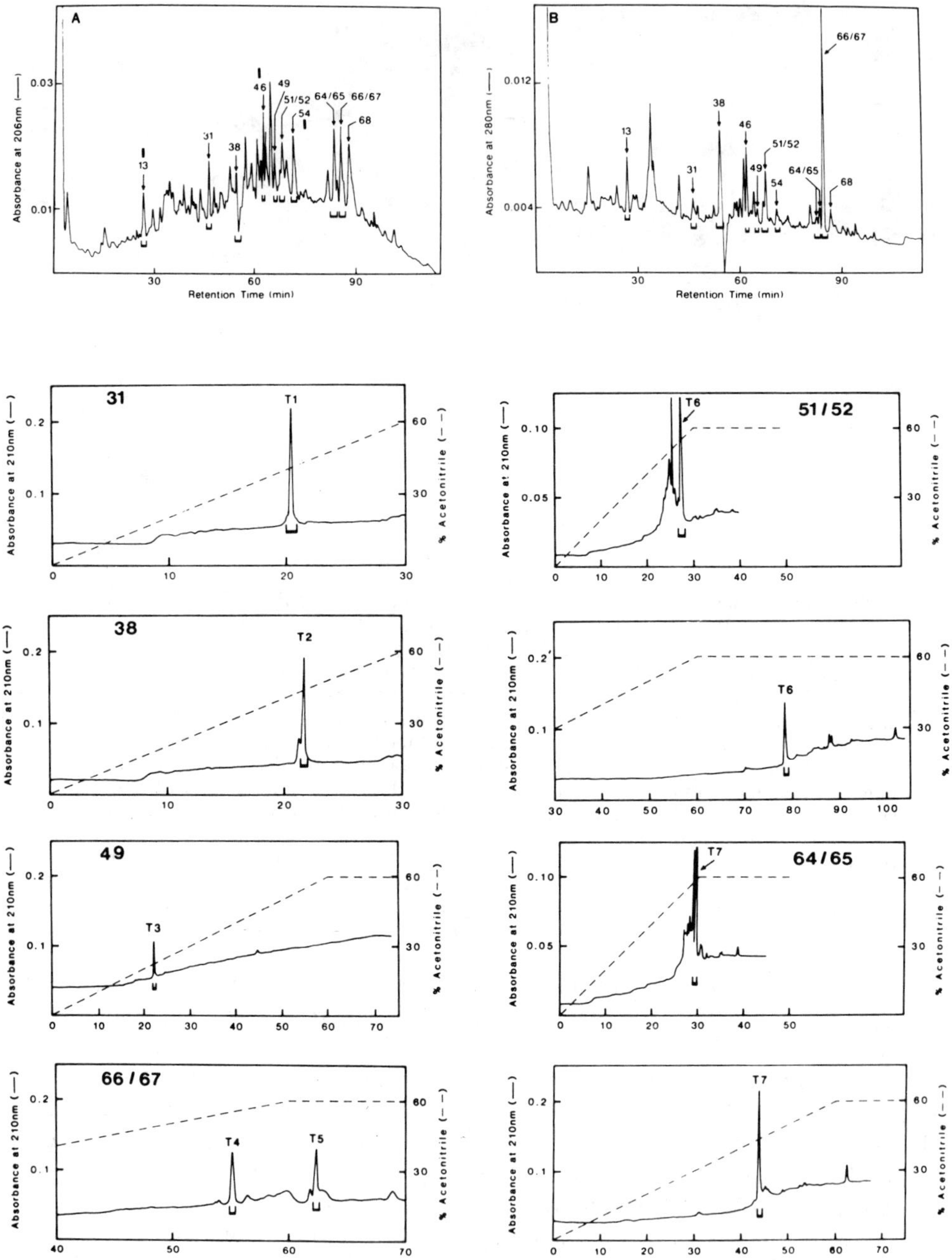

Fig. 10 — Separation of tryptic peptides of S–Cm–chicken progesterone receptor by reversed-phase HPLC, recorded at (A) 206 nm and (B) 280 nm. Purified receptor (approximately 1 nmol) was digested with TPCK–trypsin at an enzyme-to-protein ratio of 1/10 and the peptides were separated on a LiChrospher column (Chapter 3) using a linear 120 min gradient from 0% to 100% B where solvent A was 0.1% (v/v) trifluoroacetic acid–5% acetonitrile. The flow rate was 1 ml/min, and the column was operated at room temperature. All peaks were collected manually in siliconized Eppendorf tubes and stored at −70°C. Numbered peaks describe

contined next page

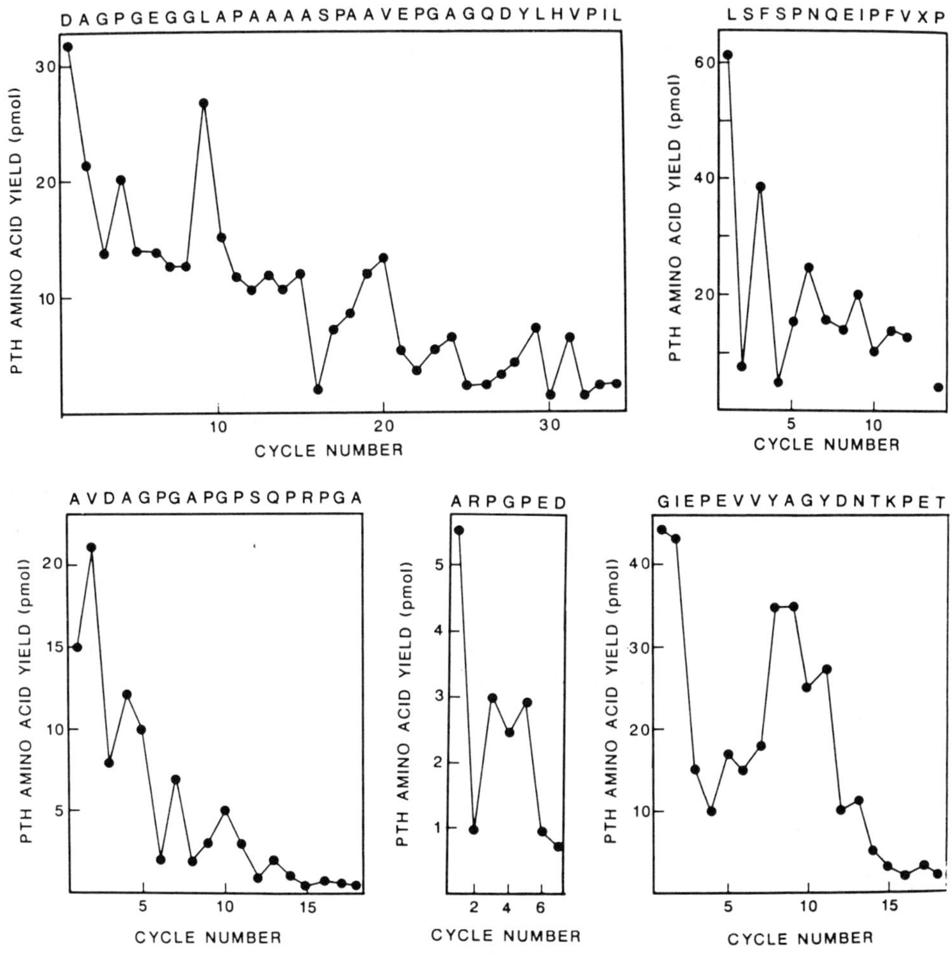

Fig. 11 — Amino acid sequence analysis of tryptic peptides T3–T7 from the chicken progesterone receptor form B. Yields of phenylthiohydantoin derivatives of amino acids are shown for each cycle of the Edman degradation. Aliquots (67%) from each cycle were analyzed by HPLC (see Chapter 3). HPLC peak heights were converted to picomoles for each derivative using values from a standard mixture of phenylthiohydantoin amino acids and were normalized to 100% injection. (From Simpson *et al.*, 1987.)

Fig. 10 — *continued*
fractions mentioned in the text, bars indicate the collected peak areas. Reversed-phase microbore HPLC purification of selected peak fractions is shown to the left. Fractions 31, 38, 49 and 66–67 were rechromatographed on MOS–Hypersil with the exception of fraction 49 where Beckmann Ultrasphere was used as the column support. Gradient was from 0% to 100% B, where solvent A was 0.1% (v/v) triofluoroacetic acid and solvent B was 0.1% (v/v) trifluoroacetic acid–60% acetonitrile. All separations were performed at a column temperature of 45°C. The tryptic peptides T1–T5 were collected manually as indicated by bars below the peaks and were used directly for sequence analysis. Second and third chromatographic steps in reversed-phase microbore columns for the purification of tryptic peptides contained in peaks 51–52 and 64–65 are shown to the right. The fractions indicated by bars were taken directly for sequence analysis. (From Simpson *et al.*, 1987.)

residues (Fig. 11A) with 30 pmol of peptide available in the first cycle. The heptapeptide ARPGPED (Fig. 11D) was sequenced starting with 5 pmol. None of the sequences obtained for peptides T3–T7 could be detected in sequence data banks. However, when the human estrogen and glucocorticoid receptor cDNAs were cloned and sequenced (see Part III of this book) it became apparent that peptide T4 exhibited considerable sequence homology to these sequences, especially that of the glucocorticoid receptor at positions 538–555. In particular, it appears that the sequence EP X_3 Y X_2 Y D X_3 P is a constant motif in the steroid receptor hormone binding doman from which T4 is derived.

VI. SUMMARY

As described above, affinity labeling has proven a highly important tool in the identification, isolation, and in the structural and immunological analysis of the chick oviduct progesterone receptor. Further, as described in Part III of this book, it provided the tools and a means of verification in the cloning of the receptor cDNA. Our current knowledge about the chicken progesterone receptor clearly shows that the conclusions from the crosslinking work, including the criticism of previous models, were apparently correct. In particular, affinity labeling was a powerful tool in the revision of the previously established concept for the roles of forms A and B. In fact, at present the only question left open is whether form A originates from form B by proteolysis, by internal initiation of translation from a common mRNA or by an as-yet unidentified mechanism. The biological question, however, concerning eventual functional differences between the two forms, which have been demonstrated by crosslinking to be clearly present in the oviduct cytosol, may now be the more challenging one.

ACKNOWLEDGEMENTS

I am grateful to all my collaborators who have contributed to the studies described in this paper. Their names are apparent from the list of references. In addition I thank A. Alberga and G. Richards for interesting discussions and P. Chambon for his continuous support. I further acknowledge the help of C. Kister, C. Werlé and B. Boulay in the preparation of this manuscript.

REFERENCES

Birnbaumer, M., Schrader, W. T. and O'Malley, B. W. (1983a) Photoaffinity labeling of the chick progesterone receptor proteins. *J. Biol. Chem.* **258**, 1637–1644.

Birnbaumer, M., Schrader, W. T. and O'Malley, B. W. (1983b) Assessment of structural similarities in chick oviduct progesterone receptor subunits by partial proteolysis of photoaffinity-labelled proteins. *J. Biol. Chem.* **258**, 7331–7337.

Cleveland, D. W., Fischer, S. G., Kirschner, M. W. and Laemmli, U. K. (1977) Peptide mapping by limited proteolysis in sodium dodecyl sulfate and analysis by gel electrophoresis. *J. Biol. Chem.* **252**, 1102–1106.

Dougherty, J. J., Puri, R. K. and Toft, D. O. (1984) Polypeptide components of two 8S forms of chicken oviduct progesterone receptors. *J. Biol. Chem.* **259**, 8004–8009.

Dure, L. S. D., Schrader, W. T. and O'Malley, B. W. (1980) Covalent attachment of a progestational steroid to chick oviduct progesterone receptor by photoaffinity labelling. *Nature* **283**, 784–786.

Formstecher, P. and Lustenberger, P. (1987) Purification of various steroid hormone receptors (glucocorticoid, progesterone, mineralocorticoid) by affinity chromatography. Design of suitable adsorbents and their applications. In: Moudgil, V. K. (ed.) *Molecular Mechanism of Steroid Hormone Action*. Walter de Gruyter, Berlin.

Grody, W., Schrader, W. T. and O'Malley, B. W. (1982) Activation, transformation, and subunit structure of steroid hormone receptors. *Endocrine Reviews* **3**, 141.

Gronemeyer, H., Harry, P. and Chambon, P. (1983) Evidence for two structurally related progesterone receptors in chick oviduct cytosol. *FEBS Lett.* **156**, 287–292.

Gronemeyer, H., Govindan, M. V. and Chambon, P. (1985) Immunological similarity between the chick oviduct progesterone receptor forms A and B. *J. Biol. Chem.* **260**, 6916–6925.

Gronemeyer, H., Turcotte, B., Quirin-Stricker, C., Bocquel, M. T., Meyer, M. E., Krozowski, Z., Jeltsch, J. M., Lerouge, T., Garnier, J. M. and and Chambon, P. (1987) The chicken progesterone receptor: sequence, expression and functional analysis. *EMBO J.* **6**, 3985–3994.

Holmes, S. D. and Smith, R. G. (1983) Identification of histidine and methionine residues in the active site of the human uterine progesterone receptor with the affinity lables 11α- and 16α-(bromoacetoxy) progesterone. *Biochemistry* **22**, 1729–1734.

Holmes, S. D., Van, N. T., Stevens, S. and Smith, R. G. (1981) Affinity labeling of the human uterine progesterone receptor with 21-, 16α- and 11α-bromoacetoxy-progesterones. *Endocrinology* **109**, 670–672.

Horwitz, K. B. and Alexander, P. S. (1983) In situ photolinked nuclear progesterone receptors of human breast cancer cells: subunit molecular weights after transformation and translocation. *Endocrinology* **113**, 2195–2201.

Joab, I., Radanyi, C., Renoir, M., Buchou, T., Catelli, M. G., Binart, N., Mester, J. and Baulieu, E. E. (1984) Common non-hormone binding component in non-transformed chick oviduct receptors of four steroid hormones. *Nature* **308**, 850–853.

Logeat, F., Pamphile, R., Loosfelt, H., Jolivet, A., Fournier, A. and Milgrom, E. (1985) One-step immunoaffinity purification of active progesterone receptor. Further evidence in favor of the existence of a single steroid binding subunit. *Biochemistry* **24**, 1029–1035.

Loosfelt, H., Atger, M., Misrahi, M., Guiochon-Mantel, A., Meriel, C., Logeat, F., Benaous, R. and Milgrom, E. (1986) Cloning and sequence analysis of rabbit progesterone receptor complementary DNA. *Proc. Natl. Acad. Sci. USA* **83**, 9045–9049.

Misrahi, M., Atger, M., D'Auriol, L., Loosfelt, H., Meriel, C., Fridlanski, F., Guiochon-Mantel, A., Galibert, F. and Milgrom, E. (1987) Complete amino acid sequence of the human progesterone receptor deduced from cloned cDNA. *Biochem. Biophys. Res. Comm.* **143**, 740–748.

Nishigori, H. and Toft, D. (1980) Inhibition of progesterone receptor activation by sodium molybdate. *Biochemistry* **19**, 77–83.

O'Malley, B. W. (1984) The molecular and cellular basis for steroid hormone action. In: Isselbacher, K. J. (ed.) *Medicine, Science, and Society*. John Wiley, New York, pp. 449–461.

Puri, R. K., Grandics, P., Dougherty, J. J. and Toft, D. (1982) Purification of "nontransformed" avian progesterone receptor and characterization. *J. Biol. Chem.* **257**, 10831–10837.

Renoir, J. M., Yang, C. R., Formstecher, P., Lustenberger, P., Wolfson, A., Redeuilh, G., Mester, G., Richard-Foy, H. and Baulieu, E. E. (1982) Progesterone receptor from chick oviduct: purification of molybdate stabilized form and preliminary characterization. *Eur. J. Biochem.* **127**, 71–79.

Schrader, W. T. (1975) Methods for extraction and quantification of receptors. *Methods Enzymol.* **36**, 187–211.

Schrader, W. T. and O'Malley, B. W. (1977) Molecular structure and analysis of progesterone receptors. In: O'Malley B. W. and Birnbaumer, L. I. (eds.) *Receptors and Hormone Action*, Vol. 11. Academic Press, New York, pp. 189–224.

Schrader, W. T., Kuhn, R. W. and O'Malley, B. W. (1977) Progesterone binding components of chick oviduct. XIII. Receptor B subunit protein purified to apparent homogeneity from laying hen oviducts. *J. Biol. Chem.* **252**, 299–307.

Schrader, W. T., Birnbaumer, M. E., Hughes, M. R., Weigel, N. L., Grody, W. W. and O'Malley, B. W. (1981) Studies on the structure and function of the chicken progesterone receptor. *Rec. Prog. Hormone Res.,* **37**, 583–633.

Sherman, M. R. and Diaz, S. C. (1977) Meroreceptor formation from a larger subcomponent of the oviduct progesterone receptor. *Ann. N.Y. Acad. Sci.* **286**, 81–86.

Sherman, M. R., Corvol. P. L. and O'Malley, B. W. (1970) Progesterone-binding components of chick oviduct. I. Preliminary characterization of cytoplasmic components. *J. Biol. Chem.* **245**, 6085–6096.

Sherman, M. R., Atienza, S. B. P., Shansky, J. R. and Hoffman, L. M. (1974) Progesterone receptors of chick oviduct. Steroid-binding "subunit" formed with divalent cations. *J. Biol. Chem.* **249**, 5351–5363.

Sherman, M. R., Tuazon, F. B., Diaz, S. C. and Miller, L. K. (1976) Multiple forms of oviduct progesterone receptors analyzed by ion exchange filtration and gel electrophoresis. *Biochemistry* **15**, 980–989.

Sherman, M. R., Pickering, L. A., Rollwagen, F. M. and Miller, L. K. (1978) Meroreceptors: proteolytic fragments of receptors containing the steroid-binding site. *Federation Proc.* **37**, 167–173.

Shyamala, G. and Leonard, L. (1980) Inhibition of uterine estrogen receptor transformation by sodium molybdate. *J. Biol. Chem.* **255**, 6028–6031.

Simpson, R. J., Grego, B., Govindan, M. V. and Gronemeyer, H. (1987) Peptide sequencing of the chick oviduct progesterone receptor form B. *Mol. Cell. Endocrinol.* **52**, 177–184.

Smyth, D. G. (1967) *Methods Enzymol.* **11**, 214–231.

Vedeckis, W. V., Schrader, W. T. and O'Malley, B. W. (1978) The chick oviduct progesterone receptor. In: Litwack, G. (ed.) *Biochemical Actions of Hormones*. Academic Press, New York, pp. 321–372.

von der Ahe, D., Renoir, J. P., Buchou, T., Baulieu, E. E. and Beato, M. (1986) Receptors for glucocorticosteroid and progesterone recognize distinct features of a DNA regulatory element. *Proc. Natl. Acad. Sci. USA* **83**, 2817–2821.

Weigel, N. L., Minghetti, P. P., Stevens, B., Schrader, W. T. and O'Malley, B. W. (1983) In: Eriksson, H. and Gustafsson, J. A. (eds.) *Steroid Hormone Receptors: Structure and Function*. Elsevier, Amsterdam, pp. 25–43.

11

Photoaffinity labeling of the human progesterone receptor

K. B. Horwitz and M. D. Francis
Department of Medicine, University of Colorado, Denver, USA

I. INTRODUCTION

Structural analysis of chick oviduct progesterone receptors (PR) in impure preparations shows the presence of two hormone binding proteins of different size (A proteins of 79 kDa, and B proteins of 105 kDa) which are hypothesized to form dimers of a larger holoreceptor (Schrader *et al.*, 1981). This proposal, as well as the origin of the smaller A protein, has been the subject of controversy. Particularly, it has been suggested that for rabbit PR very little structural information has been available for human PR. The A protein is simply a proteolytic fragment of the B protein (Loosfelt *et al.*, 1984). Until our recent studies (Estes *et al.*, 1987), human PR present in target tissues in only nanomolar concentrations were not amenable to purification. In 1982 (Horwitz *et al.*, 1982), we described a human breast cancer cell subline, T47D$_{CO}$, which constitutively contains more than 300 000 PR sites per cell. These cells seemed ideally suited for analysis of human PR structure, and the nature of the A and B proteins. Since purified PR were unavailable we selected the technique of affinity labeling to introduce a covalent radioactive tag onto PR, using the synthetic progestin 17,21-dimethyl-19-nor-pregn-4,9-diene-3,20-dione ([^{3}H]R5020; New England Nuclear). This steroid had previously been used by Dure *et al.* (1980) to photoaffinity label chick oviduct PR *in vitro*. We first improved the efficiency of *in vitro* labeling and then, in order to circumvent possible artifacts due to proteoylsis *in vitro*, we developed an *in situ* photoaffinity labeling technique using intact T47D cells. In Chapter 4 we described the technical details for *in vitro* and *in situ* photoaffinity labeling of human PRs; here we describe some studies using these methods, to analyze human PR structure.

II. ANALYSIS OF THE HUMAN PROGESTERONE RECEPTOR

A. Are photoaffinity labeled PR representative of entire receptor population?

At best, photoaffinity labeling covalently tags 10–15% of the receptor population that can be measured by conventional ligand binding assays. We have attempted to improve this efficiency by increasing UV irradiation time, or by treating cells with sequential cycles of [^{3}H]R5020 and irradiation. These measures failed to improve efficiency. Thus, until recently, we could not answer the question raised above. We have now developed a set of anti-human PR monoclonal antibodies, one of which (AB-52) cross-reacts with both B and A receptors (Estes *et al.*, 1987). We have repeated many of the studies that employed photoaffinity labeling, but measured receptors by immunoblotting instead. We find in every case that immunoblotting data parallel the results of photoaffinity labeling. Since the former method measures total protein, while the latter measures hormone binding, the coincidence of results by two independent techniques suggests that photoaffinity methods label a representative sample of PR.

B. R5020 photolysis and cellular UV damage

In an attempt to determine whether UV irradiation damages R5020, 80 nM [^{3}H)R5020 irradiated for 2 min (+UV) was compared with unirradiated control [^{3}H]R5020 (−UV) for its ability to bind to PR in intact cells and to transform receptors to the tight chromatin binding state (Fig. 1). After 5 min of hormone treatment the radioactive hormone was covalently linked to receptors in intact cells by *in situ* photoaffinity labeling as described above. Unirradiated [^{3}H]R5020 transforms >90% of cytosolic B and A receptors to the tight nuclear binding state. In contrast, with irradiated [^{3}H]R5020, no cytosolic and only minimal nuclear receptors were found; that is, <5% of the PR known to be present in these cells were photoaffinity labeled. There are two possible explanations for these results. Either (a) photolytic degradation prevented R5020 binding to PR or (b) the photolysis products bound, but could not subsequently be covalently linked to PR. To distinguish between these possiblities, we compared the ability of irradiated and unirradiated non-radioactive R5020 to compete for binding of intact [^{3}H]R5020 in cells. Unirradiated R5020 was a 100% effective competitor; irradiated R5020 was ~430% effective (data not shown). This suggested that photolysed R5020 binds PR,

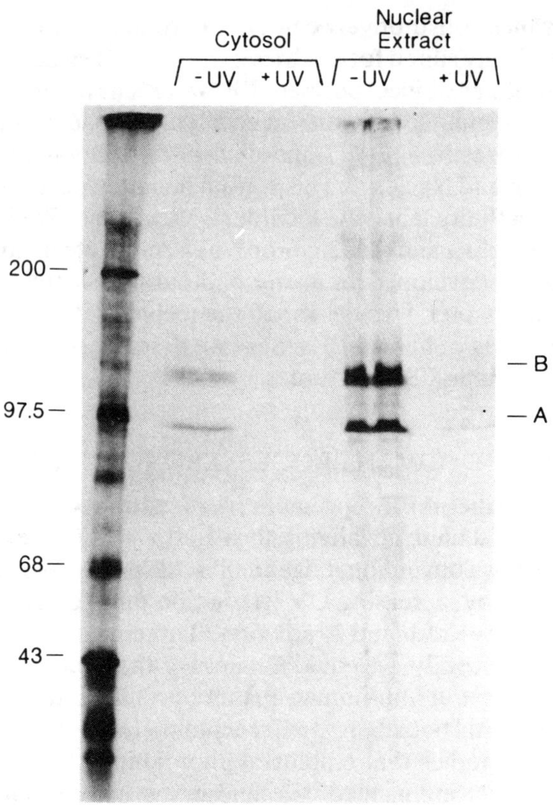

Fig. 1 — Receptor binding and translocation efficiency of UV irradiated [^{3}H]R5020 in intact cells. Serum-free medium containing 80 nM [^{3}H]R5020 was irradiated for 2 min with 300 nm UV (+UV). Unirradiated [^{3}H]R5020 served as control (−UV). Cells growing in Petri dishes were incubated for 5 min at 37°C with the [^{3}H]R5020-containing media; the dishes were washed, inverted over the transilluminator and irradiated for 2 minutes, the cells were harvested and homogenized, and cytosls and nuclear extracts were prepared. Receptors were analysed by SDS–PAGE and fluorography.

albeit with lower affinity than intact R5020. A recent study by Ilenchuk and Walters (1987) shows that R5020 photolysis products have ~10-fold lower affinity for PR than does the intact hormone. Taken together, these findings suggest that failure to demonstrate photolabeled PR using irradiated [^{3}H]R5020 in Fig. 1, is due to explanation (b) above — namely, that photolysis products of R5020 have some affinity for PR but that, after non-covalent binding, they cannot subsequently be covalently coupled to the receptors. Thus any observed photoaffinity labeling of PR must be due to the interaction of intact R5020 with receptors and not due to cross-linking of R5020 photolysis products to the proteins. These findings suggest further that photolysis damages the conjugated double bonds of the A–B rings in R5020, thereby preventing subsequent photocovalent attachment to PR, while leaving the D-ring and 17-methyl residue intact, permitting non-covalent binding of the radioactive photolysis products to PR.

As well as photolytic damage to R5020, irradiation damages the cells and alters their attachment to plastic surfaces.

(1) If cells are irradiated while growing on plastic, they can no longer be harvested by chelation of extracellular calcium with EDTA; they have to be scraped from the surface. A much greater percentage of such cells take up Trypan Blue compared with unirradiated controls. However, scraping causes some membrane damage even to unirradiated cells, and the most gentle harvesting procedure is brief incubation with Hank's–EDTA. Use of trypsin or other proteases is not recommended, since the enzymes cannot be completely washed from the cells, and their residual activity in homogenates may degrade receptors. If proteolytic enzymes must be used for harvesting cells, then specific protease inhibitors should be included in homogenization buffers.

(2) Another indication of UV-induced cellular damage comes from attempted 'pulse-chase' studies, in which cells were treated for 10 min at 37°C with [^{3}H]R5020 to transform PR to the chromatin-bound state, photoaffinity labeled by 2 min UV irradiation, then chased for 24 h with unlabeled R5020, in an attempt to distinguish between an existing and a newly synthesized receptor pool. However, we found that the cellular kinetics of receptors following irradiation differed markedly from those of unirradiated controls. We concluded that receptor dynamics cannot be studied after cells have been irradiated, and ensure in all our *in situ* studies that UV irrdiation is the last step in experimental protocols, followed immediately by cell harvesting and homogenization. We consider this cellular UV damage to be a major disadvantage of *in situ* photoaffinity labeling techniques — a problem not encountered with the use of chemically reactive steroid derivatives. Notwithstanding this, and given these limitations, when *in situ* photoaffinity labeling is used appropriately, it is a powerful tool for study of receptor structure. The structural studies discussed below employ all the methods described in detail in Chapter 4.

C. Structural analysis of human breast cancer progesterone receptors by photoaffinity labeling

The use of PR and progestins in the clinical setting (reviewed in Horwitz *et al.*, 1985a) has required an understanding of their actions at a basic molecular level. To study the structure of PR in intact cells, and to avoid the artifacts that may arise during *in vitro* hormone–receptor binding reactions, we developed an *in situ* photoaffinity labeling method for PR that permits study of the proteins with minimal incubations (Horwitz and Alexander, 1983). This method has been used to show, first, that PR consist of two equimolar hormone binding proteins that share considerable homology, second, that the two proteins form independent receptors and are not linked to one another in a dimeric complex, third, that they are covalently modified in their untransformed state, fourth, that after progestin binding and transformation — acquisition of tight nuclear binding capacity — the two PR proteins undergo further covalent modifications consistent with phosphorylation, and fifth, that nuclear receptor action may be terminated by a processing reaction that leads to the simultaneous disappearance of both proteins from the cell.

1. The 'doublet B–singlet A'

Fig. 2 illustrates the nature of the *in situ* photolabeled untransformed receptors seen after whole T47D cells have been incubated with [^{3}H]R5020 at 0°C for 2 h, UV irradiated for 2 min, and then homogenized. If the cytosol prepared from the

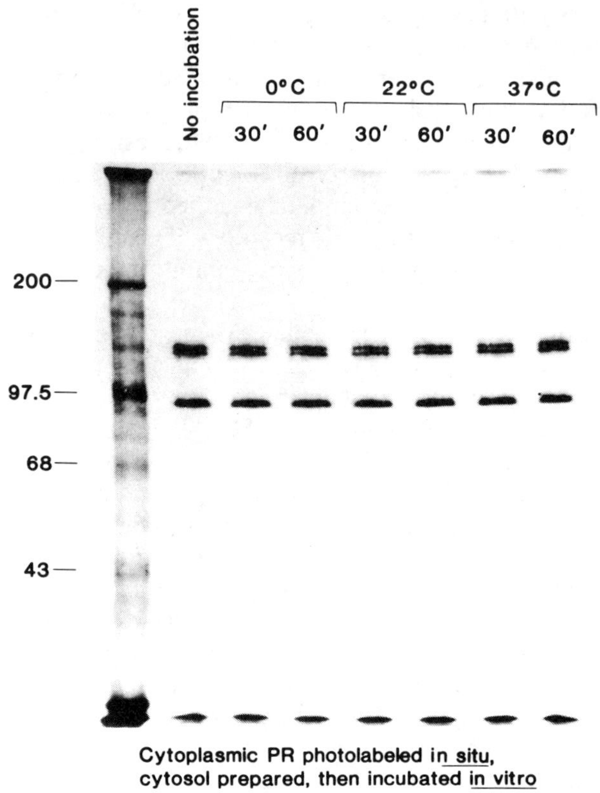

Fig. 2 — Structural stability of *in situ* photolabeled untransformed PR incubated *in vitro*. T47D cells were incubated with [^{3}H]R5020 for 3 h at 0°C, irradiated for 2 min with 300 nm of UV light, then homogenized and a cytosol was prepared. Aliquots of the cytosol were denatured immediately (no incubation), or incubated for 30 or 60 min at 0, 22, or 37°C. These cytosols were then denatured and, together with ^{14}C-labeled molecular weight standards, subjected to SDS–PAGE on a 7.5–19% acrylamide gradient gel. A fluorogram of the gel is shown.

postnuclear supernatant is subjected to electrophoresis on gradient gels under denaturing conditions, two bands are seen. The heavier B protein is a doublet of Mr~117 000–120 000 daltons. The smaller A protein is a singlet of Mr~94 000 daltons. Fig. 2 also shows the remarkable stability of these two proteins during further *in vitro* incubations that would be expected to promote proteolytic degradation. Even incubation for 1 h at 37°C fails to degrade the B protein either to A or to any other fragments. We have previously shown that proteins A and B bind only progestins specifically (Horwitz *et al.*, 1985a) and that they are present in equimolar amounts (Horwitz *et al.*, 1985b). They can be demonstrated even if homogenization and centrifugation are eliminated — that is, by *in situ* photoaffinity labeling of intact cells, solubilization of the cells in buffer containing detergent plus protease inhibitors, denaturation, and immediate electrophoresis (Horwitz *et al.*, 1985a). The 'doublet B–singlet A' is characteristic of activated but untransformed receptors, having low affinity for nuclei. This pattern is seen not only in 0°C *in situ* photolabeled receptors but also in receptors from cytosols incubated with [^{3}H]R5020 at 0°C *in vitro*.

2. A and B form independent 8S holoreceptors

We used *in situ* photoaffinity labeling, and sucrose gradient sedimentation shifts after antibody binding, to demonstrate that the two PR proteins form separate 8S holoreceptors — they are not bound together as dimers (Wei *et al.*, 1987). The antibody used was PR-6, an anti-chick oviduct PR monoclonal antibody, that binds chick B but not A receptors, and cross-reacts with human B receptors. This antibody was developed in the laboratory of D. Toft (Sullivan *et al.*, 1986).

Because it is B specific and binds to native human PR, PR-6 can be used to study the nature of the association between A and B proteins in the untransformed 8S state and in the transformed 4S state. This is of interest since there are conflicting models for the molecular interaction of the A and B proteins. One model holds that A and B dimerize (Schrader *et al.*, 1981) and the other that A and B exist as separate 8S molecules (Dougherty and Toft, 1982; Renoir and Mester, 1984). We reasoned that if the first model is correct, addition of PR-6 to receptors stabilized in the 8S state would shift both A and B proteins to the bottom of sucrose density gradients, but that A would not be shifted if the two proteins form independent 8S holoreceptor complexes. Fig. 3 shows a study in which PR, covalently labeled *in situ* with [^{3}H]R5020 by UV irradiation, were incubated with PR-6 or a control antibody and then sedimented on sucrose gradients in the continuous presence of molybdate and protease inhibitors to maintain intact and native PR conformation. Aliquots (25 μl) of every gradient fracion were countered to obtain the [^{3}H]R5020 binding profile (Fig. 3, left) and additional 175 μl aliquots of the bottom fraction (a,c) and peak 8S fractions (b,d) were analyzable by gel electrophoresis and fluorography since they were photoaffinity labeled (Fig. 3 right). The control antibody had no effect on 8S sedimentation of PR, and electrophoretic analysis showed that all B and A receptors remained in the 8S peak. In contrast, after addition of PR-6, half of the 8S radioactivity was shifted to heavier sedimenting forms. Electrophoretic analysis of the shifted photoaffinity labeled receptors at the bottom of the gradient (fraction c), as well as the residual receptors in the 8S peak (fraction d), showed only B present in the antibody-shifted fraction while A, separated from B, remained at 8S. We conclude that A and B do not dimerize but that each exists as a separate 8S receptor either because of self-association or because of association with non-hormone binding proteins. Two such proteins of 90 and 76 kDa co-purify with untransformed human B receptors (Estes *et al.*, 1987).

PR-6 can also separate B receptors from A receptors that have been transformed to the 4S species by treatment with salt. Approximately half of the radioactivity seen in the 4S peak in the presence of control antibody was shifted to heavier aggregates on addition of PR-6 and a secondary antibody (Fig. 4). The control 4S peak contained both hormone-binding proteins but only A remained at 4S after Pr-6 addition while the B receptors were shifted to heavier sedimenting fractions by the antibody (Fig. 4). It appears, then, that antibody PR-6 cross-reacts with both the native 8S as well as the tranformed 4S form of B receptors and that, like the case for

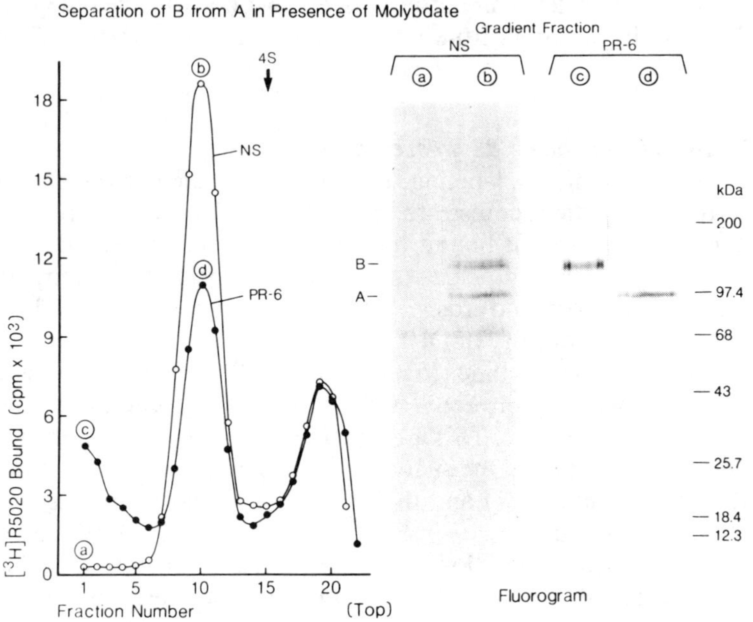

Fig. 3 — Separation of untransformed 8S B receptors from 8S A receptors by PR-6 in the presence of molybdate. Suspended T47D cells were incubated at 0°C with 80 nM [^{3}H]R5020, photoaffinity labeled *in situ,* and a labeled cytosol was prepared in buffer containing molybdate. Aliquots (200 μl) of the cytosol were incubated with antibody PR-6 or a non-specific control antibody (NS); both sets were then incubated with a second antimouse IgG, and the mixtures were layered on 5–20% molybdate-containing sucrose gradients and centrifuged. Gradient fractions (200 μl) were collected; 25 μl were counted to obtain the radioactive profiles (left panel), and 175 μl from the fractions of interest (fractions labeled a–d), were subjected to electrophoresis on SDS–PAGE, transferred to nitrocellulose, and the filter was used for immunoblotting (not shown) and fluorography (right panel). 3.28 mg protein/200 μl and 711 166 cpm specific counts /200 μl were loaded on the gradient. ^{14}C-labeled BSA was included as a 4S sedimentation marker on gradients, and its position is marked by the arrow. ^{14}C-labeled molecular weight markers were run in a parallel electrophoretic lane and their position is shown on the right of the fluorogram.

8S, there are two types of 4S species containing either A protein or B protein, but not both. The 76 kDa non-hormone binding protein co-purifies with transformed 4S B receptors (Estes *et al.*, 1987).

Since A and B are not linked in transformed PR, the two-stage nuclear binding model in which B is seen as a chromatin-specifying protein that guides A to appropriate DNA binding sites (Schrader *et al.*, 1981) is unlikely to be correct. The alternative model is that A and B proteins form independent functional receptor complexes and that, like other steroid receptors, each binds to DNA. While there appears to be consensus that A receptors bind DNA, for B receptors this point has been unsettled (Schrader *et al.*, 1981; Gronemeyer *et al.*, 1985). We have tested the ability of immunopurified transformed human B receptors to bind DNA–cellulose and find that they do so efficiently (Estes *et al.*, 1987).

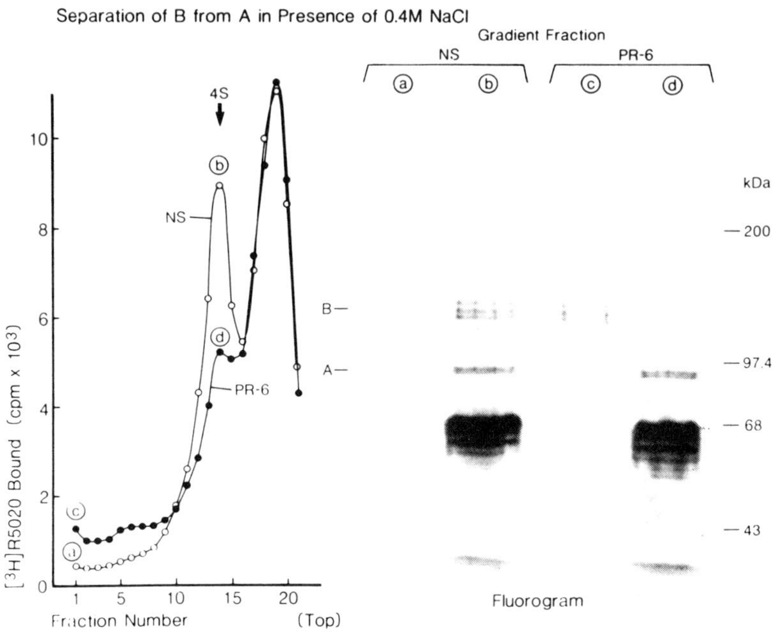

Fig. 4 — Separation of transformed 4S B receptors from 4S A-receptors in salt-containing molybdate-free sucrose gradients. T47D cytosol prepared in molybdate-free buffer was incubated with 20 nM [^{3}H]R5020 at 0°C, photoaffinity labeled *in vitro,* then brought to 0.4 M NaCl to transform receptors. Aliquots (250 μl) of labeled cytosols were incubated with antibody PR-6 or a non-specific antibody (NS), then incubated with a second antibody, layered on 0.4 M NaCl-containing sucrose gradients and fractionated as described in Fig. 3. 5.18 mg protein/250 μl and 577 000 specific cpm/250 μl were loaded on the gradient. Aliquots (25 μl) of gradient fractions were counted to obtain the sedimentation profile (left panel) and 200 μl aliquots from the fractions of interest (fractions a–d) were subjected to SDS–PAGE, electroblotting and fluorography (right panel) as described in Fig. 3. The heavily labeled protein(s) below the 68 kDa marker represent 4S sedimenting, non-specific R5020-binding fetal calf serum protein(s) previously described as a '50 kDa' binder. The protein(s) are unrelated to PR (Horwitz and Alexander, 1983).

3. Covalent modifications in the nucleus

When T47D cells are treated with hormone and also warmed, 95% of receptors become transformed and acquire tight nuclear binding capacity within 5 min (Fig. 5). In this state, receptors can no longer be recovered in cytosols, but instead must be extracted from nuclei with buffers of high ionic strength. The transformation reaction does not alter the receptors. Acutely transformed nuclear receptors (1–5 min after hormone addition) structurally resemble the untransformed cytoplasmic receptors; the 'B doublet–A singlet' structure prevails. This can also be concluded from Fig. 1, where transformation was accomplished *in vitro* by the 60 min, 37°C incubation. There also, the transformed receptors (37°C treated) resemble the untransformed receptors (0°C treated), and we would conclude that transformation itself leaves the primary structure of each of the receptor proteins unchanged, as far as can be determined by single-dimension electrophoresis.

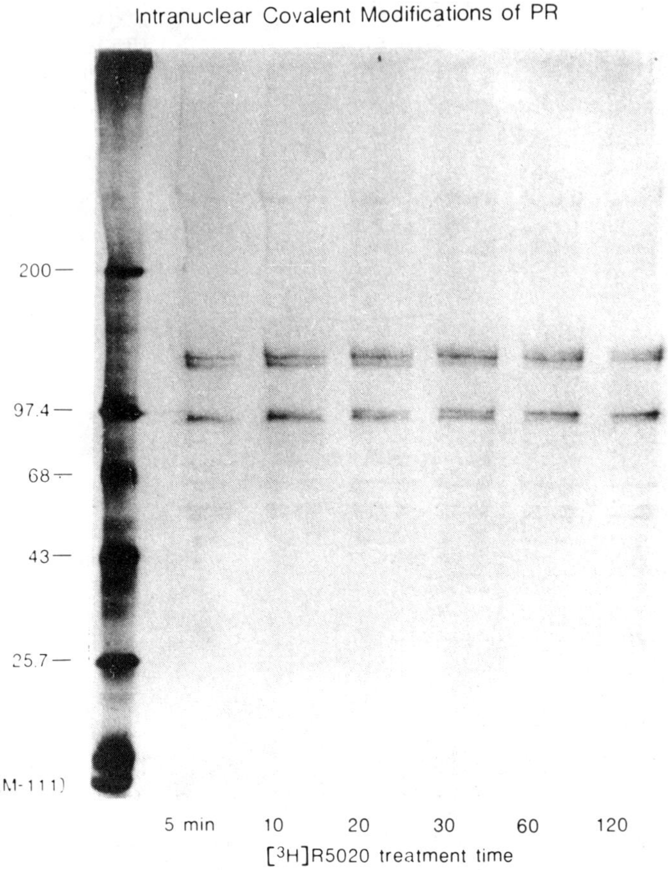

Fig. 5 — Hormone-dependent covalent modification of nuclear PR photolabeled *in situ*. T47D cells grown in Petri dishes were treated with [^{3}H]R5020 for 5 min to 2 h at 37°C. The dishes were then cooled and irradiated for 2 min with UV light. The cells were harvested and homogenized, and a nuclear extract was prepared, concentrated by ammonium sulfate precipitation, and subjected to electrophoresis followed by fluorography.

The holoreceptors, as well as the acutely transformed nuclear receptor complexes, have equal amounts of A and B, as well as equimolar levels of the two B isoforms (Fig. 5, 5 min lane). However, as nuclear residence time increases, the equimolar relationship between the two B isoforms begins to shift. Fig. 5 (10–60 min lanes) shows that the lower molecular weight band of the protein B doublet disappears with time and is entirely converted to the heavier form of the doublet approximately 30 min after the start of hormone treatment. Meanwhile protein A, which is a singlet at 5 min, splits to become a doublet by 30 min, and then the lower molecular weight form of the doublet shifts entirely to the heavier form by 60 min. Thus, both B and A display reproducible 2000–3000 dalton increases in electrophoretic mobility 30–60 min after acquisition of tight nuclear binding capacity. We have now shown that these mobility changes result from phosphorylation of both A and B receptors in nuclei (unpublished).

Longer R5020 treatment (beyond 2 h) leads to additional changes (Fig. 6). The 'double B–singlet A' condition at 5 min is replaced at 4 h by the heavier, covalently modified form of both proteins. Prolonging R5020 treatment even further leads to simultaneous down-regulation or 'processing' of both proteins; A and B are lost by a mechanism that generates no hormone-binding fragments of molecular weight greater than 10 000 daltons. The nature of this reaction is unknown, although it occurs both for estrogen receptors (Horwitz and McGuire, 1978) and PR (Horwitz *et al.*, 1982), and probably serves to terminate their action. The receptors lost from nuclear compartments do not reappear in cytoplasmic compartments after R5020 treatment so that processing represents an absolute decrease of total cellular receptors by protein mass, as measured by hormone binding and more recently by immunoblotting (Wei *et al.*, 1987).

4. Human PR contain two hormone-binding proteins

The argument has been made that A is formed *in vitro* as an artifactual proteolytic fragment of B; it is buoyed by several studies. First is the fact that despite repeated demonstration of two progestin-binding proteins in chick oviduct and human breast cancer cells, two hormone-binding proteins are usually not demonstrated for other steroid receptors (reviewed in Horwitz *et al.*, 1985a). On occasions when two proteins are seen, the proteolysis explanation is invoked (Loosfelt *et al.*, 1984). Second are studies demonstrating structural similarity between chick oviduct or human A and B receptors by peptide mapping of photolabeled receptors (Gronemeyer *et al.*, 1983; Birnbaumer *et al.*, 1983; Horwitz *et al.*, 1985a), and by immunoreactivity (Gronemeyer *et al.*, 1985). These are compelling but inconclusive, since they also fail to reveal the *ca* 25 kDa fragment that must be generated from B but not A if the proteolysis argument is true. Nor do the studies prove a proteolysis mechanism, since other explanations for the biosynthesis of partially homologous proteins are possible (see below). Third are immunoblot studies of rabbit uterine PR. These studies show that, if homogenization is carried out quickly, little or no protein A is generated. Unfortunately, the antibody used for these experiments has not been fully characterized — receptor purification data are unpublished to our knowledge (Logeat *et al.*, 1983, 1985), and the antibody used for blotting studies failed to bind to 50% of native PR in salt-containing sucrose gradients (Logeat *et al.*, 1983), suggesting that it may have very different affinities for proteins A and B, and that it belongs to the class of B-specific antibodies commonly generated from partially purified PR (Sullivan *et al.*, 1986; Estes *et al.*, 1987). Thus, the protease hypothesis has not, in our opinion, been conclusively established. A cocktail of protease inhibitors is routinely added to homogenization buffers in our studies, but A receptors levels are unaffected.

Our studies of human breast cancer PR suggest that protein B (Mr~120 000) and protein A (Mr~94 000) are integral intracellular proteins.

(1) Regardless of the site from which they are extracted (cytoplasmic or nuclear), proteins A and B are found in equimolar amounts (Horwitz and Alexander, 1983; Horwitz *et al.*, 1985a). If a protease is degrading protein B, it must be present in cytosols as well as in nuclear extracts, be equally active in buffers of different ionic strengths and pH, and its activity must stop after degrading half of the B molecules.

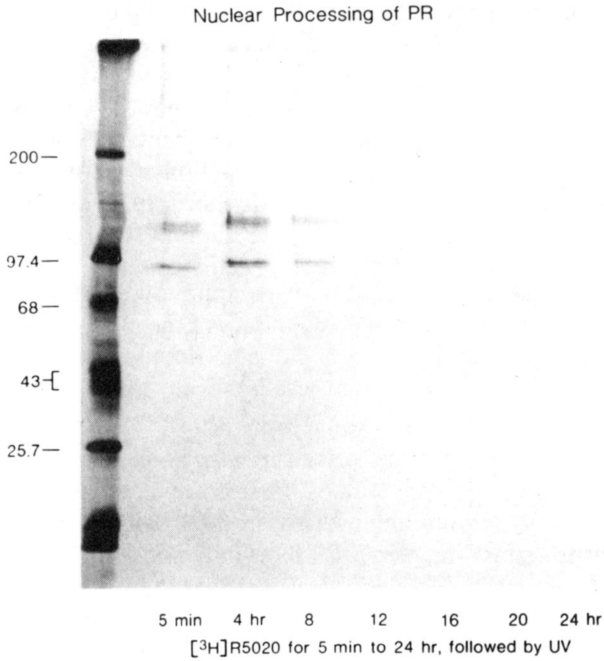

Fig. 6 — Hormone-dependent covalent modification and processing of nuclear PR photolabeled *in situ.* T47D cells growing in Petri dishes were treated with [^{3}H]R5020 for 5 min to 24 h. They were then UV irradiated before being homogenized. Nuclei were pelleted at low speed, extracted with 0.4 M KCl-containing buffer, and a soluble extract and salt-resistant pellet were obtained. The postnuclear supernatant from the low-speed run was saved. Salt-resistant nuclear pellets and cytoplasm were solubilized by 30 min of incubation at room temperature in SDS and protease inhibitors, then they and nuclear extracts were denatured by boiling in SDS–PAGE followed by fluorography. Results shown are the PR in the nuclear extracts. No receptors were seen in the other cell compartments.

(2) Degradation of B to A cannot be demonstrated during *in vitro* incubations (Fig. 2), forcing the conclusion that the putative enzyme acts instantaneously on half of the B molecules when the cells are first broken.

(3) Both receptor proteins are seen in cells photolabeled *in situ,* then lysed directly in buffer containing detergent and protease inhibitors, and immediately subjected to electrophoresis (Horwitz *et al.,* 1985a). This suggests that the putative protease acts intracellularly.

(4) The two receptor proteins are dissimilarly modified in their untransformed state. Since B is a doublet while A is a singlet in holoreceptors, one would have to postulate that, if A is a proteolytic artifact, then the offending protease clips out a domain of B containing the phosphorylation site(s). However, such a protease cannot be invoked for the 30 min resident nuclear receptors, where protein B is a singlet while protein A is a doublet (Fig. 5). Here different proteases would have to be involved, that either clip the phosphorylated B at two sites 3000 daltons apart to generate the A doublet, or clip B at a single site then dephosphorylate half of the newly generated molecules.

Thus we are left to conclude either that B is subject to degradation by a series of unusual intracellular proteases, or that A is a protein closely related to B that is formed intracellularly by other mechanisms. Possible mechanisms include gene duplication, alternate transcription from a single gene, synthesis of multiple messages by alternate processing of a single precursor RNA, or use of alternate translation start sites from a single message. Examples of all these have been described.

D. Summary

The study of human progesterone receptors is moving into a rapidly evolving and exciting state. Photoaffinity labelling, recent cloning of the cDNA for avian, rabbit and human PR, and the availability of monoclonal antibodies, pave the way for an understanding of the molecular biology of these medically important proteins.

REFERENCES

Birnbaumer, M., Schrader, W. T. and O'Malley, B. W. (1983) Assessment of structural similarities in chick oviduct progesterone receptor subunits by partial proteolysis of photoaffinity labeled proteins. *J. Biol. Chem.* **255,** 1637–1644.

Dougherty, J. J. and Toft, D. O. (1982) Characterization of two 8S forms of chick oviduct progesterone receptor. *J. Biol. Chem.* **257,** 3113–3119.

Dure, L. S., Schrader, W. T. and O'Malley, B. W. (1980) Covalent attachment of a progestational steroid to chick oviduct progesterone receptor by photoaffinity labeling. *Nature* **283,** 784–786.

Estes, P. A., Suba, E. J., Lawler-Heavner, J., El-Ashry, D., Wei, L. L., Toft, D. O., Sullivan, W. P., Horwitz, K. B. and Edwards, D. P. (1987) Immunologic analysis of human breast cancer progesterone receptors. I. Immunoaffinity purification of transformed receptors and production of monoclonal antibodies. *Biochemistry,* in press.

Gronemeyer, H., Harry, P. and Chambon, P. (1983) Evidence for two structurally related progesterone receptors in chick oviduct cytosol. *FEBS Lett.* **156,** 287–292.

Gronemeyer, H., Govindan, M. V. and Chambon, P. (1985) Immunologic similarity between the chick oviduct forms A and B. *J. Biol. Chem.* **260,** 6916–6925.

Horwitz, K. B. and Alexander, P. S. (1983) *In situ* photolinked nuclear progesterone receptors of human breast cancer cells: subunit molecular weights after transformation and translocation. *Endocrinology* **113,** 2195–2201.

Horwitz, K. B. and McGuire, W. L. (1978) Estrogen control of progesterone receptor in human breast cancer. Correlation with nuclear processing of estrogen receptor. *J. Biol. Chem.* **253,** 2223–2228.

Horwitz, K. B., Mockus, M. B. and Lessey, B. A. (1982) Variant T47D human breast cancer cells with high progesterone receptor levels despite estrogen and antiestrogen resistance. *Cell* **28,** 633–642.

Horwitz, K. B., Wei, L. L., Sedlacek, S. M. and d'Arville, C. N. (1985a) Progestin action and progesterone receptor structure in human breast cancer: A review. *Rec. Prog. Horm. Res.* **41,** 249–316.

Horwitz, K. B., Francis, M. D. and Wei, L. L. (1985) Hormone-dependent covalent modification and processing of human progesterone receptors in the nucleus. *DNA* **4,** 451–465.

Ilenchuk, T. T. and Walters, M. R. (1987) Rat uterine progesterone receptor analyzed by [^{3}H]R5020 photoaffinity labeling: evidence that the A and B subunits are not equimolar. *Endocrinology* **120,** 1449–1456.

Logeat, F., Hai, M. T. V., Fournier, A., Legrain, P., Buttin, P. and Milgrom, E. (1983) Monoclonal antibodies to rabbit progesterone receptor: cross reaction with other mammalian progesterone receptors. *Proc. Natl. Acad. Sci. USA* **80,** 6456–6459.

Logeat, F., LeCunff, M., Pamphile, R. and Milgrom, E. (1985) The nuclear-bound form of the progesterone receptor is generated through a hormone-dependent phosphorylation. *Biochem. Biophys. Res. Comm.* **131,** 421–427.

Loosfelt, H., Logeat, F., Hai, M. T. V. and Milgrom, E. (1984) The rabbit progesterone receptor. Evidence for a single steroid-binding subunit and characterization of receptor mRNA. *J. Biol. Chem.* **259,** 14196–14202.

Mockus, M. B., Lessey, B. A., Bower, M. A. and Horwitz, K. B. (1982) Estrogen insensitive progesterone receptors in a human breast cancer cell line: characterization of receptors and of a ligand exchange assay. *Endocrinology* **110,** 1564–1571.

Renoir, J. M. and Mester, J. (1984) Chick oviduct progesterone receptor: structure, immunology, function. *Mol. Cell. Endocr.* **37,** 1–13.

Schrader, W. T., Birnbaumer, M. E., Hughes, M. R., Weigel, N. L., Grody, W. W. and O'Malley, B. W. (1981) Studies on the structure and function of the chicken progesterone receptor. *Rec. Prog. Horm. Res.* **37,** 583–633.

Sullivan, W. P., Beito, T. J., Proper, J., Krco, C. J. and Toft, D. O. (1986) Preparation of monoclonal antibodies to the avian progesterone receptor. *Endocrinology* **119,** 1549–1557.

Wei, L. L., Sheridan, P. L., Krett, N. L., Francis, M. D., Toft, D. O., Edwards, D. P. and Horwitz, K. B. (1987). Immunologic analysis of human breast cancer progesterone receptors. II. Structure, phosphorylation and processing. *Biochemistry* **26,** 6262–6272.

12

Molecular weight determination of the androgen receptor by affinity labeling techniques

M. P. Johnson, D. R. Rowley, C. Y. F. Young, C. H. Chang, T. J. Lobl and D. J. Tindall
Department of Cell Biology, Baylor College of Medicine, Houston, Texas, USA, and The Upjohn Company, Kalamazoo, Michigan, USA

I. INTRODUCTION

Significant progress has been made in the last two decades in understanding the molecular basis of androgen action (for, reviews, see Liao, 1975, 1977; Mainwaring, 1978; Chan and Tindall, 1981; Tindall *et al.*, 1984; Rowley and Tindall, 1986). It appears that the mechanism of action of androgen is very similar to that of other steroid hormones. One unique feature is that in most androgen target tissues the major blood-borne androgen, testosterone, is converted intracellularly to its biologically active metabolite, 5α-dihydrotestosterone. Like other steroid hormones, dihydrotestosterone then binds with high affinity to a soluble intracellular receptor protein in a stereoscopic manner. The steroid–receptor complex subsequently acquires a greater affinity for acceptor sites on chromatin (Liao *et al.*, 1972; Mainwaring *et al.*, 1976; Klyzsejko-Stefanowicz *et al.*, 1976) and exerts an influence on the transcription of androgen-regulated genes (Mainwaring *et al.*, 1974; Heyns *et al.*, 1977; Ostrowski *et al.*, 1982; Roy *et al.*, 1983).

At present, all the molecular details of this mechanism are not fully understood. However, it is clear that the androgen receptor protein plays a critical role in the mechanism of androgen action. For example, in the genetic defect known as testicular feminizing syndrome (Tfm), androgen target organs do not respond to normal levels of testosterone (Wilson *et al.*, 1981). It has been suggested either that Tfm tissues lack androgen receptor or that there is a defect in receptor structure (Keenan *et al.*, 1974; Meyer *et al.*, 1975; Bardin and Caterall, 1981). Considering the importance of the receptor, many investigators have concentrated their efforts on studies of androgen receptor protein structure and its interaction with steroid and DNA.

Specific ligands which are commonly used to label the androgen receptor by a reversible, non-covalent interaction include [^{3}H]dihydrotestosterone, [^{3}H]methyltrienolone (also known as R1881) and [^{3}H]Mibolerone. The latter two compounds are synthetic androgens which are particularly useful because they are resistant to metabolism and show very little affinity for serum binding proteins. However, these two synthetic ligands will bind significantly to progesterone, unless excess triamcinolone acetonide is added as a competitive inhibitor (Zava *et al.*, 1979; Murthy *et al.*, 1985.) Unlabeled progestins cannot be used to block selectively binding of [^{3}H]R1881 or [^{3}H]Mibolerone to the androgen receptor because most progestins possess some affinity for the androgen receptor. Although the synthetic glucocorticoid triamcinolone acetonide is also progestogenic, it has no affinity for the androgen receptor (Zava *et al.*, 1979). Thus, inclusion of 500-fold excess non-radioactive triamcinolone acetonide is recommended when [^{3}H]R1881 0r [^{3}H]Mibolerone are used to label androgen receptors.

Binding of androgens to the androgen receptor can be quantified in vitro by the dextran-coated charcoal assay (Korenman, 1969) or the hydroxylapatite batch assay (Peck and Clark, 1977). In addition, Carroll *et al.* (1984) have established an assay for exchange labeling of the androgen receptor with R1881 as the ligand, using sodium molybdate to stabilize the receptor during incubation at 30°C for 30 min.

Recently, we (Rowley *et al.*, 1986; Johnson *et al.*, 1987) have reported that at least three forms of the androgen receptor can be identified based on their sedimentation properties in low or high ionic strength buffers. When target tissue is homogenized in low ionic buffer, there exists a 9S oligomeric form composed of steroid-binding and non-steroid-binding subunits. Under high ionic conditions, this form dissociates to a 4.5–5.0S monomeric form. When the salt concentration is lowered, the monomer associates with an unknown component containing RNA to form an intermediate 7–8S species. In addition, two proteolytic fragments of the monomer can be observed in high ionic buffer which are only 3.5–4.0S and 2.5–3.0S in size, yet still bind steroid. Various properties and relationships of these five forms are shown in Table 1.

Table 1 — Molecular forms of the androgen receptor[a]

	S	R_s (Å)	M_r (kDa)	Affinity (k_d) (nM)	DNA binding
(1) Oligomer	9	73	280	~1	No
(2) RNA bound	7–8	—	—	~1	Yes
(3) Monomer	4.5–5.0	54	109–121	~1	Yes
(4) Fragment I	3.5–4.0	33	52–59	~1	Yes
(5) Fragment II	2.5–3.0	20	22–27	~1	Yes

[a]S,=sedimentation coefficient; R_s, Stokes molecular radius; M_r, molecular weight.

II. THE MONOMERIC ANDROGEN RECEPTOR

Reports of the molecular weight for the monomeric steroid binding subunit of the androgen receptor have ranged from 25 to 167 kDa (for review, see Johnson *et al.*, 1987). The molecular weight appeared to vary among different species and target organs, as well as between different investigators.

We (Johnson *et al.*, 1987) had hypothesized that the discrepancy in receptor size was attributable to varying degrees of proteolytic breakdown of a common monomeric receptor. To test this, tissues from a diverse group of organs and species were labeled non-covalently with [^{3}H]dihydrotestosterone and analyzed by gel filtration and sucrose density gradients under non-denaturing conditions of high ionic strength (see Fig. 1). This study revealed varying amounts of three different-sized androgen receptors: 54 Å, 4.5–5.0S (109–121 kDa); 33 Å, 3.5–4.0S (52–59 kDa); 20 Å, 2.5–3.0S (22–27 kDa). Phosphocellulose chromatography diminished the appearance of the smaller forms and facilitated the appearance of the larger 54 Å form. Mixing experiments suggested that phosphocellulose was stabilizing the 54 Å form by removing putative proteases which cleave the larger form. Thus, it was concluded that the 33 and 20 Å receptor forms resulted from proteolysis of the 54 Å monomeric receptor. However, it was essential to confirm the data from these non-denaturing studies by analysis of the receptor under denaturing conditions. This was necessary to discount the possibility that the variety of receptor forms resulted from varying configurations of the folded protein. To accomplish this, partially purified androgen receptor was affinity labeled using dihydrotestosterone 17β-bromoacetate and the covalently labeled sample was analyzed by SDS gel electrophoresis.

III. ELECTROPHILIC AFFINITY LABELS

Electrophilic steroid affinity labels have been used extensively to characterize a number of steroid-metabolizing enzymes (Ganguly and Warren, 1971; Sweet *et al.*, 1972; Thomas and Strickler, 1983) and, more recently, steroid receptor proteins (see Katzenellenbogen and Katzenellenbogen, 1984; Gronemeyer and Govindan, 1986; or other chapters in this volume). These compounds possess a good 'leaving' group such as a bromine atom or a methylsulfonyloxy (mesylate) moiety which is displaced,

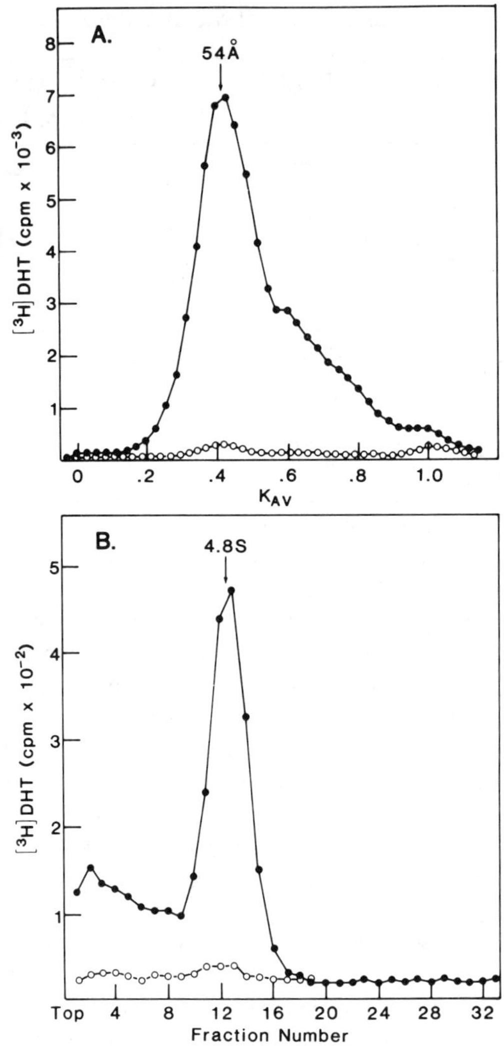

Fig. 1 — Hydrodynamic properties of R3327 rat tumor cytosol androgen receptor. A: gel filtration chromatography of bound [^{3}H]dihydrotestosterone (DHT) from R3327G tumor homogenized in TTES buffer and labeled with 6 nM [^{3}H]DHT in the presence (○) or absence (●) of 600 nM non-radioactive DHT during centrifugation for 1 h at 105 000*g* and 2°C to prepare cytosol. Cytosol was treated using a pellet of dextran-coated charcoal to remove free steroid. Aliquots of 1.0 ml were applied to parallel columns of agarose A1.5m, which were equilibrated in $TTEGK_{400}$ buffer. The void volume is indicated by a distribution coefficient (K_{av}) of 0.0, whereas a K_{av} of 1.0 refers to the total column volume. B: sucrose density gradient sedimentation of androgen receptor from an aliquot of the peak fraction from the gel filtration analysis shown in A. The 300 μl sample was layered onto 6–20% sucrose gradients (5 ml) in $TTEGK_{400}$ buffer and centrifuged for 20 h in an SW50.1 rotor at 190 000*g* and 2°C (Reprinted with permission from *Biochemistry*, **26**(2). Copyright 1987 American Chemical Society.)

thus allowing the steroid to bind covalently to the active site of the protein. For the androgen receptor protein, 17β-[(bromoacetyl)oxy]-[1,2,4,5,6,7,16,17-$^{3}H_{8}$]15α-androstan-3-one ([^{3}H]dihydrotestosterone 17β-bromoacetate) has proven to be an effective affinity label (see Fig. 2). A similar steroid, testosterone bromoacetate, was

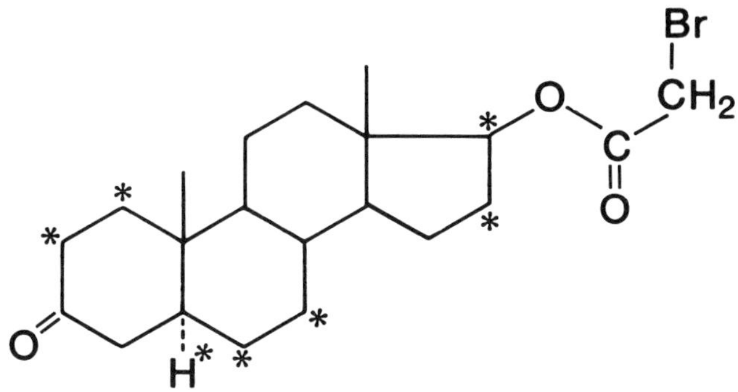

Fig. 2 — Chemical structure of dihydrotestosterone 17β-bromoacetate: *, incorporation of tritium.

first used in its ^{14}C-radiolabeled form (14 mCi/mmol) by Mainwaring and Johnson (1980) to label androgen receptors. [^{3}H]DHT-bromoacetate has been synthesized with a much higher specific radioactivity (170 Ci/mmol) by Dr. Thomas J. Lobl of the Upjohn Company, Kalamazoo, Michigan (Change *et al.*, 1982).

A. Synthesis of dihydrotestosterone 17β-bromoacetate

Bromoacetate derivatives of 5α-dihydrotestosterone were synthesized according to LeGaillard and Dantrevaux (1977) and reported in Chang *et al.* (1984). The unlabeled dihydrotestosterone 17β-bromoacetate which we synthesized was pure as judged by nuclear magnetic resonance spectroscopy, mass spectroscopy, infrared spectroscopy, and elemental analysis. A similar protocol was used to synthesize the tritiated compound from [^{3}]dihydrotestosterone. Chromatography of the tritium-labeled compound on a thin-layer plate showed only one peak of radioactivity, which co-migrated with the non-radioactive material.

B. Binding of [^{3}H]dihydrotestosterone 17β-bromoacetate to the androgen receptor

1. Binding affinity

In order to determine whether the non-radioactive affinity label would compete with dihydrotestosterone for the active binding site on the androgen receptor, we incubated rat prostate cytosol with [^{3}H]dihydrotestosterone or dihydrotestosterone 17β-bromoacetate. Fig. 3 shows the copmpetition by both steroids. Dihydrotestos-

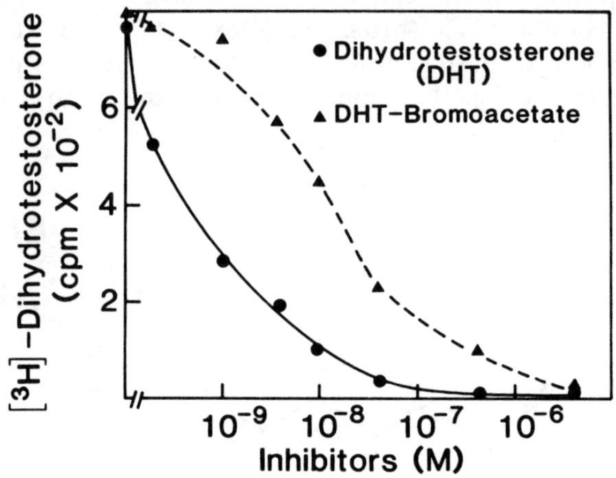

Fig. 3 — Competition of binding in rat prostate cytosol. Aliquots (250 μl) of prostate cytosol were incubated for 18 h at 0°C with 1 nM [^{3}H]dihydrotestosterone in the absence or presence of different concentrations of either dihydrotestosterone or dihydrotestosterone 17β-bromoacetate. Bound activity was determined by the charcoal binding assay. (Reprinted with permission frm *Biochemistry* **23**, 2527. Copyright 1984 American Chemical Society.)

terone competed with [^{3}H]dihydrotestosterone for the binding site in a dose-dependent manner. 50% of the binding activity was competed with 0.5 nM dihydrotestosterone, which is consistent with the dissociation constant reported by Liao *et al.*, (1972). Dihydrotestosterone 17β-bromoacetate also competed with [^{3}H]dihydrotestosterone in a dose-dependent manner. Approximately 15 nM of dihydrotestosterone 17β-bromoacetate was required to compete for 50% of the binding sites, indicating that dihydrotestosterone 17β-bromoacetate has less affinity than dihydrotestosterone for the binding site. Nonetheless, 100% of the [^{3}H]dihydrotestosterone was displaced by dihydrotestosterone 17β-bromoacetate at higher concentrations (4 μM).

2. Steroid specificity

The steroid specificity of the [^{3}H]dihydrotestosterone 17β-bromoacetate binding was investigated (Table 2). Both dihydrotestosterone 17β-bromoacetate and dihydrotestosterone competed with 100% of the radioactive dihydrotestosterone 17β-bromoacetate for the receptor binding site at the concentrations tested. Testosterone competed less well (69%). Progesterone, 17β-estradiol, and cortisol were unable to compete. These data suggest that the affinity label binds to the active site of the androgen receptor in a structure-specific manner.

3. Covalent attachment

Experiments were performed to determine whether specific covalent binding remains after trichloroacetate precipitation and treatment with organic solvents. Partially purified rat prostate cytosol was incubated with either [^{3}]dihydrotestoster-

Table 2 — Binding specificity of [^{3}H]dihydrotestosterone 17β-bromoacetate in rat prostate cytosol

Steroid	Inhibition[a] (%)
Dihydrotestosterone 17β-bromoacetate	100
Dihydrotestosterone	100
Testosterone	69
Progesterone	<1
17β-estradiol	<1
Cortisol	<1

[a]Rat prostate cytosol was incubated with or without 125 nM unlabeled competing steroids. After 4 h incubation at 0°C, 12.5 nM [^{3}H]dihydrotestosterone 17β-bromoacetate was added to all tubes. The incubation was continued for 3 h. Bound radioactivity was determined by charcoal binding assay. (Reprinted with permission from *Biochemistry* **23**, 2527. Copyright 1984 American Chemical Society.)

one or the bromoacetate-derivatized affinity label. After incubation, half of the samples were subjected to charcoal binding assay to determine the total binding activity to non-denatured receptor, whereas the remaining samples were precipitated with trichloroacetate and extracted with ether to determine the amount of radioactivity covalently bound to the denatured protein. The results presented in Table 3 show that 18% of the receptor-bound affinity label became covalently

Table 3 — Covalent binding of [^{3}H]dihydrotestosterone 17β-bromoacetate–receptor complexes in rat prostate cytosol.

	Specific binding (cpm)		%
	Charcoal assay	Organic solvent	
Dihydrotestosterone 17β-bromoacetate[a]	50483	9149	18
Dihydrotestosterone[b]	36000	257	0.7

[a]Rat prostate cytosol was saturated with unlabeled testosterone (16 nM) for 2 h at 0°C before being applied to a DNA–Sepharose column. The flow-through fractions were pooled and precipitated with 40% ammonium sulfate. The pellet was resuspended in phosphate buffer. The sample was incubated with 16 nM [^{3}H]dihydrotestosterone 17β-bromoacetate$\pm$1.6 μM dihydrotestosterone 17β-bromoacetate at 0°C for 3 h. At the end of the incubation, samples were either treated with a solution containing 1% charcoal or precipitated with 10% trichloroacetate followed by extraction with ether.

[b]The pellet from 40% ammonium sulfate was resuspended in phosphate buffer and was incubated with 16 nM [^{3}H]dihydrotestosterone$\pm$1.6 μM dihydrotestosterone at 0°C for 3 h. The samples were assayed as described in footnote *a*. (Reprinted with permission from *Biochemistry* **23**, 2527. Copyright 1984 American Chemical Society.)

attached. In contrast, only 0.7% of receptor-bound [^{3}H]dihydrotestosterone remained associated with the denatured receptor. Covalent binding was linear up to 64 nM [^{3}H]dihydrotestosterone 17β-bromoacetate (data not shown), indicating that binding of the affinity label was concentration dependent as well. This agrees with the binding study described in Fig. 3, which showed that approximately 4 μM of the affinity label is needed to saturate all the binding sites. These results clearly suggest that dihydrotestosterone 17β-bromoacetate is binding covalently to the active site of the androgen receptor.

C. Molecular weight of affinity labeled androgen receptor

Denaturing gel electrophoresis of affinity labeled receptor was used to confirm the estimate of molecular weight (110–120 kDa), as determined by non-denaturing studies using gel filtration and sucrose gradients. The receptor was partially purified using a combination of steroid affinity chromatography and differential DNA chromatography as described previously (Chang *et al.*, 1982; Rowley *et al.*, 1984) and as shown in Table 4. The samples were labeled for 3 h at 4°C using 10 nM [^{3}H]dihydrotestosterone 17β-bromoacetate, with or without 1000-fold excess of the unlabeled ligand. The partially purified receptor preparation was incubated at 22°C for 30 min to effect exchange of ligands and covalent attachment, then precipitated with 10% trichloroacetate. The pellets were dissolved in SDS sample buffer and electrophoresed through SDS–polyacrylamide gels. The protein profile of the gel, stained with silver nitrate, revealed a band corresponding to a molecular weight of 118 kDa (Fig. 4A) which coincided with a band at 118 kDa, revealed by fluorography of a parallel gel lane (Fig. 4B). This band of radioactivity was displaced by a 1000-fold excess of the non-radioactive steroid ligand (Fig. 4C). It is interesting to note that the molecular weight of the androgen receptor monomer is similar to that of some of the other steroid receptors, as determined by affinity labeling (see other chapters in this volume). Also present were bands at approximately 30 kDa, which correspond to proteolytic fragments of the androgen receptor.

IV. PHOTOAFFINITY LABELING OF THE ANDROGEN RECEPTOR

In 1982, Chang *et al.*, reported that both dihydrotestosterone 17β-bromoacetate and the synthetic androgen methyltrienolone (R1881) were used to affinity label highly purified androgen receptor. The latter compound (see Fig. 5) is a photoactivatable ligand whose affinity for the androgen receptor is similar to that of the natural ligand, dihydrotestosterone (1 nM) and is substantially greater than that of dihydrotestosterone 17β-bromoacetate (4 nM). Both of these affinity labels identified a 60 kDa protein in SDS polyacrylamide gels of androgen receptor from steer seminal vesicle. It is apparent now that this protein was a proteolytic fragment of the receptor monomer (Johnson *et al.*, 1987). Nonetheless, R1881 appeared to be a useful photoaffinity ligand for highly purified androgen receptor.

In 1984, Mainwaring and Randall reported that methyltrienolone was not very efficient or useful as an affinity label with crude cytosol preparations of androgen receptor in rat ventral prostate. These authors observed that only 5–6% of the receptor-bound ligand became covalently attached. Considering the low abundance of androgen receptor in crude prostate cytosols, this represented very little radioacti-

Table 4 — Purification of androgen receptor from Dunning H tumor

Step	Protein (mg)	Bound [^{3}H]DHT (pmol)	Spectroscopic activity (pmol/mg)	K_d[a] (nM)	Yield (%)	*x*-fold
Cytosol (from 35 g tumor)	1530	81.9	0.054	2.7	100	1
DNA I flow-through	1290	62.7	0.049	2.1	77	0.9
40% $(NH_4)_2SO_4$ precipitate	275	38.5	0.140	7.0	47	2.62
Affinity chromatography	2.8	31.0	11.1	n.d.[b]	38	207
DNA II salt eluate	0.002	9.38	4920	3.9	12	91,900

[a]Determined by Scatchard analysis.
[b]Not determined.

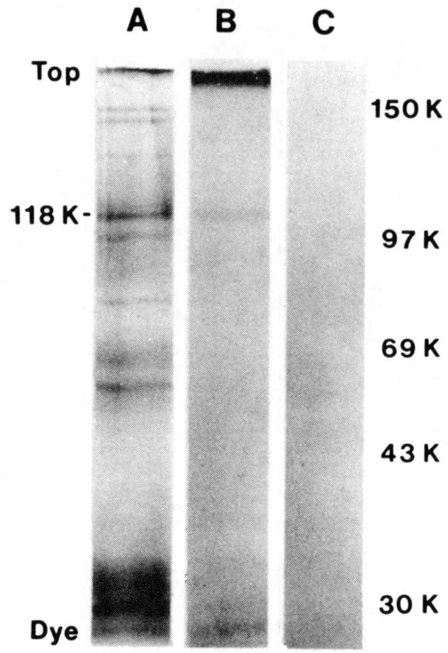

Fig. 4 — SDS gel electrophoresis of affinity labeled androgen receptor. The androgen receptor was partially purified and labeled covalently with [^{3}H]dihydrotestosterone 17β-bromoacetate. Gel electrophoresis was performed according to Laemmli (1969) procedure. The gels were stained by the silver nitrate technique (Wray *et al.*, 1970). Parallel gels were treated with EN^{3}HANCE solution and dried. The dried gel was exposed to Kodak X-omat AR film for fluorographic demonstration of radioactivity. Lane A is the protein profile stained by silver nitrate. Lanes B and C are the fluorographic profiles. Lane B represents the sample labeled with [^{3}H]dihydrotestosterone 17β-bromoacetate, and lane C represents sample labeled additionally with a 1000-fold excess of unlabeled dihydrotestosterone 17β-bromoacetate. (Reprinted with permission from *Endocrinology* **114**, 1776. Copyright 1984 The Endocrine Society.)

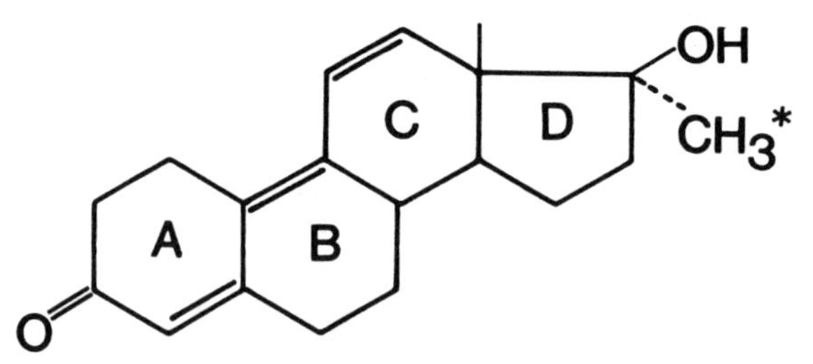

Fig. 5 — Chemical structure of methyltrienolone (R1881): *, incorporation of tritium.

vity. Moreover, this study revealed that photoillumination of [^{3}H]R1881 could lead to its association with unsaturable, non-receptor proteins. Consequently, these authors did not recommend the use of [^{3}H]R1881 to photoaffinity label androgen receptors of crude cytosols. Similarly, Brinkmann *et al.* (1985, 1986) reported that a purification of at least 40-fold appeared to be essential for the use of [^{3}H]R1881 in photoaffinity labeling studies of androgen receptors by SDS polyacrylamide gel electrophoresis.

V. SUMMARY AND CONCLUSIONS

Significant progress is being made in understanding the molecular basis of androgen action. The androgen receptor protein will continue to be the focus of intense study because of its important role in this mechanism. Methods have been developed to stabilize androgen receptor protein structure during purification and analysis. Affinity labeling of the receptor by [^{3}H]dihydrotestosterone 17β-bromoacetate has helped to resolve the discrepancy regarding the molecular weight of the monomeric androgen receptor (110–120 kDa). [^{3}H]methyltrienolone (R1881) was found to be less useful because of its low percentage of covalent attachment on photoactivation. Successful covalent labeling with either of these affinity ligands depends on the use of extensively purified samples; the greater the purity, the better. This is a somewhat unfortunate requirement, since one of the potential uses of affinity labeling is as a tool to selectively label receptors of crude samples during the purification process.

REFERENCES

Bardin, C. W. and Catterall, J. F. (1981) Testosterone: A major determinant of extragenital sexual dimorphism. *Science* **211,** 1285–1294.

Brinkmann, A. O., Kuiper, G. G. J. M., de Boer, W., Mulder, E. and van der Molen, H. J. (1985) Photoaffinity labeling of androgen receptors with 17β-hydroxy-17α-[^{3}H]methyl-4,9,11-estratrien-3-one. *Biochem. Biophys. Res. Commun.* **126,** 163–169.

Brinkmann, A. O., Kuiper, G. G. J. M., de Boer, W., Mulder, E., Bolt, J., van Steenbrugge, G. J. and van der Molen, H. J. (1986) Characterization of androgen receptors after photoaffinity labeling with [^{3}H]methyltrienolone (R1881). *J. Steroid Biochem.* **24,** 245–249.

Carroll, S. L., Rowley, D. R., Chang, C. H. and Tindall, D. J. (1984) Exchange assay for androgen receptors in the presence of molybdate. *J. Steroid Biochem.* **21,** 353–359.

Chan, L. and Tindall, D. J. (1981) Steroid hormone action. In: Collu, R., Ducharme, J. R. and Cuyda, H. (eds.), *Pediatric Endocrinology*.Raven Press, New York, pp. 63–97.

Chang, C. H., Rowley, D. R. and Tindall, D. J. (1982) Purification and characterization of the androgen receptor from steer seminal vesicle. *Biochemistry* **21,** 4102–4109.

Chang, C. H., Lobl, T. J., Rowley, D. R. and Tindall, D. J. (1984) Affinity labeling of the androgen receptor in rat prostate cytosol with 17β[(bromoacetyl)oxy]-5α-androstan-3-one. *Biochemistry* **23,** 2527–2533.

Ganguly, M. and Warren, J. C. (1971) Affinity labeling of steroid binding sites. Synthesis of cortisone 21-iodoacetate and study of 20β-hydroxysteroid dehydrogenase. *J. Biol. Chem.* **246,** 3646–3652.

Gronemeyer, H. and Govindan, M. V. (1986) Affinity labeling of steroid hormone receptors. *Mol. Cell. Endocrinol.* **46,** 1–19.

Heyns, W., Peeters, B. and Mous, J. (1977) Influence of androgens on the concentration of prostatic binding protein (PBP) and its mRNA in rat prostate. *Biochem. Biophys. Res. Commun.* **77,** 1492–1499.

Johnson, M. P., Young, C. Y. F., Rowley, D. R. and Tindall, D. J. (1987) Common molecular weight of androgen receptor monomers from various target tissues. *Biochemistry* **26**(2), in press.

Katzenellenbogen, J. A. and Katzenellenbogen, B. S. (1984) Affinity labeling of receptors for steroid and thyroid hormones. *Vitamins Hormones* **41,** 213–274.

Keenan, B. S., Meyer, W. J., Hadjain, A. J., Jones, H. W. and Migeon, C. J. (1974) Syndrome of androgen insensitivity in man: Absence of 5α-dihydrotestosterone binding protein in skin fibroplasts. *J. Clin. Endocrinol. Metab.* **38,** 1143–1146.

Klyzejko-Stefanowicz, L., Chui, J. F., Tsai, Y. H. and Hnilica, L. S. (1976) Acceptor proteins in rat androgenic tissue chromatin. *Proc. Natl. Acad. Sci. USA* **73** 1954–1963.

Korenman, S. G. (1969) Comparative binding affinity of estrogens and its relation to estrogenic potency. *Steroids* **13,** 163–177.

Laemmli, U. K. (1970) Cleavage of structural proteins during the assembly of the head of bacteriophage T4. *Nature* **227,** 680–685.

LeGaillard, F. and Dantrevaux, M. (1977). Affinity labeling of human transcortin. *Biochim. Biophys. Acta* **495,** 312–323.

Liao, S. (1975) Cellular receptors and mechanisms of action of steroid hormones. *Int. Rev. Cytol.* **41,** 87–172.

Liao, S. (1977) Molecular actions of androgens. In: Litwack, G. (ed.) *Biochemical Actions of Hormones*. Academic Press, New York, pp. 351–406.

Liao, S., Liang, T. and Tymockzo, J. L. (1972) Structural recognitions in the interactions of androgens and recptor proteins in their association with nuclear acceptor components. *J. Steroid Biochem.* **3,** 401–419.

Mainwaring, W. I. P. (1978) Androgen receptors and biologic actions: A survey. In: O'Malley, B. W. and Birnbaumer, L. (eds.) *Receptors and Hormone Action II*. Academic Press, New York, pp. 105–120.

Mainwaring, W. I. P. and Johnson, A. D. (1980) Use of the affinity label 17β-bromoacetoxytestosterone in the purification of androgen receptor proteins. In: Bresciani, F. (ed.) *Perspectives in Steroid Receptor Research*. Raven Press, New York, pp. 89–97.

Mainwaring, W. I. P. and Randall, V. A. (1984) Limitations in the use of tritiated methyltrienolone for the photoaffinity labeling of androgen receptor proteins. *J. Steroid Biochem.* **21,** 209–216.

Mainwaring, W. I. P., Mangan, F. R., Irving, R. A. and Jones, D. A. (1974) Specific changes in the mRNA content of the rat ventral prostate gland after androgenic stimulation. Evidence from the synthesis of aldolase mRNA. *Biochem J.* **144,** 413–426.

Mainwaring, W. I. P., Symes, E. K. and Higgins, S. J. (1976) Nuclear components responsible for the retention of steroid–receptor complexes especially from the standpoint of the specificity of hormone responses. *Biochem. J.* **156,** 129–141.

Meyer, W. J., Midgeon, B. R. and Migeon, C. J. (1975) Locus on human X chromosome for DHT receptor and androgen insensitivity. *Proc. Natl. Acad. Sci USA* **72,** 1469–1472.

Murthy, L. R., Johnson, M. P., Rowley, D. R., Young, C. Y. F., Scardino, P. T. and Tindall, D. J. (1985). Characterization of steroid receptors in human prostate using mibolerone. *Prostate* **8,** 241–253.

Ostrowski, M. C., Kistler, M. K. and Kistler, W. S. (1982) Effect of castration on the synthesis of seminal vesicle secretory protein IV in the rat. *Biochemistry* **21,** 3525–3529.

Peck, E. J. and Clark, J. H. (1977) Effect of ionic strength on charcoal adsorption assays of receptor–estradiol complexes. *Endocrinology* **101,** 1034–1043.

Rowley, D. R. and Tindall, D. J. (1986) In: Litwack, G. (ed.) *Biochemical Actions of Hormones,* Vol. 13, Academic Press, New York, pp. 305–324.

Rowley, D. R., Chang, C. H. and Tindall, D. J. (1984) Effects of sodium molybdate on the androgen receptor from the R3327 prostatic tumor. *Endocrinology* **114,** 1776–1783.

Rowley, D. R., Premont, R. T., Johnson, M. P., Young, C. Y. F. and Tindall, D. J. (1986) Properties of an intermediate-sized androgen receptor: Association with RNA. *Biochemistry* **25,** 6988–6995.

Roy, A. K., Chatterjee, B., Demyan, W. F., Milin, B. S., Motwani, N. M., Nath, T. S. and Schiop, M. J. (1983) Hormone and age-dependent regulation of $\alpha_{2\mu}$-globulin gene expression. *Rec. Prog. Horm. Res.* **39,** 425–459.

Sweet, F., Arias, F. and Warren, J. C. (1972) Affinity labeling of steroid binding sites. Synthesis of 16α-bromoacetoxyprogesterone and its use for affinity labeling of 20β-hydroxysteroid dehydrogenase. *J. Biol. Chem.* **247,** 3424–3433.

Thomas, J. L. and Strickler, R. C. (1983) Human placental 17β-estradiol dehydrogenase and 20α-hydroxysteroid dehydrogenase. Studies with 6β-bromoacetoxyprogesterone. *J. Biol. Chem.* **258,** 1587–1590.

Tindall, D. J., Chang, C. H., Lobl, T. J. and Cunningham, G. R. (1984) Androgen antagonists in androgen target tissues. *Pharmacol. Ther.* **24,** 367–400.

Wilson, J. D., George, F. W. and Griffin, J. E. (1981) The hormonal control of sexual development. *Science* **211,** 1278–1284.

Wray, W., Boulikas, T., Wray, V. P. and Hancock, R. (1981) Silver staining of proteins in polyacrylamide gels. *Anal. Biochem.* **118,** 197–203.

Zava, D. T., Landrum, B., Horwitz, K. B. and McGuire, W. L. (1979) Androgen receptor assay with [^{3}H]methyltrienolone (R1881) in the presence of progesterone receptors. *Endocrinology* **104,** 1007–1012.

13

Photoaffinity labeling of thyroid hormone receptors

Z. D. Horowitz and H. H. Samuels
Department of Medicine, New York University Medical Center, New York, USA

I. INTRODUCTION

Thyroid hormones have an important role in the growth, development, and metabolism of animal tissues (Oppenheimer and Samuels, 1983; Wolff and Wolff, 1964) and appear to play an important role in the neoplastic transformation of C3H/10T$^1/_2$ cells induced by X-irradiation (Guernsey *et al.*, 1981) and chemical carcinogens (Borek *et al.*, 1983). Thyroid hormone also plays a role in the modulation of adenovirus transformation of Fischer rat embryo fibroblast (CREF) cells (Fisher *et al.*, 1983) and has been related to the induction of hepatocellular carcinoma by hepatitis B virus, because the viral integration site is adjacent to a liver cell DNA sequence highly homologous with c-erb-A (thyroid hormone receptor) (Dejean *et al.*, 1986). Studies in intact animals and in cultured cells show that the regulation of specific genes by 3,5,3′5′-tetraiodo-L-thyronine (L-thyroxine, L-T4) and 3,5,3′-triiodo-L-thyronine (L-triiodothyronine, L-T3) occurs through the stimulation of mRNA accumulation which codes for specific proteins. The thyroid hormone regulated effects which have been studied in detail include the following: the stimulation of growth hormone synthesis in rat anterior pituitary *in vivo* (Hervas *et al.*, 1975); stimulation of growth hormone gene expression in cultured rat pituitary cells (Dobner *et al.*, 1981; Evans *et al.*, 1982; Martial *et al.*, 1977; Samuels and Shapiro, 1976; Samuels *et al.*, 1979a,b; Seo *et al.*, 1977; Shapiro *et al.*, 1978; Spindler *et al.*, 1982; Yaffe and Samuels, 1984; Flug *et al.*, 1987); stimulation of malic enzyme mRNA (Magnuson and Nikodem, 1983; Siddiqui *et al.*, 1981; Towle *et al.*, 1981), as well as several other genes which encode for hepatic proteins with unknown functions; Spot 14 and Spot 11 (Carr *et al.*, 1984; Liaw and Towle, 1984; Tao and Towle, 1986).

Several clonal strains of growth hormone producing rate pituitary cells (GH_1, GH_3, and GC) have been shown to be excellent model systems for thyroid hormone action (Evans *et al.*, 1982; Martial *et al.*, 1977; Samuels and Shapiro, 1976; Samuels *et al.*, 1979a,b; Seo *et al.*, 1977; Shapiro *et al.*, 1978; Spindler *et al.*, 1982; Yaffe and Samuels, 1984). In these cells physiologic concentrations of L-T3 and L-T4 stimulate growth hormone mRNA accumulation (Martial *et al.*, 1977; Samuels *et al.*, 1979a; Seo *et al.*, 1977; Shapiro *et al.*, 1978) with kinetics of stimulation similar to that described in anterior pituitary after hormone administration (Hervas *et al.*, 1975; Martial *et al.*, 1977; Samuels and Shapiro, 1976; Samuels 1979a). These, and essentially all other cellular responses to thyroid hormone in intact animals (Oppenheimer and Samuels, 1983) and in cultured cells (Samuels *et al.*, 1983a,b), appear to be mediated by a nuclear associated receptor.

Thyroid hormone receptors associate with nuclear components in the absence of ligand (Samuels *et al.*, 1974; Oppenheimer *et al.*, 1974) and, unlike the steroid hormone receptors, have not been identified in the extra-nuclear fraction of cells. GH_1 cells have approximately 15 000 receptor molecules per cell nucleus, with equal binding capacity for L-T3 and L-T4 (Samuels *et al.*, 1979a, 1983a,b), but with 10-fold lower affinity for L-T4. Photoaffinity labeling studies of the thyroid hormone receptor in intact cells using *N*-2-diazo-3,3,3-trifluoropropionyl-L-triiodothyronine (L-T3-PAL) (Fig. 1) have documented that the receptor is present in several molecular weight forms (Casanova *et al.*, 1984; Pascual *et al.*, 1982). In intact GH_1 cells, L-[^{125}I]T3-PAL labels an abundant 47 000 molecular weight (M_r) receptor species (75% of total receptor) and a less abundant M_r~57 000 doublet form (approximately 25% of receptor) (Fig. 2), which show identical affinities for a variety of thyroid hormone analogues. This observation and the fact that both photoaffinity labeled species generate identical size ^{125}I-peptide fragments with *Straphylococcus*

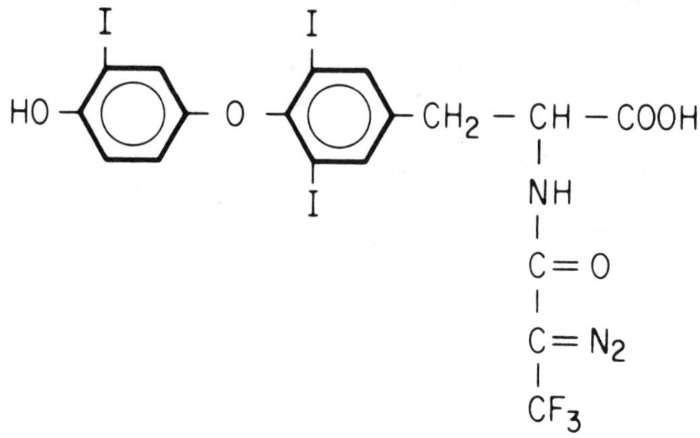

Fig. 1 — Structure of L-T3-PAL (*N*-2-diazo-3,3,3-trifluoropropionyl-L-T3). (From Pascual *et al.* (1982), with permission.)

aureus V8 protease digestion suggest that the ligand-binding domains of the two receptor forms are very similar, if not identical (Casanova *et al.*, 1984).

The precise relationship between the two receptor forms is unknown. Whether or not they represent transcripts from different receptor genes, alternate transcripts of the same gene, alternate mRNA processing, alternate translation initiation sites of the same mRNA, alternate post synthetic modification, or a combination of interacting processes, remains very much open to question. Each of these possibilities raises important questions concerning points at which the expression of the thyroid hormone receptor gene may be influenced. How the receptor interacts with target genes or with other transcriptional regulators, whether or not the different receptor forms affect the same or different thyroid hormone responsive genes or different thyroid hormone responsive elements within a single target gene, are all questions which require investigation.

In this chapter we will discuss these and other studies which utilize the photoaffinity label derivatives (PAL) of L-T3, L-T4 and L-rT3 (3,3′,5′-triiodo-L-thyronine, L-reverse-T3). We also review more general information relating to the role of thyroid hormone receptor as the mediator of the molecular action of thyroid hormone and show how photoaffinity labeling can be coupled with other techniques to shed additional light on the structure and function of this gene regulatory protein.

II. GENE REGULATION BY THYROID HORMONE

The stimulation of growth hormone gene transcription in rat pituitary tumor cell lines, GC and GH_1 cells, is mediated by thyroid hormone receptor (Samuels *et al.*, 1977; Yaffe and Samuels, 1984; Samuels *et al.*, 1983a,b). Previously, dexamethasone had been shown to increase independently the production of growth hormone in GC cells only slightly, but to act synergistically with thyroid hormone to augment greatly the growth hormone response (Shapiro *et al.*, 1978; Samuels *et al.*, 1979b; Samuels *et*

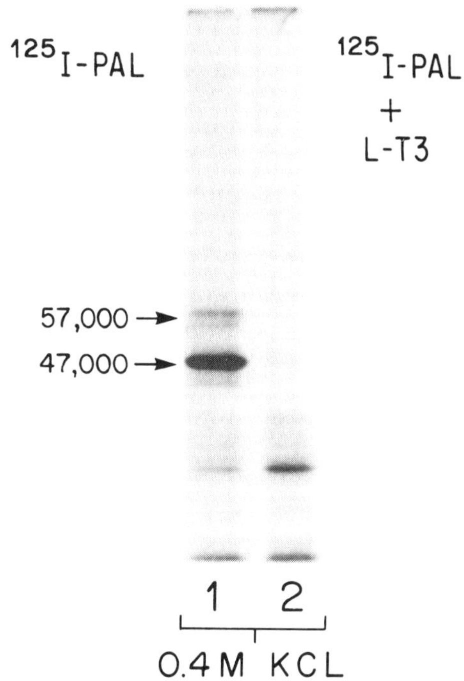

Fig. 2 — Receptor forms identified by photoaffinity labeling in intact cells using L-[^{125}I]T3-PAL. GH_1 cells were incubated for 1.5 h with 2 nM L-[^{125}I]T3-PAL (^{125}I-PAL) without (lane 1) or with (lane 2) 5 μM L-T3. Cell monolayers were irradiated at 254 nm for 2 min at 4°C. Nuclear proteins were extracted with 0.4 M KCl buffer, precipitated with 80% ethanol, and electrophoresed in 6–15% polyacrylamide–SDS gels and autoradiographed for 36 h at −80°C. 300 μg of protein was applied to each gel lane. The following standard proteins were electrophoresed in a parallel lane (not shown here): conalbumin, M_r=77 000; bovine serum albumin, M_r=68 000; ovalbumin, M_r=45 000; deoxyribonuclease I, M_r=31 000; chymotrypsinogen, M_r=25 700.

al., 1977). More recently (Yaffe and Samuels, 1984), the abundance of nuclear L-T3–receptor complexes has been correlated with the rate of growth hormone mRNA synthesis and gene transcription stimulated by L-T3 and/or dexamethasone. The *in vitro* transcription rate was determined by [^{32}P]UTP labeling of nuclei and the mRNA synthetic rate was determined by [^{3}H]uridine labeling of intact cells. Radiolabeled growth hormone mRNA sequences were quantitated by hybridzation to immobilized plasmid containing growth hormone cDNA. Thyroid hormone caused a reduction in nuclear receptor levels and a parallel decrease in growth hormone mRNA synthesis. Incubation with dexamethasone alone resulted in a minimal increase or decrease in growth hormone mRNA synthesis, whereas L-T3 plus dexamethasone yielded a synergistic increase which, like L-T3 alone, decreased with time in parallel with a reduction in L-T3–receptor levels. These results indicate that thyroid hormone is a primary regulator of the growth hormone gene. No lag period occurred between the kinetics of L-T3 binding to receptor and stimulation of growth hormone gene transcription. At 30 min, 74% of the maximum level of L-T3–receptor complexes were formed and 71% of the maximum growth hormone gene transcription occurred. [^{3}H]uridine pulse-chase studies showed that the growth hormone mRNA had a half-life of 50 h, which is in agreement with the half-life of growth hormone synthesis in de-induced cells, suggesting that the enhancement of the growth hormone response is directly caused by thyroid hormone (Yaffe and Samuels, 1984).

Thyroid hormone regulation of the growth hormone gene has been found to be entirely dependent on sequences within the 5′-flanking region (Casanova *et al.*, 1985). Using a chimeric plasmid (pGH-cat) in which a reporter gene, chloramphenicol acetyl transferase (CAT) (Gorman *et al.*, 1982), is linked to various fragments of the 5′-flanking sequence of the rat growth hormone gene, and transiently expressed in GC cells, it has been shown that L-T3 stimulates CAT expression 10- to 15-fold (Flug *et al.*, 1987). In this, and in other studies (Crew and Spindler, 1986), the growth hormone DNA sequence between −236 and the cap site has been shown to be necessary for thyroid hormone regulated expression. Transfection studies in GC cells (Flug *et al.*, 1987; Casanova *et al.*, 1985) have documented that DNA sequences between −210 and −181 bp are essential for regulated expression by thyroid hormone and that these sequences function most efficiently with the homologous rat growth hormone promotor (−104 to +7). Interestingly, a sequence between −209 and −198 in the rat growth hormone gene shows a high degree of homology with a sequence in the 5′-flanking region of the α-myosin heavy chain (Mahdavi *et al.*, 1984) and hepatic Spot 14 genes (Liaw and Towle, 1984), although their involvement in thyroid hormone regulated expression of these two genes has not yet been reported.

Basal expression of the rat growth hormone gene requires DNA sequences between −104 and the cap site (Flug *et al.*, 1987). Elements between −145 and −104 were found to enhance basal expression about 3- to 5-fold. Interestingly, the −145 to −105 region is very highly conserved among rat, bovine, and human growth hormone genes (Barta *et al.*, 1981; Page *et al.*, 1981; Seeburg, 1982; Gordon *et al.*, 1983), even though this region is not necessary for thyroid hormone regulated expression (Flug *et al.*, 1987). However, DNA between −104 to +7 appears to contain a 3′ element which, along with sequences between −210 to −181, is important for regulated expression by thyroid hormone since sequences between −210 and −181 elicit L-T3 stimulation when ligated to the homologues but not to several viral promotors. This implies that some type of protein–protein interaction between proteins which bind between −210 to −181 and −104 to +7 occurs in thyroid hormone regulated expression.

Recently the DNA sequences of the growth hormone gene which specifically bind nuclear proteins have been identified (Ye and Samuels, 1987). These proteins may be cell type specific or not, e.g. thyroid hormone receptor. To study this question, nuclear extracts derived from cells which endogenously express growth hormone (GC, GH, GH_4Cl), as well as from cells which do not express the growth hormone gene (H4TG rat hepatoma cells and Rat 2 fibroblasts), were screened for specific interaction with DNA fragments of the 5′-flanking region of the rat growth hormone gene, as defined in transient expression studies (Flug *et al.*, 1987). Non-denaturing polyacrylamide gel electrophoresis of ^{32}P-DNA–protein complexes (Garner and Revzin, 1981; Fried and Crothers, 1981, 1984) was used to detect any specific DNA–nuclear protein interactions. (When protein is bound to DNA, the complex has decreased electrophoretic mobility relative to the free ^{32}P-DNA fragments.)

Nuclear proteins only from cells which endogenously express the growth hormone gene were found to bind the −104 to +7 DNA fragment (Ye and Samuels, 1987) which has been shown to be required for basal expression (Flug *et al.*, 1987). DNase I footprinting (Dynan and Tjian, 1983), performed under the same binding conditions as for DNA–protein binding in the gel mobility shift experiments, further resolved the sequence specificity to between −95 and −68 bp upstream of the cap site (Ye and Samuels, 1987). No footprint was formed by nuclear proteins of rat cells or tissues which do not express the growth hormone gene, e.g. H4TG, Rat 2, liver, testis (Ye and Samuels, 1987). A less abundant protein (or proteins), which was not cell type specific, bound to the thyroid hormone responsive region within −236 to −146. However, it has not been documented that this protein is thyroid hormone receptor.

III. THYROID HORMONE RECEPTOR IS A PROTO-ONCOGENE PRODUCT

The avian erythroblastosis virus (AEV), a defective leukemia retrovirus, induces sarcomas and erythroblastosis *in vivo* and induces transformation of fibroblasts and erthyroblasts to neoplastic phenotypes *in vitro* (Graf and Beug, 1983). When the molecular cloning of the AEV genome was achieved (Vennstrom *et al.*, 1980) it became possible to isolate the cellular homologue of the retroviral gene, c-erb-A (Vennstrom and Bishop, 1982; Gonda *et al.*, 1982). Two domains of the AEV genome were described: the v-erb-A region and the v-erb-B region, which was eventually shown to encode a truncated epidermal growth factor (EGF) receptor (Hayman *et al.*, 1983; Jansson *et al.*, 1983; Debuire *et al.*, 1984; Nilsen *et al.*, 1985; Goodwin *et al.*, 1986). Although it had been shown, by deletion mutations, that v-erb-B is responsible for the transforming potential of AEV, this property requires the presence of v-erb-A for full expression. v-erb-A by itself is incapable of inducing transformation (Graf and Beug, 1983; Frykberg *et al.*, 1983; Raines *et al.*, 1985; Damm *et al.*, 1987). The chromosomal location of the human c-erb-A proto-oncogene was identified on chromosome 17 near the breakpoint found typically in human pro-myelocytic leukemia (Spurr *et al.*, 1984). c-erb-B was localized to chromosome 7 (Raines *et al.*, 1985). Although these data provide a basis for the association of the v-erb-A DNA and a site of chromosomal translocation in neoplastic cells, they do not explain how v-erb-A is mechanistically related to the transforming event.

Recently it has been shown that the translation products of the chicken (Sap *et al.*, 1986) and human (Weinberger *et al.*, 1986) c-erb-A homologue of AEV are thyroid hormone binding proteins with properties very characteristic of thyroid hormone receptor. Whereas the chick c-erb-A product (M_r~46 000) is a high affinity thyroid hormone binding protein, the v-erb-A product (M_r~75 000), containing viral gag sequences at the 5′ end, appears not to bind thyroid hormone, but is a DNA binding protein (Sap *et al.*, 1986). The molecular weights of the products of the human c-erb-A cDNA, synthesized by *in vitro* translation, were 55 000, 52 000, and 35 000. The two larger proteins were thought to represent products of alternate translation initiation sites, whereas the M_r~35 000 product was ascribed to proteolysis of the larger forms. The product of the chicken c-erb-A cDNA was of M_r~46 000.

Our own data derived from L-/[^{125}I]T4-PAL labeled receptor from normal human fibroblasts (Fig. 3) (Horowitz *et al.*, 1987) demonstrate a receptor molecular weight

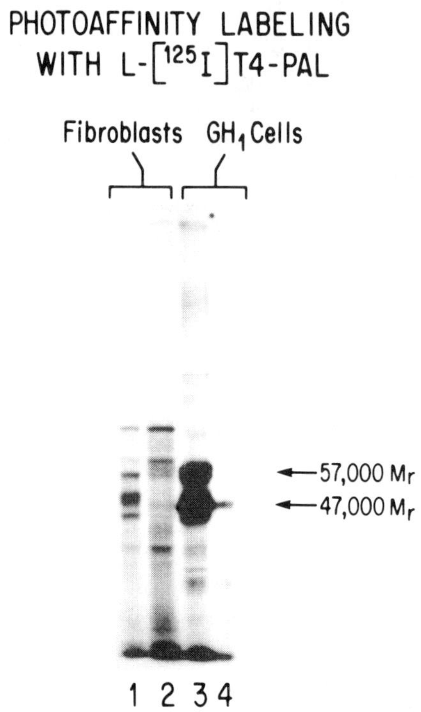

Fig. 3 — Intact GH_1 and B144 (normal human fibroblast) cells were photoaffinity labeled with L-[^{125}I]T4-PAL, harvested, the nuclei isolated, and extracted with micrococcal nuclease and 0.4 M KCl, as described in Chapter 6. Electrophoresis of the nuclear extracts and autoradiography (48 h exposure) were as described in the text. Molecular weight standards were conalbumin (76 000 kd) BSA (66 000 kd) and ovalbumin (45 000 kd). Lanes 1 and 2 contain approximately 100 μg of protein derived from B144 cells incubated in the presence of 2 nM L-[^{125}I]T4-PAL, without (lane 1), or with (lane 2), 1 μM L-T3. Lanes 3 and 4 contain approximately 250 μg of protein derived from GH_1 cells incubated in the presence of 2 nM L-[^{125}I]T4-PAL without (lane 3), or with (lane 4), 1 μm L-T3. Thyroid hormone receptor identified by L-[^{125}I]T4-PAL in GH_1 cells is indicated by the arrows labeled '47 000 M_r' and '57 000 M_r. (From Horowitz *et al.* (1987).)

of about 57 000 and 52 000 (human) and, in GH_1 cells (rat), 57 000 and 47 000, also indicating species differences in receptor molecular weight. Although the authors (Weinberger *et al.*, 1986) ascribe all thyroid hormone binding to the two higher molecular weight species, the evidence supporting this conclusion is not presented. We found (Fig. 4) that L-[^{125}I]rT3-PAL identifies an M_r~35 000 species in nuclear extracts from GC and GH_1 cells, but this labeling is not inhibited by non-radioactive L-T3 or L-rT3, suggesting that the M_r~35 000 species is not a functional receptor protein (Horowitz *et al.*, 1987).

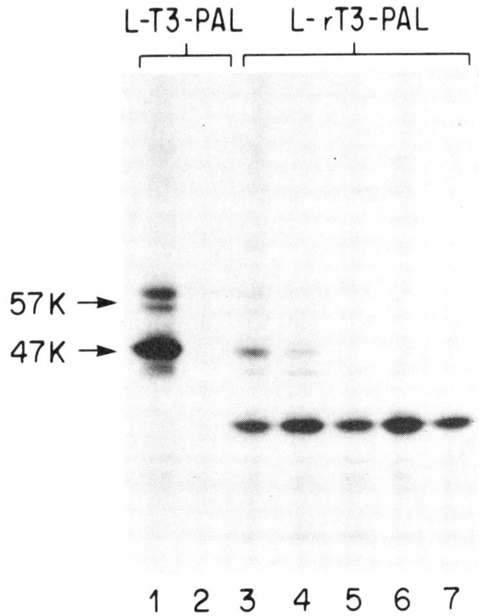

Fig. 4 — Receptor forms were identified in intact cells by photoaffinity labeling using L-[^{125}I]T3-PAL and L-[^{125}I]rT3-PAL. GH_1 cells were incubated for 1.5 h with 2 nM L-[^{125}I]T3-PAL without (lane 1), or with (lane 2) 2 μM L-T3, or with 10 nM L-[^{125}I]rT3-PAL in the absence (lane 3) or in the presence of 100 nm L-rT3 (lane 4), 2 μM L-rT3 (lane 5), or in the presence of 100 nM L-T3 (lane 6) or 1 μM L-T3 (lane 7). Following incubation, cell monolayers were irradiated at 254 nm for 2 min at 4°C. Nuclear preparations were digested with micrococcal nuclease, extracted with 0.4 M KCl buffer, precipitated with 80% ethanol, electrophoresed, and autoradiographed. 300 μg of protein, derived from two 25 cm^2 culture flasks, as applied per lane. Exposure time was seven days. The following standard proteins were electrophoresed in a parallel lane (not shown): conalbumin, M_r=77 000; bovine serum albumin, M_r=68 000; ovalbumin, M_r=45 000. (From Horowitz *et al.* (1987).)

IV. THYROID HORMONE RECEPTOR AS PART OF A GENE FAMILY

The recent cloning of the cDNAs encoding the human glucocorticoid receptor (Hollenberg *et al.*, 1985), the chicken (Krust *et al.*, 1986) and human (Green *et al.*, 1986) estrogen receptors, the human progesterone receptor (Connelly *et al.*, 1986; Jeltsch *et al.*, 1986), the human aldosterone receptor (Arriza *et al.*, 1987), and the avian receptor for vitamin D (McDonnell *et al.*, 1987) has led to the discovery that these steroid receptors, and possibly others, are members of a family of related genes including also the c-erb-A gene, all of which are potentially oncogenic (Green and Chambon, 1986). Each of these genes demonstrates a high degree of homology, 50–60%, within the putative DNA-binding domain, a 66 amino acid domain, termed the 'C' region. No significant homology exists between v-erb-A and the steroid hormone receptors within the 'B' region, the hydrophilic hinge, or 'D' region, or the ligand-binding domain, region 'E'.

Conservation of amino acid sequence homology among the chicken and human c-erb-A cDNAs in the putative DNA-binding (91%) and the ligand-binding (88%)

domains of the v-erb-A and c-erb-A is very high, while other regions of the proteins are more variable. Using a chimeric receptor, it has been shown that replacement of the human estrogen receptor C region with that of the human glucocorticoid receptor causes estrogen to activate glucocorticoid responsive elements of target genes (Green and Chambon, 1987), thus showing that it is the C region which determines the DNA target specificity of the steroid hormone receptors. Furthermore, these hormone receptors, and other proteins, contain within the putative DNA-binding domain a highly conserved cysteine–histidine-rich sequence with metal-binding properties (Berg, 1986a,b; Connelly *et al.*, 1986; Krust *et al.*, 1986; Jeltsch *et al.*, 1986; Sabbah *et al.*, 1987) crucial for DNA binding.

Unlike steroid hormone receptors, thyroid hormone receptors associate with nuclear components in the absence of ligand (Samuels *et al.*, 1974; Oppenheimer and Surks, 1975; Spindler *et al.*, 1982). GH_1 cells have approximately 15 000 receptor molecules per cell nucleus, with an equal binding capacity for L-T3 and L-T4 (Samuels *et al.*, 1979c, 1983a,b), but with a 10-fold lower affinity for L-T4. The receptor can be efficiently extracted from nuclei with 0.4M KCl (Samuels *et al.*, 1974; Perlman *et al.*, 1982) and has a sedimentation coefficient of 3.8 S.

V. REGULATION OF THYROID HORMONE RECEPTOR ABUNDANCE BY THYROID HORMONE

The thyroid hormone receptor half-life and synthetic rate in cultured GH_1 cells have been quantitated by using a dense amino acid labeling technique (Raaka and Samuels, 1981). These studies demonstrated that the receptor has a half-life of about 4.5 h and a production rate of 1800 molecules per hour per cell. These studies also documented that the nuclear receptor concentration is determined only by the synthetic rate and half-life. No significant cytoplasmic receptor pool has been identified in cultured GH_1 cells, or other cells in culture, or in intact animals (Samuels *et al.*, 1974; Oppenheimer and Surks, 1975) even though many cytoplasmic proteins non-specifically bind thyroid hormone. In GC and GH_1 cells, L-T3 decreases the level of its own receptor in a time- and dose-related fashion. Dense amino acid labeling has shown that the decrease in thyroid hormone receptor level is found within 24 h and occurs, specifically, because of a decrease in receptor synthetic rate (Raaka and Samuels, 1981).

A. Estimation of receptor half-life by dense amino acid labeling

Nuclear associated receptor is in a dynamic steady state. In the absence of thyroid hormone, receptor has a half-life of approximately 4–5 h and a synthetic rate of 1500–2000 molecules per cell per hour (Raaka and Samuels, 1981). In GH_1 and GC cells, L-T3 or other iodothyronines elicit a reduction in nuclear receptor levels which is directly related to the extent of occupancy of receptor by hormone (Raaka and Samuels, 1981; Samuels *et al.*, 1979c). In the absence of hormone, receptor levels remain relatively constant at about 240 fmol/100 μg cell DNA (15 000 molecules/cell nucleus). However, cells which receive 5 nM L-[^{125}I]T3 from the beginning of the study (a concentration of L-T3 at which greater than 95% of the receptor population is occupied) show a rapid decrease in nuclear receptor levels which attains a new steady state within 24 h of incubation.

To investigate the mechanism of L-T3 regulation of receptor levels, our laboratory developed a dense amino acid labeling technique which quantitates receptor half-life and synthetic rates and measures the influence of hormone on these processes (Raaka and Samuels, 1981). To estimate receptor half-life and the rate of the receptor synthesis, cells are incubated with amino acids uniformly labeled with the non-radioactive dense isotopes ^{15}N, ^{13}C, and ^{2}H. Newly synthesized proteins will be of higher density than pre-existing proteins and can be separated using velocity gradient centrifugation. The density of most proteins is about 1.36 g/cm^3. Therefore, the total substitution of amino acids containing all three heavy isotopes results in a maximal density increase of approximately 8% (0.109 g/cm^3), yielding protein with a density of about 1.47 g/cm^3.

The velocity of sedimentation of a particle is directly related to the difference between the particle density (d_p) and the density of the gradient (d_g). If a gradient is constructed such that the average density is 1.2 g/ml, the difference in density between the dense protein and the gradient ($d_p - d_g$) is 0.27 (1.47−1.2), while the value for the protein of normal density is 0.16 (1.36−1.2). Under the correct gradient conditions, therefore, an 8% increase in the density of protein can result in a 1.5-fold increase (0.27/0.16) in the velocity of sedimentation. To achieve an average gradient density of 1.2 g/ml, sucrose gradients are constructed with D_2O instead of H_2O. The newly synthesized protein of high density can be separated from pre-existing protein of normal density by velocity sedimentation and the corresponding populations of dense and normal receptor quantitated using L-[^{125}I]T3 (Raaka and Samuels, 1981).

Fig. 5 illustrates the time course of dense amino acid labeling of thyroid hormone receptor in GH_1 cells. Cells were first incubated without hormone in medium containing dense amino acids for 1.5, 5, or 20 h (Raaka and Samuels, 1981). 1 h before harvesting the cells, 5 nM L-[^{125}I]T3 was added to the medium to quantitate receptor levels. Labeled dense and normal receptors were then extracted from isolated nuclei with 0.4 M KCl and separated by sedimentation in 17–32% D_2O–sucrose gradients (Fig. 5). After 1.5 h in dense amino acid medium approximately 90% of the receptors are of normal density and a more rapidly sedimenting species corresponding to newly synthesized dense receptor is identified (fraction 12). After 5 h in dense medium approximately equal amounts of normal and dense receptor are present. After 20 h, approximately 95% of the receptor is of the dense form. By measuring the kinetics of disappearance of receptor of normal density, the receptor half-life can be determined. In cells cultured without thyroid hormone exposure, except for the 1 h incubation with L-[^{125}I]T3 to occupy receptor, the half-life ($t_{1/2}$) of the receptor was estimated to be 4.7±0.5 h (Raaka and Samuels, 1981).

B. Estimation of receptor synthetic rates

Under steady state conditions the synthetic rate of receptor (k_s) can be calculated from the degradation rate constant (k_d) ($k_d = 0.693/t_{1/2}$) and the steady state amount of receptor (R) using the formula $k_s = k_d R$ (Raaka and Samuels, 1981; Schimke, 1975). This value is about 30 fmol of receptor/(h 100 μg DNA), which is equivalent to the synthesis of 1700 molecules of receptor/(h cell). The synthetic rate can also be assessed by determining the kinetics of appearance of newly synthesized dense receptor using the formula $k_s = k_d R_d/(1 - e^{-k_d t})$, where R_d is the amount of newly synthesized dense receptor present in the nucleus at any time (t) after the addition of

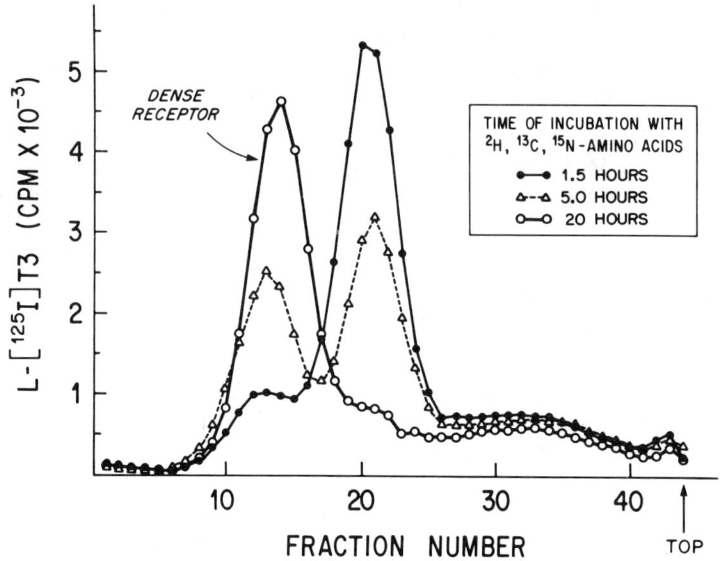

Fig. 5 — Separation of normal and dense thyroid hormone receptors in sucrose–D_2O gradients. GH_1 cells were cultured in medium without thyroid hormones for a total of 43 h. Normal medium was replaced with dense medium 1.5, 5, or 20 h prior to harvesting the cells. The dense medium was supplemented with 5 nM L-[^{125}I]T3 1 h before harvesting the cells to identify the receptor. Nuclei were isolated and hormone–receptor complexes were extracted with 0.4 M KCl. The extracts were sedimented in sucrose–D_2O gradients. The direction of sedimentation is from right to left. (From Raaka and Samuels (1981), with permission.)

dense amino acids to cells (Raaka and Samuels, 1981; Schimke, 1975). As shown in Fig. 6 a significant reduction in nuclear associated dense receptor occurs during the first 6 h of incubation with L-T3. Dense amino acid labeling studies showed no alteration in receptor half-life during this period (Raaka and Samuels, 1981). As shown in Fig. 6 the decrease in the amount of receptor occurring at early times of incubation of cells with L-T3 is primarily due to a decrease in receptor synthesis.

The effect of L-T3 on reducing receptor levels during the first 15 h of incubation appears to be solely a result of a rapid decrease in the synthetic rate of receptor. At incubation times greater than 24 h, a 30% reduction in receptor half-life also occurs (Raaka and Samuels, 1981). Although the mechanism of L-T3 regulation of receptor levels has not yet been defined, all studies indicate that the reduction of receptor is mediated by an interaction of hormone with nuclear associated receptor (Raaka and Samuels, 1981; Samuels and Shapiro, 1976; Samuels 1979a). This observation suggests that the thyroid hormone-receptor complex lowers receptor levels by decreasing the expression of the receptor gene, resulting in a decrease in steady state amounts of thyroid hormone receptor mRNA.

Thyroid hormone-mediated reduction of its own receptor is rapidly reversed after removal of hormone by cells (Raaka and Samuels, 1981). Over a 20 h period receptor re-populates the nucleus to a level found in cells cultured chronically without hormone. The recovery of receptor is due entirely to an increase in the level

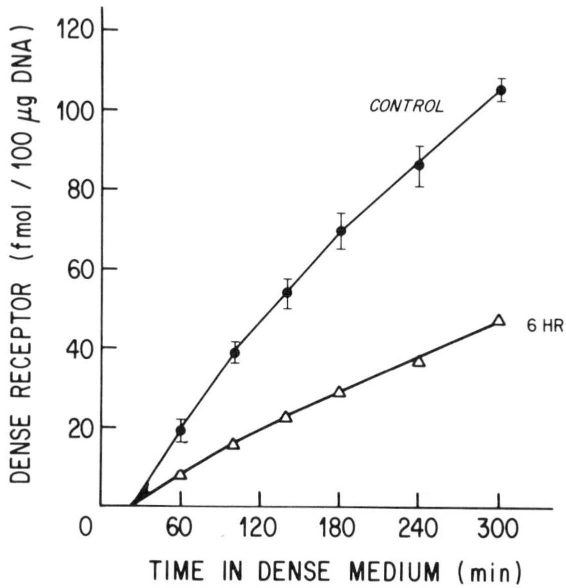

Fig. 6 — Accumulation of newly synthesized dense receptor in cells first incubated with 5 nM L-[^{125}I]T3 for 6 h (△). Control cells (●) received 5 nM L-[^{125}I]T3 for 1 h prior to harvesting the cells. The rates of appearance of newly synthesized dense receptor (k_s) can be calculated from the accumulation results using the formula $k_s = k_d R_d/(1 - e^{-k_d t})$, where R_d is the amount of newly synthesized dense receptor present in the nucleus at any time (t) after addition of dense amino acids to cells. The degradation rate constant ($k_d = 0.693/t_{1/2}$) was calculated using a half-life of 4.7 h. (From Raaka and Samuels (1981), with permission.)

of receptor synthesis, matching the rate which occurs in cells cultured in the absence of hormone. Therefore, during the onset of the recovery from thyroid hormone-mediated receptor level reduction, changes in nuclear receptor levels are primarily determined by rapid alterations in receptor synthesis, which is controlled by thyroid hormone.

VI. PHOTOAFFINITY LABELING OF THYROID HORMONE RECEPTORS

A. Photoaffinity labeling studies in intact cells

Although the hydrodynamic properties of thyroid hormone receptors have been well characterized in GH_1 cells (Perlman *et al.*, 1982) and in rat liver (Latham *et al.*, 1976), this type of analysis provides limited information regarding the structural features of receptor in cells. To examine the structure of the receptor in more detail, a photoaffinity label derivative of L-[^{125}I]T3 was developed (Casanova *et al.*, 1984; Pascual *et al.*, 1982). The structure of *N*-2-diazo-3,3,3-trifluoropropionyl-L-T3 (L-T3-PAL) is shown in Fig. 1. This compound has approximately 1/10 of the affinity of L-T3 for the receptor and binds to receptor in intact cells and in nuclear extracts.

B. Different receptor forms

Fig. 2 illustrates the electrophoretic profile of photoaffinity labeled receptor isolated from GH_1 cell nuclei by extraction with 0.4 M KCl. Lane 1 represents cells incubated with 2 nM L-[^{125}-I]T3-PAL while the cells represented in lane 2 also received a 1000-fold molar excess of non-radioactive L-T3. L-[^{125}I]T3-PAL labeled an abundant M_r~47 000 component as well as a less abundant M_r~57 000 doublet species and labeling of both components was inhibited by non-radioactive L-T3. Several lines of evidence indicate that the radiolabeled components each have properties characteristic for receptor. First, a variety of iodothyronine analogues differentially inhibit the extent of coupling to all forms as predicted by their known relative affinities for receptor (Pascual *et al.*, 1982). Secondly, the components show amino acid sequence homology based on peptide mapping with *S. aureus* V8 protease, trypsin, and chymotrypsin (Pascual *et al.*, 1982).

Both M_r~47 000 and M_r~57 000 receptor forms, labeled with L-[^{125}I]T4-PAL or with L-[^{125}I]T3-PAL, bind equally well to DNA–cellulose under conditions of low ionic strength (0.05 M KCl) and elute identically over a concentration range of 0.15–0.4 M KCl (Casanova *et al.*, 1984). Similarly, stepwise salt extraction of nuclei derived from L-[^{125}I]T3-PAL labeled cells, followed by DNA–cellulose chromatography, reveals no preferential extraction of one of the receptor forms. This suggests that the DNA-binding region of the two receptor species has similar affinity for non-specific (calf thymus) DNA.

A number of possibilities could account for the identification of the different M_r receptor forms: (1) zero-length 254 nm UV enhanced cross-linking of the M_r~47 000 receptor to a non-hormone-binding subunit (or subunits) or an M_r~10 000 peptide (or peptides) with which it is in contact in chromatin; (2) cleavage of an M_r~10 000 fragment from the M_r~57 000 form by 254 nm irradiation; (3) the artifactual proteolytic cleavage of a native M_r~57 000 species to the M_r~47 000 form by proteases released as a result of cell lyses; (4) post-synthetic modification of the 47 000 component (e.g. glycosylation, phosphorylation, etc.) which could result in anomalous migration in sodium dodecylsulfate (SDS)–polyacrylamide gels; (5) the M_r~57 000 form is converted or processed *in situ* to the M_r~47 000 species in a functional precursor–product type relationship; (6) the different receptor forms may represent the products of different genes which have diverged in evolution or represent alternative splicing of a single receptor gene.

We have provided evidence which tends to exclude several of the above possibilities. It was reported by van der Walt *et al.* (1982) that underivatized L-[^{125}I]T4 has the capacity to couple to proteins at wavelengths greater than 300 nm, a wavelength at which UV damage to proteins does not occur (Dure *et al.*, 1980). To exclude UV peptide cleavage we compared the receptor forms derivatized by L-[^{125}I]T3-PAL at 254 nm with those formed with un-derivatized L-[^{125}I]T4 using irradiation at 310 nm through a Pyrex glass plate which eliminates wavelengths below 300 nm where peptide cleavage may occur (Dure *et al.*, 1980). Fig. 7 shows that L-[^{125}I]T3-PAL and L-[^{125}I]T4, irradiated at 254 nm and 310 nm respectively, yield identical M_r~57 000 and M_r~47 000 components, thereby confirming the absence of UV-induced peptide cleavage. As described later, the same molecular weight species are identified by photocrosslinking of underivatized L-[^{125}I]T3 or L-[^{125}I]T4 at 254 nm.

Although the coupling efficiency of L-[^{125}I]T4 at 310 nm is about 20-fold lower than that of L-[^{125}I]T3-PAL (Casanova *et al.*, 1984), the results indicate that the

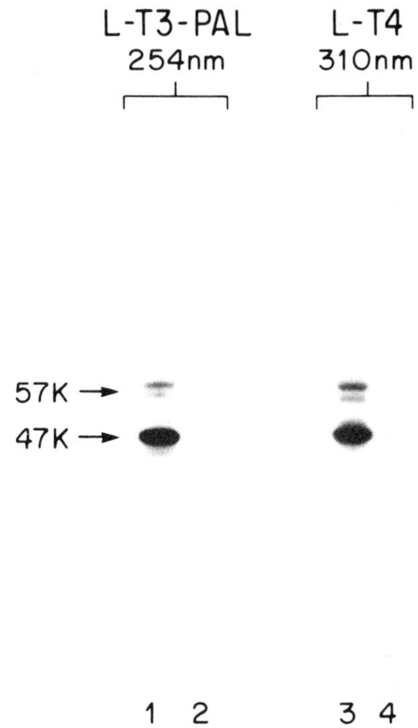

Fig. 7 — Receptor forms identified by photoaffinity labeling using L-[^{125}I]T3-PAL and underivatized L-[^{125}I]T4. GH_1 cells were incubated for 1.5 h with 2 nM L-[^{125}I]T3-PAL without (lane 1) or with (lane 2) 5 μM L-T3 or for 1.5 h with 2 nM L-[^{125}I]T4 without (lane 3) or with (lane 4) 5 μM L-T3. Cell monolayers incubated with L-[^{125}I]T3-PAL were irradiated at 254 nm for 2 min at 4°C. Cells incubated with L-[^{125}I]T4 were irradiated for 40 min at 4°C using a Westinghouse FS-20 UV B lamp having an emission peak at 310 nm. UV irradiation below 300 nm was absorbed with a 2 mm thick Pyrex plate (Corning). Nuclear proteins were extracted with 0.4 M KCl buffer, precipitated with 80% ethanol, and electrophoresed. 400 μg of protein was applied to each gel lane. Autoradiographic exposure time was 36 h for the L-[^{125}I]T3-PAL samples (lanes 1 and 2) and 23 days for the L-[^{125}I]T4 samples (lanes 3 and 4). (From Casanova *et al.* (1984), with permission.)

different receptor forms do not reflect a unique property of L-[^{125}I]T3-PAL or result from peptide cleavage secondary to irradiation at 254 nm. This study also indicates that it is unlikely that the M_r~57 000 species contains the M_r~47 000 component crosslinked to protein(s) having an M_r of approximately 10 000. Although this could occur at 254 nm, which can act as a zero-length protein–protein crosslinking agent, this does not occur at wavelengths above 300 nm (Kunkel *et al.*, 1981). Studies using a wide variety of protease inhibitors, at each stage of receptor isolation beginning from the time of cell lyses, suggest that the different receptor forms do not reflect artifactual proteolysis for proteases released from broken cells. This is also supported by studies which showed an identical abundance of receptor forms whether nuclei are rapidly processed at 4°C and extracted with 0.4 M KCl or first incubated at 25°C for 30 min prior to extraction (Pascual *et al.*, 1982).

C. Estimation of the half-life and relative synthetic rates of the two photoaffinity labeled receptor forms

To explore further the properties and the possible relationship of the two M_r receptor forms we coupled the technique of dense amino acid and photoaffinity labeling to assess the half-lives and the relative synthetic rates of the various species (Fig. 8) (Casanova *et al.*, 1984). Cells were incubated either with normal medium

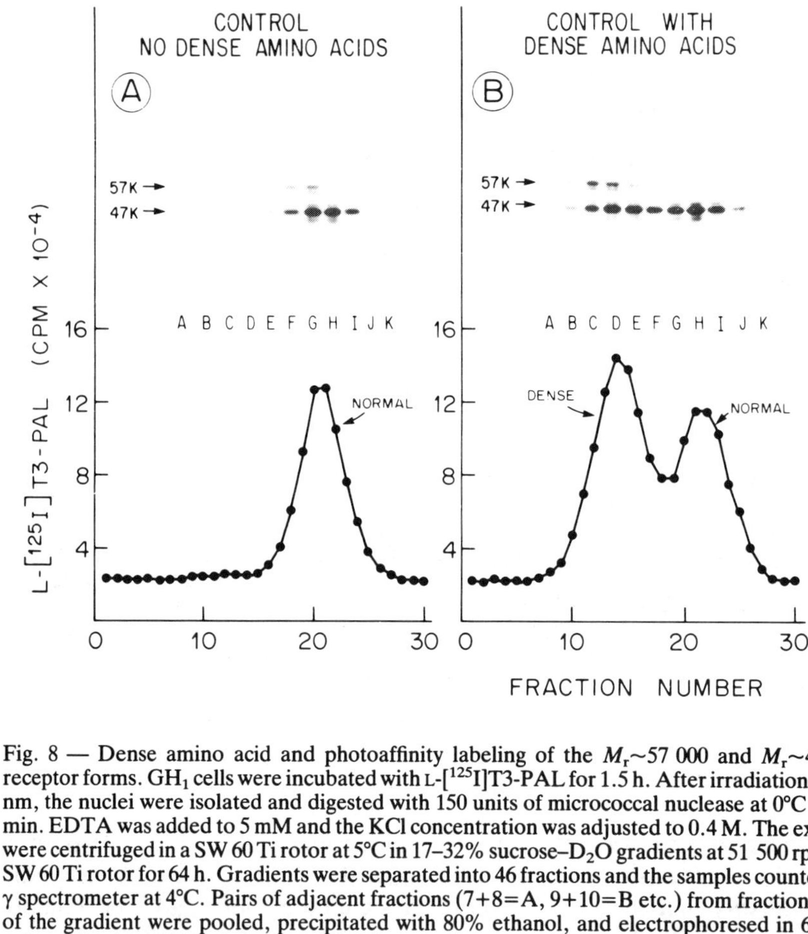

Fig. 8 — Dense amino acid and photoaffinity labeling of the M_r~57 000 and M_r~47 000 receptor forms. GH_1 cells were incubated with L-[^{125}I]T3-PAL for 1.5 h. After irradiation at 254 nm, the nuclei were isolated and digested with 150 units of micrococcal nuclease at 0°C for 45 min. EDTA was added to 5 mM and the KCl concentration was adjusted to 0.4 M. The extracts were centrifuged in a SW 60 Ti rotor at 5°C in 17–32% sucrose–D_2O gradients at 51 500 rpm in a SW 60 Ti rotor for 64 h. Gradients were separated into 46 fractions and the samples counted in a γ spectrometer at 4°C. Pairs of adjacent fractions (7+8=A, 9+10=B etc.) from fractions 7–28 of the gradient were pooled, precipitated with 80% ethanol, and electrophoresed in 6–15% polyacrylamide–SDS gels. The autoradiographs (72 h at −80°C) are illustrated as insets to the figure. A, cells incubated with L-[^{125}I]T3-PAL for 1.5 h; B, cells incubated with dense amino acids for a total of 5 h and with L-[^{125}I]T3-PAL for the last 1.5 h. Sedimentation is illustrated from right to left: A utilized four 25 cm^2 flasks of GH_1 cells while B utilized eight 25 cm^2 culture flasks. (From Casanova *et al.* (1984), with permission.)

(Fig. 8A) or with normal medium followed by a 5 h incubation with dense amino acid medium (Fig. 8B). The experiment in Fig. 8B used twice the number of flasks as the control cells (Fig. 8A). The control cells as well as the cells labeled with dense amino acids received L-[^{125}I]T3-PAL for the last 1.5 h of incubation. After irradiation, nuclei were extracted with 0.4M KCl and the samples were centrifuged in 17–32% sucrose–D_2O gradients. These results are illustrated together with the SDS–gel electrophoretic pattern derived from precipitating two adjacent fractions (7+8=A, 9+10=B, 11+12=C etc.), from fractions 7–28 of the gradients.

Without dense amino acids the labeled receptor sediments as a single peak of normal density containing both the M_r~47 000 and 57 000 receptor forms. Gradient analysis of receptor from cells incubated with dense amino acid media for 5 h demonstrates two peaks of newly synthesized dense and pre-existing receptor of normal density. The corresponding SDS–gell shows that the M_r~47 000 form appears equally abundant in the dense and normal receptor fractions with the M_r~57 000 doublet is highly abundant in the newly synthesized dense receptor pool. The radioactive bands were excised from the gel and quantitated with a γ spectrophotometer. These results confirmed the autoradiographic studies and showed that the M_r~47 000 form was equally divided between both pools, whereas the M_r~57 000 form was highly enriched in the newly synthesized receptor pool (Casanova *et al.*, 1984).

The results of Fig. 8 further suggest that the two M_r receptor forms do not derive from aritifactual proteolysis of the M_r~57 000 form or post-synthetic modification of the M_r~47 000 species to generate the M_r~57 000 component(s). If an irreversible postsynthetic modification occurred, the amount of newly synthesized M_r~57 000 receptor would be more abundant in the pre-existing pool and the reverse is found. If a rapidly reversible postsynthetic modification occurred, the ratio of the 57 000 to 47 000 forms would be similar in both pools. In addition, random artifactual proteolysis also cannot explain these findings since the ratio of the two M_r receptor forms would be equal in both the dense and normal receptor pool.

On the basis of the relative amounts of dense and normal receptor at the 5 h dense amino acid incubation time, it is possible to calculate the half-life and the relative synthetic rate of each form using the formula $R_n = R_t e^{-k_d t}$ in which R_n is the receptor of normal density, R_t represents the total labeled receptor (dense plus normal), t equals the time of labeling (5 h), and k_d is the degradation rate constant. k_d was used to calculate the half-life using the formula $t_{1/2} = 0.693/k_d$. The $t_{1/2}$ of total receptor (4.6 h), calculated by summing the dense M_r ~ 57 000 and 47 000 components and the normal density forms, is in good agreement with that determined for total receptor based on L-[^{125}I]T3 binding (4.7 h) (Raaka and Samuels, 1981). The $t_{1/2}$ of the individual forms was 2.4 h for the M_r ~ 57 000 species and 5.6 h for the M_r ~ 47 000 component. The 'relative synthetic rates' of the M_r ~ 47 000 and 57 000 forms were calculated from the amounts of these photoaffinity labeled ^{125}I-forms and the estimated coupling efficiencies. These are not absolute synthetic rates since the coupling efficiency of each form may not be identical. However, both M_r receptor forms showed similar relative synthetic rates, which raises the possibility that the M_r ~ 57 000 form is an *in situ* precursor of the M_r ~ 47 000 component which is more highly abundant because of its longer half-life (Casanova *et al.*, 1984).

D. Comparison of photoaffinity and chemical affinity labeling studies using ^{125}I-iodothyronine analogues

In recent studies we have examined the coupling of various ^{125}I-iodothyronine analogues with receptors in intact cells and in isolated receptor preparations in nuclear extracts. We examined whether L-[^{125}I]rT3-PAL couples to the same receptor as L-[^{125}I]T3-PAL. In addition, we have also compared the coupling efficiencies at 254 nm of L-[^{125}I]T3, L-[^{125}I]T3-PAL, L-[^{125}I]T4, and L-[^{125}I]T4-PAL. The coupling efficiency of underivatized hormone was found to be higher at 254 nm than in previous studies at 310 nm (Casanova *et al.*, 1984; van der Walt *et al.*, 1982). Furthermore, although L-[^{125}I]T4-PAL has a lower affinity for the receptor than L-[^{125}I]T3-PAL it has a higher coupling efficiency. In addition, we have also compared the M_r species labeled with L-[^{125}I]T3-PAL with those formed with *N*-bromoacetyl-L-T3, a chemical affinity label, which has been reported to couple covalently to a component of M_r~56 000 in rat liver nuclear extracts (Nikodem *et al.*, 1980; van der Walt *et al.*, 1982).

1. *L-[^{125}I]rT3-PAL couples to the same receptor species as identified with L-[^{125}I]T3-PAL*

Since several reports have suggested that L-rT3 interacts with a separate receptor species (Weirsinga and Broenink, 1982; Weirsinga *et al.*, 1982), we investigated labeling of receptor in intact cells using a L-[^{125}I]rT3-PAL derivative (Horowitz *et al.*, manuscript submitted). We found (Fig. 4) that covalent labeling by L-[^{125}I]rT3-PAL was partially inhibited by co-incubation with 10-fold molar excess of unlabeled L-rT3 and was completely inhibited by 10-fold excess of unlabeled L-T3 or 1000-fold excess of either unlabeled L-rT3 or L-T3. The relative abundances of the two molecular weight receptor species labeled by L-[^{125}I]T3-PAL and L-[^{125}I]rT3-PAL are the same. As with L-[^{125}I]T3-PAL, L-[^{125}I]rT3-PAL was not found to label specifically any protein in the extranuclear fraction of GH_1 cells (not shown).

2. *N-bromoacetyl-L-[^{125}I]T3 does not identify thyroid hormone receptor in GH_1 cell nuclear extracts*

Nikodem *et al.* (1980) reported having labeled an M_r~56 000 protein in rat liver nuclear extracts with *N*-bromoacetyl-L-[^{125}I]T3 (BrAc[^{125}I]T3), a chemical affinity label. This species was reported to represent thyroid hormone receptor even though no inhibition of coupling with an excess of unlabeled L-T3 was performed. To assess whether BrAc[^{125}I]T3 labeled receptor, we compared labeling of L-[^{125}I]T3-PAL and BrAc[^{125}I]T3 using GH_1 nuclear extracts. The synthesis of *N*-hydroxysuccinimide and BrAc[^{125}I]T3 was carried out as described (Nikodem *et al.*, 1980). In order to ensure that receptor was fully occupied in the competition studies, 10 μM L-T3 was added to the extract before or simultaneously with BrAc[^{125}I]T3. Fig. 9 demonstrates that, when nuclear extract was incubated with BrAc[^{125}I]T3 in the presence or absence of 10 μM non-radioactive L-T3, no specifically labeled protein bands were found (lanes 1 and 2) that corresponded to L-[^{125}I]T3-PAL labeled receptor (lanes 3 and 4) in the same GH_1 cell nuclear extract. BrAc[^{125}I]T3 does non-specifically modify a major component (M_r~53 000) which is also non-specifically labeled by L-[^{125}I]T3-PAL and can be seen just above the M_r~47 000 receptor band (lanes 3 and 4).

3. *Estimation of the coupling efficiencies of different photoaffinity probes*

The relative coupling efficiencies of L-[^{125}I]T3, L-[^{125}I]T4, L-[^{125}I]T3-PAL, and L-[^{125}I]T4-PAL were determined by comparing the total amount of radioactivity

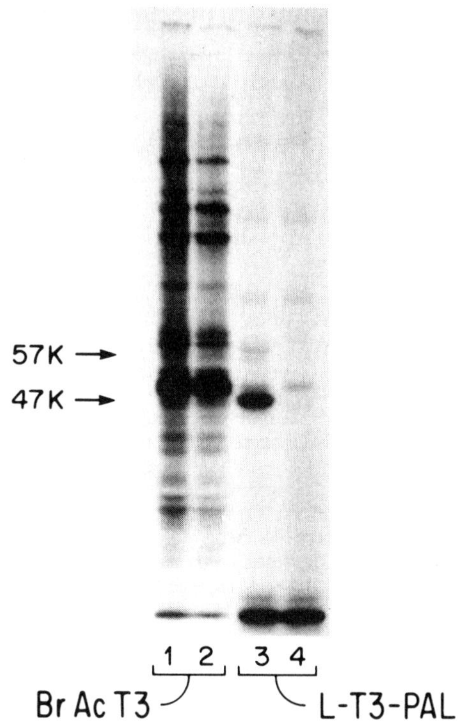

Fig. 9 — Comparison of receptor forms identified by L-[^{125}I]T3-PAL and nuclear proteins labeled with bromoacetyl L-[^{125}I]T3. GH_1 cell nuclei were digested with micrococcal nuclease and then extracted with 0.4 M KCl. The extract was divided into four aliquots, each of which was incubated for 1.5 h at 0°C, in the dark, with 8 nM bromoacetyl L-[^{125}I]T3 (lanes 1 and 2) or 4 nM L-[^{125}I]T3-PAL (lanes 3 and 4) and in the presence of 10 μM non-radioactive L-T3 (lanes 2 and 4). Following the incubation, the samples containing L-[^{125}I]T3-PAL were irradiated at 254 nm for 2 min at 0°C. The samples were precipitated with 80% ethanol, electrophoresed, and autoradiographed. Equal amounts of protein (approximately 400 μg) were applied per lane. (From Horowitz *et al.* (1987).)

associated with the thyroid hormone receptor with that amount of radioactivity covalently coupled to receptor (Horowitz *et al.*, manuscript submitted). In this analysis intact cells were incubated with the various compounds. After UV irradiation, the nuclear extracts were sedimented in sucrose gradients. The gradients were fractionated and the region below the 3.8 S receptor peak, the receptor peak, and fractions above the receptor peak were precipitated and electrophoresed in SDS–gels. After autoradiography, the individual labeled bands were excised and quantitated in a γ counter. Table 1 shows that L-[^{125}I]T4-PAL appears to be at least 9-fold more efficient than L-[^{125}I]T3-PAL. In a number of experiments we have observed a coupling efficiency for L-[^{125}I]T4-PAL which varied from 13.5% to 50% with about a 5- to 10-fold increase above that found with L-[^{125}I]T3-PAL. The enhanced coupling efficiency of L-[^{125}I]T4-PAL could result from both crosslinking of the photoactivated carbene intermediate (Pascual *et al.*, 1982) and crosslinking due to a mechanistically unrelated UV-induced phenyl radical on the outer phenolic

Table 1 — Determination of coupling efficiencies

Compound	Density gradient (cpm)	SDS–PAGE (cpm)	Coupling efficiency (%)
L-[^{125}I]T4-PAL	13 668	1849	13.5
L-[^{125}I]T3-PAL	25 926	398	1.5
L-[^{125}I]T4	24 500	134	0.5
L-[^{125}I]T3	63 692	—	—

Photoaffinity labeling of intact cells with each of the ^{125}I-labeled hormones (0.5 nM) or [^{125}I]PAL probes (2 nM), and isolation of receptor in nuclear extract was carried out as described in the text. Equal amounts of protein were layered on a discontinuous sucrose density gradient (10%, 20%, 30%) and the samples were centrifuged in a SW 60 Ti rotor at 58 000 rpm for 20 h. 46 fractions (70 μl each) were collected at 4°C and the radioactivity in each fraction was determined. Fractions 1–15, 16–30, and 31–46 of each gradient were pooled and precipitated with ethanol and prepared for and electrophoresed as described in the text. The gel was fixed, stained and destained, and autoradiographed as described above. The bands of interest containing receptor were excised from the gel and the radioactivity was determined in a γ counter. The extent of background radioactivity in the gel was low and uniform at approximately 50 cpm per gel section. The value listed for the 'Density gradient' represents the radioactivity in fractions 16–30, the only fractions from each gradient which contained receptor identified on SDS–PAGE, as an M_r~47 000 and M_r~57 000 doublet species. Both molecular weight forms were excised and counted together. (From Horowitz *et al.* (1987).)

ring (van der Walt *et al.*, 1982). However, the coupling efficiency of underivatized L-[^{125}I]T4 is significantly less than that of L-[^{125}I]T3-PAL and 25-fold less than that of L-[^{125}I]T4-PAL. An alternative possibility is that L-[^{125}I]T4-PAL may be positioned in the receptor ligand-binding site in a sterically favored position relative to amino acid residues capable of undergoing covalent modification.

VII. SUMMARY AND CONCLUSIONS

Photoaffinity label probes of iodothyronines can specifically identify thyroid hormone receptor in intact cells and in solubilized receptor preparations. These probes have certain advantages over a chemical affinity label in analyzing receptor structure. First, a photoaffinity label probe covalently crosslinks only after photoactivation so it is possible to demonstrate, with appropriate competitive inhibition studies, that the photoaffinity label probe associates with the receptor specifically. Secondly, since crosslinking only occurs after photolysis, it is possible to adjust the concentration of the photoaffinity label to maximize association with 'specific' binding sites relative to 'non-specific' associations prior to covalent linkage by photoactivation.

The different [^{125}I]iodothyronine-PAL analogues may be useful as probes of the thyroid hormone receptor ligand binding domain since PAL compounds with different affinities for receptor may photocouple to different receptor residues within, or proximate to, the ligand binding region. Because of the recent documentation that the c-erb-A cDNA encodes a thyroid hormone receptor, photoaffinity label probes may also prove useful in studying the structure of normal and mutated receptor made by *in vitro* translation as well as in probing the organization of receptor in chromatin. Lastly, they may provide insights into possible alterations of receptor structure in patients with partial end organ resistance to thyroid hormone (Refetoff *et al.*, 1967; Eil *et al.*, 1982) and in transformed cells possessing the v-erb-A oncogene.

ACKNOWLEDGMENT

The research from the authors' laboratory described in this chapter was supported by Grants AM16636 and AM32352 (H.H.S.) and AM01372 (Z.D.H.) from the National Institutes of Health.

REFERENCES

Arriza, J. L., Weinberger, C., Cerelli, G., Glaser, T. M., Handelin, B. L., Housman, D. E. and Evans, R. M. (1987) Cloning of human mineralicoticoid receptor complementary DNA: Structural and functional kinship with the glucocorticoid receptor. *Science* **237,** 268–275.

Barta, A., Richards, R. I., Baxter, J. D. and Shire, J. (1981) Primary structure and evolution of rat growth hormone gene. *Proc. Natl. Acad. Sci. USA* **78,** 4867–4871.

Berg, J. M. (1986a) Potential metal-binding domains in nucleic acid binding proteins. *Science* **232,** 485–487.

Berg, J. M. (1986b) More metal-binding fingers. *Nature* **319,** 264–265.

Borek, C., Guernsey, D. L., Ong, A. and Edelman, I. S. (1983) Critical role played by thyroid hormone in induction of neoplastic transformation by chemical carcinogens in tissue culture. *Proc. Natl. Acad. Sci. USA* **80,** 5749–5752.

Carr, F. E., Jump, D. B. and Oppenheimer, J. H. (1984) Distribution of thyroid hormone-responsive translation products in rat liver polysome and postribosomal ribonucleoprotein populations. *Endocrinology* **115,** 1737–1745.

Casanova, J., Horowitz, Z. D., Copp, R. P., McIntyre, W. R., Pascual, A. and Samuels, H. H. (1984) Photoaffinity labeling of thyroid hormone nuclear receptors: influence of *n*-butyrate and analysis of the half-lives of the 47 000 and 57 000 molecular weight receptor forms. *J. Biol. Chem.* **259,** 12084–12091.

Casanova, J., Copp, R. P., Janocko, L. and Samuels, H. H. (1985) 5′-flanking DNA of the rat growth hormone gene mediates regulated expression by thyroid hormone. *J. Biol. Chem.* **260,** 11744–11748.

Connelly, D. M., Sullivan, W. P., Toft, D. O., Birnbaumer, M., Cook, R. G., Maxwell, B. L., Schrader, W. T. and O'Malley, B. W. (1986) Molecular cloning of the chicken progesterone receptor. *Science* **233,** 767–770.

Crew, M. D. and Spindler, S. R. (1986) Thyroid hormone regulation of the transfected rat growth hormone promotor. *J. Biol. Chem.* **261,** 5018–5022.

Damm, K., Beug, H., Graf, T. and Vennstrom, B. (1987) A single point mutation in *erbA* restores the erythroid transforming potential of a mutant avian erythroblastosis virus (AEV) defective in both *erbA* and *erbB* oncogenes. *EMBO J.* **6,** 375–382.

Dayton, A. I., Selden, J. R., Laws, G., Dorney, D. J., Finan, J., Tripputi, P., Emanuel, B. S., Rovera, G., Nowell, P. C. and Croce, C. M. (1984) A human c-erb-A oncogene homologue is closely proximal to the chromosome 17 breakpoint in acute promyelocytic leukemia. *Proc. Natl. Acad. Sci. USA* **81,** 4495–4499.

Debuire, B., Henry, C., Benaissa, M., Biserte, G., Claverie, J. M., Saule, S., Martin, P. and Stehelin, D. (1984) Sequencing the erb-A gene of avian erythroblastosis virus reveals a new type of oncogene. *Science* **224,** 1456–1459.

Dejean, A., Bouguelerat, L., Grzesclnik, K. H. and Tiollais, P. (1986) Hepatitis B virus DNA integration in a sequence homologous to v-erb-A and steroid receptor genes in a hepatocellular carcinoma. *Nature* **322,** 70–72.

Dobner, P. R., Kawasaki, E. W., Yu, L.-Y. and Bancroft, F. C. (1981) Thyroid hormone or glucocorticoid hormone induces pre-growth hormone mRNA and its probable nuclear precursor in rat pituitary cells. *Proc. Natl. Acad. Sci. USA* **78,** 2230–2234.

Dynan, W. S. and Tjian, R. (1983) The promotor-specific transcription factor SP 1 binds to upstream sequences in the SV40 early promotor. *Cell.* **35,** 79–87.

Eil, C., Fein, H. G., Smith, T. J., Furlanetto, R. W., Bourgeois, M., Sterling, M. W. and Weintraub, B. D. (1982) Nuclear binding of L-[^{125}I]triiodothyronine in dispersed cultured skin fibroblasts from patients with resistance to thyroid hormone. *J. Clin. Endo. Metabol.* **55,** 502–510.

Evans, R. M., Birnberg, N. C. and Rosenfeld, M. G. (1982) Glucocorticoid and thyroid hormones transcriptionally regulate growth hormone gene expression. *Proc. Natl. Acad. Sci. USA* **79,** 7659–7663.

Fisher, P. B., Guernsey, D. L., Weinstein, I. B. and Edelman, I. S. (1983) Modulation of adenovirus transformation by thyroid hormone. *Proc. Natl. Acad. Sci. USA* **80,** 196–200.

Flug, F., Copp, R. P., Horowitz, Z. D., Janocko, L., Plotnick, M. and Samuels, H. H. (1987) *cis*-acting elements of the rat growth hormone gene which mediate basal and regulated expression by thyroid hormone. *J. Biol. Chem.* **262.**

Fried, M. and Crothers, D. M. (1981) Equilibria and kinetics of lac repressor–operator interactions by polyacrylamide gel electrophoresis. *Nucleic Acids Res.* **9,** 6505–6525.

Fried, M. G. and Crothers, D. M. (1984) Equilibrium studies of the cyclic AMP receptor protein–DNA interaction. *J. Mol. Biol.* **172,** 241–262.

Frykberg, L., Palmieri, S., Beug, H., Graf, T., Hayman, M. J. and Vennstrom, B. (1983) Transforming capacities of avian erythroblastosis virus mutants deleted in the erb A or erb B oncogenes. *Cell* **32,** 227–238.

Garner, M. M. and Revzin, A. (1981) A gel electrophoresis method for quantifying the binding of proteins to specific DNA regions: application to components of the *Escherichia coli* lactose operon regulator system. *Nucleic Acids Res.* **9,** 3047–3060.

Gonda, T. J., Sheiness, D. K. and Bishop, J. M. (1982) Transcripts from the cellular homologs of retroviral oncogenes: distribution among chicken tissues. *Mol. Cell. Endocrinol.* **2,** 617–624.

Goodwin, R. G., Rottman, F. M., Calloghan, T., Kung, H. J., Maroney, P. A. and Nilsen, T. W. (1986) C-erb B activation in avian leukosis virus-induced erythroblastosi: multiple epidermal growth factor receptor mRNA's are generated by alternative mRNA processing. *Mol. Cell. Endocrinol.* **6,** 3128–3133.

Gordon, D. F., Quick, D. P., Erwin, C. R., Donebon, J. E. and Maurer, R. A.

(1983) Nucleotide sequence of the bovine growth hormone chromosomal gene. *Mol. Cell. Endocrinol.* **3,** 81–95.

Gorman, C. M., Moffat, L. F. and Howard, B. H. (1982) Recombinant genomes which express chloramphenical acetyltransferase in mammalian cells. *Mol. Cell. Biol.* **2,** 1044–1051.

Graf, T. and Beug, H. (1983) Role of v-erb-A and v-erb-B oncogenes of avian erythroblastosis virus in erythroid cell transformation. *Cell* **34,** 7–9.

Green, S. and Chambon, P. (1986) A superfamily of potentially oncogenic hormone receptors. *Nature* **324,** 615–617.

Green, S. and Chambon, P. (1987) Oestradiol induction of a glucocorticoid-responsive gene by a chimaeric receptor. *Nature* **325,** 75–78.

Green, S., Walter, P., Kumar, V., Krust, A., Nornert, J. M., Argos, P. and Chambon, P. (1986) Human oestrogen receptor cDNA: sequence, expression, and homology to v-erb-A. *Nature* **320,** 134–139.

Groul, D. J. (1980) 3,5,3′-triiodo-L-thyronine receptor-containing chromatin fragments: production by nuclease digestion. *Endocrinology* **107,** 994–999.

Guernsey, D. L., Borek, C. and Edelman, I. S. (1981) Crucial role of thyroid hormone in X-ray-induced neoplastic transformation in cell culture. *Proc. Natl. Acad. Sci. USA* **78,** 5708–5711.

Hayman, M. J., Ramsay, G. M., Savin, K. and Kitchener, G. (1983) Identification and characterization of the avian erythroblastosis virus erb B gene product as a membrane glycoprotein. *Cell* **32,** 579–588.

Hervas, Morreale de Escobar, G. and Escobar Del Ray, F. (1975) Rapid effects of single small doses of L-thyroxine and triiodo-L-thyronine on growth hormone, as studied in the rat by radioimmunoassay. *Enocrinology* **97,** 91–101.

Hollenberg, S. M., Weinberger, C., Ong, E. S., Cerelli, G., Oro, A., Lebo, R., Thompson, E. B., Rosenfeld, M. G. and Evans, R. M. (1985) Primary structure and expression of a functional glucocorticoid receptor cDNA. *Nature* **318,** 635–641.

Horowitz, Z. D., Casanova, J., Pascual, A., Sahnoun, H. and Samuels, H. H. (1987) Photoaffinity labeled iodoythronine derivatives identify identical receptor molecular weight forms with coupling efficiency inversely proportional to binding affinity. Manuscript submitted.

Jansson, M., Philipson, L. and Vennstrom, B. (1983) Isolation and characterization of multiple human genes homologous to the oncogenes of avian erythroblastosis virus. *EMBO J.* **2,** 561–565.

Jeltsch, J. M., Krozowski, Z., Quirin-Stricker, C., Gronemeyer, H., Simpson, R. J., Garnier, J. M., Krust, A., Jacob, F. and Chambon, P. (1986) Cloning of the chicken progesterone receptor. *Proc. Natl. Acad. Sci. USA* **83,** 5424–5428.

Krust, A., Green, S., Argos, P., Kumar, V., Walter, P., Bornert, J. M. and Chambon, P. (1986) The chicken oestrogen receptor sequence: homology with v-erb-A and the human oestrogen and glucocorticoid receptors. *EMBO J.* **5,** 891–897.

Kunkel, G. R., Mehrabian, M. and Martinson, H. G. (1981) Contact-site cross-linking agents. *Mol. Cell. Endocrinol.* **34,** 3–13.

Liaw, C. W. and Towle, H. C. (1984) Characterization of a thyroid hormone-responsive gene from rat. *J. Biol. Chem.* **259,** 7253–7260.

Magnuson, M. A. and Nikodem, V. M. (1983) Molecular cloning of a cDNA sequence for rat malic enzyme: direct evidence for induction *in vivo* of rat liver malic enzyme mRNA by thyroid hormone. *J. Biol. Chem.* **258,** 12712–12717.

Mahdavi, V., Chambers, A. P. and Nadal-Ginard, B. (1984) Cardiac alpha- and beta-myosin heavy chain genes are organized in tandem. *Proc. Natl. Acad. Sci. USA* **81,** 2626–2630.

Martial, J. A., Baxter, J. D., Goodman, H. M., Seeburg, P. H. (1977) Regulation of growth hormone messenger RNA by thyroid hormone and glucocorticoid hormones. *Proc. Natl. Acad. Sci. USA* **74,** 1816–1820.

McDonnell, D. P., Mangelsdorf, D. J., Pike, J. W., Haussler, M. R. and O'Malley, B. W. (1987) Molecular cloning of complementary DNA encoding the avian receptor for vitamin D. *Science* **235,** 1214–1217.

Nikodem, V. M., Cheng, S. Y. and Rall, J. E. (1980) Affinity labeling of rat liver thyroid hormone receptor. *Proc. Natl. Acad. Sci. USA* **77,** 7064–7068.

Nilsen, T. W., Maroney, P. A., Goodwin, R. G., Rottman, F. M., Crittenden, L. B., Raines, M. A. and Kung, H. J. (1985) C-erb B activation in ALV-induced erythroblastosis: novel mRNA processing and promotor insertion result in expression of an amino-truncated EGF receptor. *Cell* **41,** 719–728.

Oppenheimer, J. H. and Samuels, H. H. (eds.) (1983) *Molecular Basis of Thyroid Hormone Action.* Academic Press, New York.

Oppenheimer, J. H. and Surks, M. I. (1975) Biochemical basis of thyroid hormone action. In: Litwack, G. (ed.) *Biochemical Actions of Hormones,* Vol. 3. Academic Press, New York, pp. 119–157.

Oppenheimer, J. H., Schwartz, H. L. and Surks, M. I. (1974) Tissue differences in the concentration of triiodothyronine nuclear binding sites in the rat: liver, kidney, pituitary, heart, brain, spleen, testis. *Endocrinology* **95,** 897–903.

Page, G. S., Smith, S. and Goodman, H. M. (1981) DNA sequences of the rat growth hormone gene: location of the 5′ terminus of the growth hormone mRNA and identification of an internal transposon-like element. *Nuc. Acids. Res.* **9,** 2087–2104.

Pascual, A., Casanova, J. and Samuels, H. H. (1982) Photoaffinity labeling of thyroid hormone nuclear receptors in intact cells. *J. Biol. Chem.* **257,** 9640–9647.

Perlman, A. J., Stanley, F. and Samuels, H. H. (1982) Thyroid hormone nuclear receptor: evidence for multimeric organization in chromatin. *J. Biol. Chem.* **257,** 930–938.

Raaka, B. M. and Samuels, H. H. (1981) Regulation of nuclear thyroid hormone receptor levels in GH_1 cells by 3,5,3′-triiodo-L-thyronine: use of dense amino acid labeling to determine the influence of hormone on the receptor half-life and the rate of appearance of newly synthesized receptor. *J. Biol. Chem.* **256,** 6883–6889.

Raines, M. A., Lewis, W. G., Crittenden, L. B. and King, H. J. (1985) C-erb B activation in avian leukosis virus-induced erythroblastosis: clustered integration sites and the arrangement of provirus in the c-erb-B alleles. *Proc. Natl. Acad. Sci. USA* **82,** 2287–2291.

Refetoff, S. L., DeWind, T. and DeGroot, L. J. (1967) Familial syndrome combining deaf-mutism, stippled epiphyses, goiter, and abnormally high PBI: possible target organ refractoriness to thyroid hormone. *J. Clin. Endo. Metabol.* **27,** 279–294.

Sabbah, M., Redeuih, G., Secco, C. and Baulieu, E. E. (1987) The binding activity of estrogen receptor to DNA and heat shock protein (M_r 90 000) is dependent on receptor-bound metal. *J. Biol. Chem.* **262,** 8631–8635.

Samuels, H. H. and Shapiro, L. E. (1976) Thyroid hormone stimulates *de novo* growth hormone synthesis in cultured GH_1 cells: evidence for the accumulation of a rate limiting RNA species in the induction process. *Proc. Natl. Acad. Sci. USA* **73,** 3369–3373.

Samuels, H. H., Tsai, J. S., Casanova, J. and Stanley, F. (1974) *In vitro* characterization of solubilized nuclear receptors from rat liver and cultured GH_1 cells. *J. Clin. Invest.* **54,** 853–865.

Samuels, H. H., Horowitz, Z. D., Stanley, F., Casanova, J. and Shapiro, L. E. (1977) Thyroid hormone controls glucocorticoid hormone action in cultured GH_1 cells. *Nature* **268,** 254–257.

Samuels, H. H., Stanley, F. and Casanova, J. (1979a) Relationship of receptor affinity to the modulation of thyroid hormone nuclear receptor levels and growth hormone synthesis by L-triiodothyronine and iodothyronine analogs in cultured GH_1 cells. *J. Clin. Invest.* **63,** 1229–1240.

Samuels, H. H., Stanley, F. and Shapiro, L. E. (1979b) Control of growth hormone synthesis in cultured GH_1 cells by 3,5,3′-triiodo-L-thyronine and glucocorticoid agonists and antagonists: studies on the independent and synergistic regulation of growth hormone response. *Biochemistry* **18,** 715–721.

Samuels, H. H., Stanley, F. and Casanova, J. (1979c) Depletion of L-3,5,3′-triiodothyronine and L-thyroxine in euthyroid calf serum for use in cell culture studies of the action of thyroid hormone. *Endocrinology* **105,** 80–85.

Samuels, H. H., Stanley, F., Casanova, J. and Shao, T. C. (1980) Thyroid hormone nuclear receptor levels are influenced by the acetylation of chromatin-associated proteins. *J. Biol. Chem.* **255,** 2499–2508.

Samuels, H. H., Casanova, J., Horowitz, Z. D., Raaka, B. M., Shapiro, L. E., Stanley, F., Tsai, J. S. and Yaffe, B. M. (1983a) Hormonal regulation of growth hormone synthesis in cell cultures: role of thyroid and glucocorticoid hormone receptors. In: Litwack, G. (ed.) *Biochemical Actions of Hormones,* Vol. 10. Academic Press, New York, Chap. 4, pp. 116–162.

Samuels, H. H., Perlman, A. J., Raaka, B. M. and Stanley, F. (1983b) Thyroid hormone receptor synthesis and degradation and interaction with chromatin components. In: Oppenheimer, J. H. and Samuels, H. H. (eds.) *Molecular Basis of Thyroid Hormone Action.* Academic Press, New York, Chap. 4, pp. 99–137.

Sap, J., Munoz, A., Damm, K., Goldberg, Y., Ghysdael, J., Leutz, A., Beug, H. and Vennstrom, B. (1986) The c-erb-A protein is a high-affinity receptor for thyroid hormone. *Nature* **324,** 635–640.

Schimke, R. T. (1975) Methods for analysis of enzyme synthesis and degradation in animal tissues. *Methods Enzymol.* **40,** 241–266.

Seeburg, P. H. (1982) The human growth hormone gene family: nucleotide sequences show recent divergence and predict a new polypeptide hormone. *DNA* **1,** 239–249.

Seo, H., Vassart, G., Brocas, H. and Refetoff, S. (1977) Triiodothyronine stimulates specifically growth hormone mRNA in rat pitutitary tumor cells. *Proc. Natl. Acad. Sci. USA* **74,** 2054–2058.

Shapiro, L. E., Samuels, H. H. and Yaffe, B. M. (1978) Thyroid and glucocorticoid hormones synergistically control growth hormone mRNA in cultured GH_1 cells. *Proc. Natl. Acad. Sci. USA* **75,** 45–49.

Siddiqui, U. A., Goldflam, T. and Goodridge, A. G. (1981) Nutritional and hormonal regulation of the translational levels of malic enzyme and albumin mRNA's in avian liver cells *in vivo* and in culture. *J. Biol. Chem.* **256,** 4544–4550.

Spindler, S. R., Mellon, S. H. and Baxter, J. D. (1982) Growth hormone gene transcription is regulated by thyroid and glucocorticoid hormones in cultured rat pituitary tumor cells. *J. Biol. Chem.* **257,** 11627–11623.

Spurr, N. K., Solomon, S., Jansson, M., Sheer, D., Goodfellow, P. N., Bodmer, W. F. and Vennstrom, B. (1984) Chromosomal localization of the human homologues to the oncogenes erb A and erb B. *EMBO J.* **3,** 159–163.

Tao, T. Y. and Towle, H. C. (1986) Coordinate regulation of rat liver genes by thyroid hormone and dietary carbohydrates. *Annals NY Acad. Sci.* **478,** 20–30.

Towel, H. C., Mariash, C. N., Schwartz, H. L. and Oppenheimer, J. H. (1981) Quantitation of rat liver messenger ribonucleic acid for malic enzyme induction by thyroid hormone. *Biochemistry* **20,** 3486–3492.

van der Walt, B., Nikodem, V. M. and Cahnmann, H. J. (1982) Use of underivatized thyroid hormones for photoaffinity labeling of binding proteins. *Proc. Natl. Acad. Sci. USA* **79,** 3508–3512.

Vennstrom, B. and Bishop, J. M. (1982) Isolation and characterization of chicken DNA homologues to the putative oncogenes of avian erythroblastosis virus. *Cell* **28,** 135–143.

Vennstrom, B., Fanshier, L., Moscovici, C. and Bishop, J. M. (1980) Molecular cloning of the avian erythroblastosis virus genome and recovery of oncogenic virus by transfection of chicken cells. *J. Virol.* **36,** 575–585.

Weinberger, C., Hollenberg, S. M., Ong, E. S., Harmon, J. M., Brower, S. T., Cidlowski, J., Thompson, E. B., Rosenfeld, M. G. and Evans, R. M. (1985) Identification of human glucocorticoid receptor complementary DNA clones by epitope selection. *Science* **228,** 740–772.

Weinberger, C., Thompson, C. C., Ong, E. S., Lebo, R., Groul, D. J. and Evans, R. M. (1986) The c-erb-A gene encodes a thyroid hormone receptor. *Nature* **324,** 641–646.

Weirsinga, W. M. and Broenink, M. (1982) Differential release of nuclear reverse T3 and T3 receptors into medium: further evidence for two distinct nuclear binding sites for thyroid hormones. *58th Annual Meeting of the American Thyroid Association.* Abstract T-26.

Wolff, E. C. and Wolff, J. (1964) The mechanism of action of the thyroid hormones. In: Pitt-Rivers, R. and Trotter, W. R. (eds.). *The Thyroid Gland.* Butterworth, London, pp. 237–282.

Yaffe, B. M. and Samuels, H. H. (1984) Hormonal regulation of the growth hormone gene: relationship of the rate of transcription to the level of nuclear thyroid hormone–receptor complexes. *J. Biol. Chem.* **259,** 6284–6291.

Ye, Z.-S. and Samuels, H. H. (1987) Cell and sequence specific binding of nuclear proteins to 5′-flanking DNA of the rat growth hormone gene. *J. Biol. Chem.* **262,** 6313–6317.

Part III
Cloning and functional analysis of steroid and thyroid hormone receptors

14

The superfamily of nuclear receptor genes: DNA cloning strategies

J. M. Jeltsch
Institut de Chimie Biologique, Faculté de Médecine, Strasbourg, France

I. TOWARDS A POWERFUL cDNA CLONING SYSTEM

Fifteen years after the first reports on the synthesis of DNA complementary to RNA (cDNA) in 1972 (Verma *et al.*, 1972; Bank *et al.*, 1972; Ross *et al.*, 1972), double-stranded (ds) cDNA cloning (usually called cDNA cloning) has become a routine technology in molecular biology laboratories. The generation of new cloning vectors and of the corresponding bacterial hosts, the commercial availability of a battery of highly specific enzymes and other biochemicals, and various improvements of previous protocols led to the development of two powerful cloning systems based on bacteriophage λgt10 and λgt11 vectors (Huynh *et al.*, 1985).

In cDNA cloning, three major problems have to be faced. If the mRNA of interest is of low abundance and available only in small amounts — as is the case for nuclear receptors — a large number of independent recombinants (from 100 000 to several millions) may need to be screened in order to detect a positive clone. Secondly, these have to be analyzed in one experiment and therefore a system is required in which the recombinants occupy minimal space during the screening procedure. A third aspect is the conservation of the libraries which have been constructed from precious mRNA — usually with some difficulty — and which may have to be rescreened several times. At present, the λ systems give the most satisfying solutions to these three problems because (i) high numbers of recombinants are generated, (ii) high density screening is possible and (iii) stability of conserved libraries is guaranteed, provided that enough care is taken over storage.

Both λgt10 and λgt11 vectors accept DNA inserts up to 8 kb in length. λgt10 is a 'simple' cloning vector with the advantage that recombinants can be positively selected using an *Escherichia coli hfl* (high frequency lysogen) strain (Huynh *et al.*, 1985). The large and regular plaques formed on lysis and the high yield of phage DNA are important properties of this cloning vehicle, which is preferentially used for screening procedures involving nucleic acids as probes.

λgt11 is an expression vector which, in the appropriate host, produces the bacterial enzyme β-galactosidase (blue plaques on X-gal plates). Foreign DNA fragments are inserted (via EcoRI linkers if necessary) at the unique EcoRI site located close to the 3′ end of the β-galactosidase coding sequence. Following induction, recombinant λgt11 clones produce a fused protein composed of a 114 kDa β-galactosidase fragment linked to the amino acid sequence encoded by the DNA insert. The fused protein has lost the β-galactosidase activity (resulting in white plaques in X-gal plates) but, in principle, can show functional properties of the insert-encoded polypeptide such as enzymatic activity or ligand binding. The λgt11 system is designed mainly for the screening of recombinants with antibodies, although nucleic acid probes have been used under certain conditions (see below). Theoretically, a disadvantage of the λgt11 expression system is that only 1 out of 6 clones generated from a given DNA will be expressed as the corresponding protein (two orientations and three reading frames are possible for the DNA insert relative to the β-galactosidase coding region). This is, however, not a problem in practise because high numbers of recombinants can be screened in one experiment. In consequence, hundreds of DNA libraries have been built with these two systems within the 1983–1987 period. DNA inserts have been generated either from various mRNA sources (e.g. plants and mammals) to yield cDNA libraries or from size-selected genomic DNA fragments isolated after restriction endonuclease digestion. In the case of prokaryotes such as bacteria (Young *et al.*, 1985; Sommer *et al.*, 1987) and lower eukaryotes such as yeast (Young and Davis, 1983) partial restriction enzyme digests could be used in the construction of genomic libraries bearing in mind their lower genomic complexity. Additionally, because introns are either absent or rare in these organisms, the genomic fragments can be screened by expression, thus avoiding the cDNA cloning step. In mammals, genomic sublibraries, constructed in λgt10 after a complete restriction enzyme digest and fragment enrichment (based on Southern analysis), may be used, as demonstrated in the cloning of the human mineralocorticoid receptor (Arriza *et al.*, 1987).

In addition to the λgt10 and λgt11 systems, the Okayama–Berg cloning technique based on plasmid vectors (Okayama and Berg, 1983) can be used efficiently in cases where enriched cDNA sublibraries have to be constructed. This is of special importance if the screening is based on differential hybridization (see Table 1).

The cloning of the nuclear receptor superfamily is an example where the λgt cloning systems provided a major breakthrough. With the exception of the mouse and rat glucocorticoid receptor cloning, which has been done, at least in part, with plasmid vectors, the λgt systems have been used in every case (see Table 1). The development of these efficient cDNA cloning systems coincided with progress in steroid hormone receptor protein purification techniques. These led to the gen-

Table 1 — Overview summarizing the materials, methods and results sections from the publications presenting nuclear receptor cDNA cloning

Receptor	Cloning system			ds cDNA synthesis		Way of screening		Recombinants screened			Number of positives obtained			Percentage of the mRNA length		References
	λgt11	λgt10	Other (plasmid)	Enrichment[a]	Priming[b]	Antibodies[c]	Nucleic acid probes[d]	λgt11	λgt10	Other (plasmid)	λgt11	λgt10	Other (plasmid)	Coding	Total	
Glucocorticoid																
Human	+			M	T	P		2×10^5			3			82	37	Weinberger *et al.* (1985)
Mouse	+		+	No	T,R		C*	10^7		4.1×10^5	1		1	100	~67	Danielsen *et al.* (1986)
Rat			+	M	T		O**			540			6	57	86	Miesfeld *et al.* (1984)
Mineralocorticoid																
Human		+		?	?		C***		3×10^6			11		100	100	Arriza *et al.* (1987)
Estrogen																
Human	+	+		M	R	M	S	1.5×10^6	3×10^5		10	5		100	34	Walter *et al.* (1985)
Rat		+		M+C	T,R		S*		3×10^5			15		100	~85	Koike *et al.* (1987)
Chicken		+		M	R		C*		8.6×10^6			33		100	28	Krust *et al.* (1986)
X. laevis		+		?	?		C*		2×10^5			1		100	~37	Weiler *et al.* (1987)
Progesterone																
Rabbit	+	+		M	R	M,P	C	3×10^6	3×10^5		2	25		100	90	Loosfelt *et al.* (1986)
Human		+		M	R		C*		5×10^5			115		100	81	Misrahi *et al.* (1987)
Chicken																
(a)	+			M	?	M		3×10^6			3			64	31	Conneely *et al.* (1986)
(b)	+	+		M	T,R	P	C,S	10^6	10^6		2	3		87	45	Jeltsch *et al.* (1986)
Retinoic acid																
Human	+	+		No	R		S*	5×10^5	5×10^5		3	2		100	~38	Petkovich *et al.* (1987)
Thyroid (T3)																
Human																
Placenta		+		No	T		C*		2×10^6			2		100	100	Weinberger *et al.* (1986)
Testis	+			No	T		C*	3×10^5			2			100	?	Benbrook and Pfahl (1987)
Chicken	+			No	T		C*	10^5			10			100	~60	Sap *et al.* (1986)
Vitamin D3																
Chicken	+			No	T,R	M	C	10^7			2			~49	~28	McDonnell *et al.* (1987)

?, no information in the reference cited; C, S, probes derived from the same receptor sequences (either cDNA (C and S) or protein (S)); C*, probes for homologous screening to get the same receptor cDNA from other species; S*, synthetic probes derived from a cloned receptor cDNA or from compilation of conserved sequences in various cloned cDNAs to achieve both homologous and heterologous screening; O**, cDNA probes for differential hybridization; C***, probes for heterologous screening using a cloned receptor cDNA.

[a] M, mRNA; C, cDNA.

[b] T, oligo-dT; R, random.

[c] M, monoclonal; P, polyclonal.

[d] S, synthetic; C, cloned DNA; O, other.

eration of antibodies and the microsequencing of receptor fragments whose identification was facilitated by affinity labeling techniques. Thus the tools were available for trying new systems after the failure of conventional plasmid cDNA cloning technology. We know now that the main reason why the plasmid technique was unsuccessful for cloning hormone receptor cDNAs is the very low abundance of the corresponding mRNAs. In fact, the need to generate and screen 10^6 independent recombinants in receptor cloning projects is common (see Table 1). The cloning, screening and easy manipulation of a similar number of recombinants with plasmid cDNA (expression) systems is extremely difficult.

Despite the successful application of the conventional λgt cDNA cloning system, several improvements (in particular the directional cDNA cloning (Han and Rutter, 1988)) have been incorporated in order to overcome particular cloning problems. These are mainly concerned with the artefactual cloning of unrelated cDNAs together in the same molecule (with or without an EcoRI site at the junction) (for a discussion of this problem see Huynh *et al.* (1985)), the random orientation in the vector with respect to the 5′-to-3′ gene organization and the problem of the recovery of a whole cDNA insert when genuine internal EcoRI sites are present (for details see Han and Rutter, 1988).

II. STRATEGIES FOR RECEPTOR cDNA AND GENE CLONING

A. cDNA cloning

A general scheme is presented in Fig. 1. In this field of research the main emphasis has been put on cDNA cloning and less on genomic cloning. Since the cDNAs of various receptors have now been identified, this situation may change and the study of their gene promoter regions, for example, will become attractive. The original route for cDNA expression cloning was to purify the protein and to raise polyclonal and/or monoclonal antibodies (A1–5/6 in Fig. 1). For constructing cDNA libraries more than half of the reports listed in Table 1 used a combination of mRNA enrichment and random priming cDNA synthesis techniques made necessary by a low mRNA abundance and long 3′-untranslated mRNA sequences (see Chapter 15). In parallel, further work at the protein level yielded partial amino acid sequences (A1–3, 10, 11, in Fig. 1), which provided the basis for the design of synthetic oligodesoxynucleotides (A12 in Fig. 1), used as probes for the screening of DNA libraries. An example of this double approach, i.e. antibody screening with λgt11 and synthetic oligonucleotide screening with λgt10, is the cloning of the human estrogen receptor cDNA (Walter *et al.*, 1985; see Table 1).

A very different approach was based on the initial purification of a specific mRNA by the immunoprecipitation of polysomes (A1–6, B1 in Fig. 1) which allowed the group of Keith Yamamoto to identify clones of the rat glucocorticoid receptor cDNA in a plasmid vector using a differential hybridization technique to find 6 positives in as few as 540 clones (C1–3, 4^2, 5^2, 6 in Fig. 1). This was actually the first report of a steroid hormone receptor cDNA cloning (Miesfeld *et al.*, 1984).

These approaches exemplify two opposite strategies: on the one hand the purification of mRNA before screening (glucocorticoid receptors) — here the most difficult and hazardous steps are at the RNA level — and on the other hand the generation of a large library with less emphasis on the purification of the starting generation of a large library with less emphasis on the purification of the starting material (estrogen receptor) — here the difficulties are at the screening level. The second strategy was feasible only by using λgt techniques; today it is by far the preferred one.

As depicted in Table 1, the successful isolation of a first cDNA clone was based most often on immunological screenings. Later crosshybridization studies (Northern blots) with mRNA isolated from various species showed that the nucleotide sequence conservation between homologous receptors allowed the screening of cDNA libraries with nucleic acid probes, thus avoiding the more tedious cloning by expression (see Table 1). In the estrogen receptor subfamily, for example, a cloned human receptor cDNA fragment or a synthetic oligonucleotide based on human receptor cDNA sequences was used as a probe to isolate the rat, chicken and *Xenopus laevis* estrogen receptor cDNAs (see Table 1 and $C8^1$, 9 in Fig. 1).

Note that special care is needed when screening a λgt11 library with a DNA probe which has been purified from a plasmid vector after subcloning. If the vector contains pBR322 sequences, the strains Y1088, Y1089 and Y1090 have to be avoided because their use will result in a high background owing to their episome, which is a pBR322-based recombinant. A strain which does not contain an episome should be used for the screening. We use currently any simple $hsdR^-$ $hsdM^+$ SupE SupF strain. After isolation, positive clones can be amplified in Y1088 as usual for λgt11 recombinants. A similar route ($C4^1$ or 4^3, 5^2–8^1, 9 in Fig. 1) was used to clone the DNA corresponding to the c-erbA protein. This was particularly interesting because it had been noticed that the product of v-erbA contained sequences homologous to the human estrogen and glucocorticoid receptors (see Chapter 15). Therefore, v-erbA derived probes were used to detect c-erbA cDNA in human λgt10 or chicken λgt11 libraries (see Table 1 and Chapter 17). To identify the ligand of this putative receptor, cDNA containing the open reading frame was expressed *in vitro* and *in vivo* ($C8^1$ in Fig. 1). From the molecular weight of the protein bands observed, only one known hormone receptor was a potential candidate but did not belong to the class of steroid receptors: the thyroid hormone receptor. In fact, it was demonstrated by ligand binding of transiently expressed c-erbA protein that it was identical with a thyroid hormone receptor (see Chapter 17). Using a similar approach, a second distinct thyroid receptor was cloned from human testis (see Table 1) and it is likely that further members of the erbA–thyroid hormone receptor family will be identified soon.

The cloning of the human mineralocorticoid and retinoic acid receptors (Table 1) represents two important strategies in the isolation of more distantly related members of a superfamily of genes. The cloning of the mineralocorticoid receptor was based on the observation that additional bands appeared on Southern blots hybridized and washed under low stringency conditions when using a glucocorticoid receptor fragment corresponding to a conserved amino acid region in the gene family. These bands were not seen under high stringency conditions, indicating the existence of a distantly related gene. A genomic sublibrary was constructed in λgt10 according to the Southern blotting data D1–3, 4^1, 5^2–7, 8^1 in Fig. 1) which was screened under low stringency conditions with the glucocorticoid receptor cDNA fragment. The insert of one positive clone was then used to rescreen λgt10 cDNA libraries. By expression the cDNA obtained was shown to encode the human mineralocorticoid receptor.

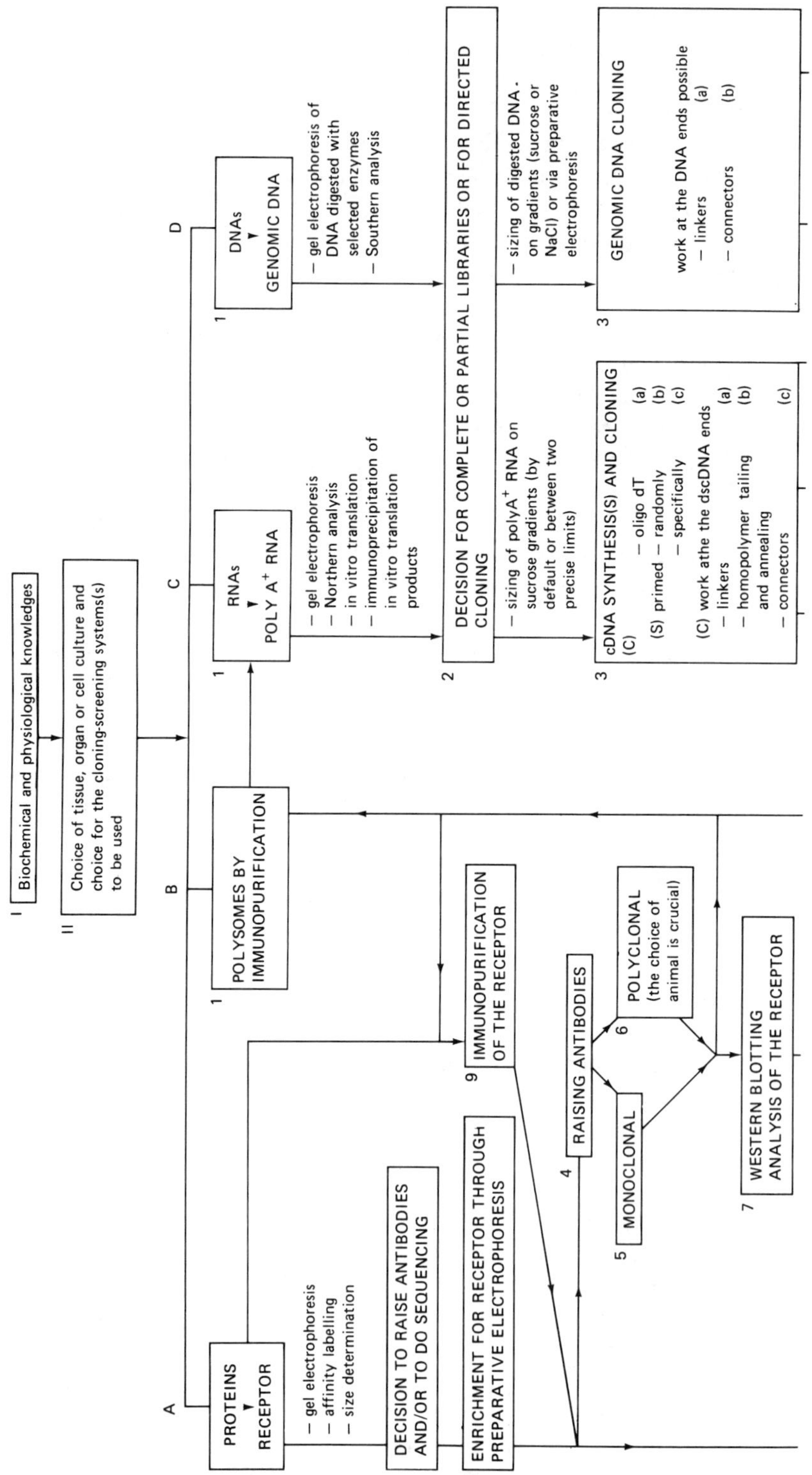
I Biochemical and physiological knowledges
II Choice of tissue, organ or cell culture and choice for the cloning-screening systems(s) to be used
A
B
C
D
1 PROTEINS ▼ RECEPTOR
— gel electrophoresis
— affinity labelling
— size determination
2 DECISION TO RAISE ANTIBODIES AND/OR TO DO SEQUENCING
3 ENRICHMENT FOR RECEPTOR THROUGH PREPARATIVE ELECTROPHORESIS
4 RAISING ANTIBODIES
5 MONOCLONAL
6 POLYCLONAL (the choice of animal is crucial)
7 WESTERN BLOTTING ANALYSIS OF THE RECEPTOR
9 IMMUNOPURIFICATION OF THE RECEPTOR
1 POLYSOMES BY IMMUNOPURIFICATION
1 RNAs ▼ POLY A+ RNA
— gel electrophoresis
— Northern analysis
— in vitro translation
— immunoprecipitation of in vitro translation products
1 DNAs ▼ GENOMIC DNA
— gel electrophoresis of DNA digested with selected enzymes
— Southern analysis
2 DECISION FOR COMPLETE OR PARTIAL LIBRARIES OR FOR DIRECTED CLONING
— sizing of polyA+ RNA on sucrose gradients (by default or between two precise limits)
— sizing of digested DNA. on gradients (sucrose or NaCl) or via preparative electrophoresis
3 cDNA SYNTHESIS(S) AND CLONING (C)
(S) primed — oligo dT (a)
— randomly (b)
— specifically (c)
(C) work athe the dscDNA ends
— linkers (a)
— homopolymer tailing and annealing (b)
— connectors (c)
3 GENOMIC DNA CLONING
work at the DNA ends possible
— linkers (a)
— connectors (b)

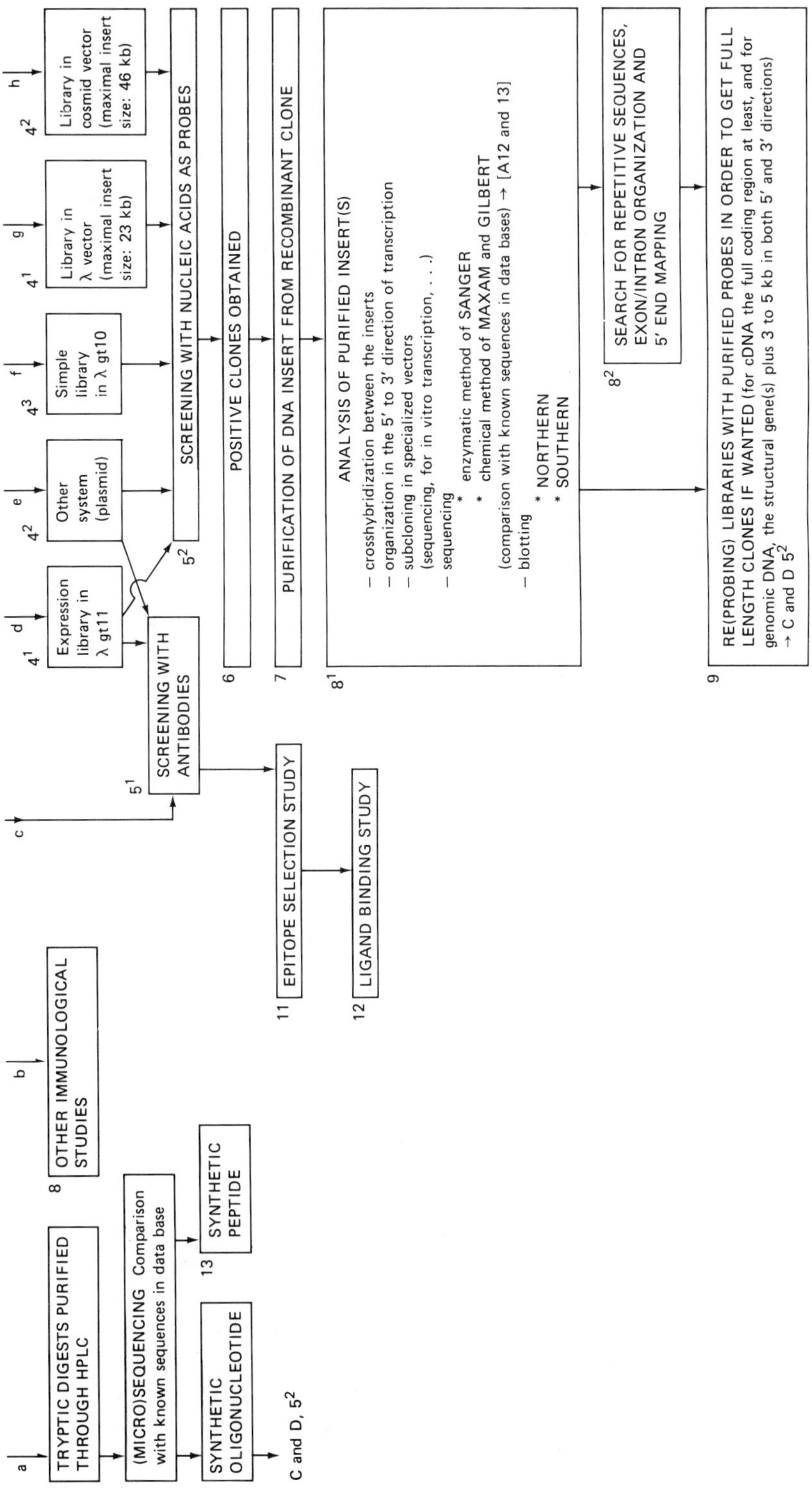

Fig. 1 — A general detailed strategy for steroid and thyroid hormone receptor cDNA and gene cloning.

By the screening of human cDNA libraries with a consensus synthetic oligonucleotide probe corresponding to the DNA binding domain of steroid hormone receptors, a novel human receptor cDNA was cloned which exhibited some sequence homology to the c-erbA–thyroid hormone receptor subfamily. It could be shown, however, that the corresponding protein did not bind thyroids. As both ligand and target genes were unknown, Petkovich *et al.* (1987) took advantage of the functional segmentation of the supergene family. Using the 'finger-swapping' technique (see Chapter 15 for details and references) they established an assay system in which individual or mixtures of candidate ligands could be tested. As confirmed by ligand binding experiments this approach identified the cloned cDNA as corresponding to the human retinoic acid receptor.

B. Gene cloning

Extensive reviews have been published on this subject to which the reader is referred for detailed information (Williams and Blattner, 1980; Dahl *et al.*, 1981; Kaiser and Murray, 1985). What should be kept in mind? Briefly, a good cloning system is able to select for recombinants (examples: the spi phenotype in λ vectors and the inability of cosmid vectors to concatenate and be packaged *in vitro*) often requiring appropriate bacterial hosts. The DNA, which will be digested and size fractionated before cloning, must be long enough for efficient cloning. This requirement is more important for cosmid than for λ vectors because the mean DNA size which can be inserted in a cosmid vector is roughly 2.5 times higher than that for λ. The λ EMBL 3 and 4 replacement vectors (Frischauf *et al.*, 1983) have been used extensively to construct genomic libraries with size-fractionated genomic DNA produced by either MboI or Sau3AI partial digestion. These two enzymes generate compatible cohesive ends with the BamHI cloning sites of the cloning vectors.

Once a genomic library has been successfully established it may be used for years without any problems. For example, for the cloning of the chicken progesterone receptor gene in our laboratory we used a λ Charon 4A genomic library (Dodgson *et al.*, 1979) that had been stored since 1979. Cosmid cloning vectors are more diverse than λ vectors, mainly because any plasmid can be converted to a cosmid vector provided that the 'cos' region of λ has been inserted correctly. Thus there is a wide heterogeneity among cosmid vectors and for details the reader is referred to the review by Dahl *et al.* (1981).

As far as gene cloning strategies are concerned, both λ- and cosmid-based genomic libraries should be constructed in order to generate a reservoir of overlapping clones. Southern blotting analyses, performed with cloned cDNA probes, are used to select genomic DNA fragments for enrichment and directed cloning.

After the cloning of a receptor gene its structural organization and the regulation of transcription will be studied. In the case of the receptor genes it may be interesting to look for exon–protein domain relationships and for the exon–intron pattern in related receptors. These results may be relevant for phylogenetic studies. At present, no complete receptor gene cloning has been reported. However, initial results have been obtained for the chicken progesterone and human estrogen receptor genes. These have revealed some exciting aspects of gene organization. For example, the DNA binding domain C is encoded by two small exons (of approximately 120 bases), separated by an intron of more than 10 kb. Interestingly, each of these exons corresponds to a so-called 'DNA-binding finger' previously characterized in yeast (see also Chapter 15).

In four years the approaches described above, initially concentrated on steroid hormone receptors, have led to the detection of the existence of an entire family of receptor genes. This family comprises known (steroid and thyroid hormone) and unknown (retinoic acid) receptors and further members will be detected in the future. The strategy for this approach is $C1 \rightarrow 4^3 \rightarrow 9$ in Fig. 1, which corresponds to an iterative procedure.

REFERENCES

Arriza, J. L., Weinberger, C., Cerelli, G., Glaser, T. M., Mandelin, B. L., Houseman, D. E. and Evans, R. M. (1987) Cloning of human mineralocorticoid receptor complementary DNA: structural and functional kinship with glucocorticoid receptor. *Science* **237,** 268–275.

Bank, A., Terada, M., Metafora, S., Dow, L. and Marks, P. A. (1972) In vitro synthesis of DNA components of human genes for globins. *Nature New Biology* **235,** 167–169.

Benbrook, D. and Pfahl, M. (1987) A novel thyroid hormone receptor encoded by a cDNA clone from a human testis library. *Science* **238,** 788–791.

Conneely, O. M., Sullivan, W. P., Toft, D. O., Birnbaumer, M., Cook, R. G., Maxwell, B. L., Zarucki-Schulz, T., Greene, G. L., Schrader, W. T. and O'Malley, B. W. (1986) Molecular cloning of the chicken progesterone receptor. *Science* **233,** 767–770.

Dahl, H. H., Flavell, R. A. and Grosveld, F. G. (1981) The use of genomic libraries for the isolation and study of eukaryotic genes. In: Williamson, R. (ed.) *Genetic Engineering*, Vol. 2. Academic Press, London, pp. 49–127.

Danielsen, M., Northrop, J. P. and Ringold, G. (1986) The mouse glucocorticoid receptor: mapping of functional domains by cloning, sequencing and expression of wild-type and mutant receptor protein. *EMBO J.* **5,** 2513–2522.

Dodgson, J. B., Strommer, J. and Engel, J. D. (1979) Isolation of the chicken β-globin gene and a linked embryonic β-like globin gene from a chicken DNA recombinant library. *cell* **17,** 879–887.

Frischauf, A. M., Lehrach, H., Poustka, A. and Murray, N. (1983) Lambda replacement vectors carrying polylinker sequences. *J. Mol. Biol.* **170,** 827–842.

Han, J. H. and Rutter, W. J. (1988) Isolation of intact mRNA and construction of full length cDNA libraries: use of a new vector, λgt22, and primer–adapters for directional cDNA cloning. In: Setlow, J. K. (ed.) *Genetic Engineering*, Vol. 10. Plenum, New York, in press.

Huynh, T. V., Young, R. A. and Davis, R. W. (1985) Constructing and screening cDNA libraries in λgt10 and λgt11. In: Glover, D. M. (ed.) *DNA Cloning*, Vol. 1. IRL Press, Oxford, pp. 49–78.

Jeltsch, J. M., Krozowski, Z., Quirin-Stricker, C., Gronemeyer, H., Simpson, R. J., Garnier, J. M., Krust, A., Jacob, F. and Chambon, P. (1986) Cloning of the chicken progesterone receptor. *Proc. Natl. Acad. Sci. USA* **83,** 5424–5428.

Kaiser, K. and Murray, N. (1985) The use of phage lambda replacement vectors in the construction of representative genomic DNA libraries. In: Glover, D. M. (ed.) *DNA Cloning*, Vol. 1. IRL Press, Oxford, pp. 1–47.

Koike, S., Sakai, M. and Muramatsu, M. (1987) Molecular cloning and characterization of rat estrogen receptor cDNA. *Nucleic Acids Res.* **15,** 2499–2513.

Krust, A., Green, S., Argos, P., Kumar, V., Walter, P., Bornert, J. M. and Chambon, P. (1986) The chicken oestrogen receptor sequence: homology with v-erbA and the human oestrogen and glucocorticoid receptors. *EMBO J.* **5,** 891–897.

Loosfelt, H., Atger, M., Misrahi, M., Guiochon-Mantel, A., Meriel, C., Logeat, F., Benarous, R. and Milgrom, E. (1986) Cloning and sequence analysis of rabbit progesterone-receptor complementary DNA. *Proc. Natl. Acad. Sci. USA* **83,** 9045–9049.

McDonnell, D. P., Mangelsdorf, D. J., Pike, J. W., Haussler, M. R. and O'Malley, B. W. (1987) Molecular cloning of complementary DNA encoding the avian receptor for vitamin D. *Science* **235,** 1214–1217.

Miesfeld, R., Okret, S., Wikstrom, A. C., Wrange, O., Gustafsson, J. A. and Yamamoto, K. R. (1984) Characterization of a steroid hormone receptor gene and mRNA in wild-type and mutant cells. *Nature* **312,** 779–781.

Misrahi, M., Atger, M., D'Auriol, L., Loosfelt, M., Meriel, C., Fridlandsky, F., Guiochon-Mantel, A., Galibert, F. and Milgrom, E. (1987) Complete amino acid sequence of the human progesterone receptor deduced from cloned cDNA. *Biochem. Biophys. Res. Comm.* **143,** 740–748.

Okayama, H. and Berg, P. (1983) A cDNA cloning vector that permits expression of cDNA inserts in mammalian cells. *Mol. Cell. Biol.* **3,** 280–289.

Petkovich, M., Brand, N. J., Krust, A. and Chambon, P. (1987) A human retinoic acid receptor which belongs to the family of nuclear receptor. *Nature* **330,** 444–450.

Ross, J., Aviv, H., Scolnick, E. and Leder, P. (1972) In vitro synthesis of DNA complementary to purified rabbit globin mRNA. *Proc. Natl. Acad. Sci. USA* **69,** 264–268.

Sap, J., Munoz, A., Damm, K., Goldberg, Y., Ghysdael, J., Leutz, A., Beug, H. and Vennström, B. (1986) The c-erbA protein is a high-affinity receptor for thyroid hormone. *Nature* **324,** 635–640.

Sommer, P., Bruyere, T., Ogier, J. A., Garnier, J. M., Jeltsch, J. M. and Klein, J. P. (1987) Cloning of the saliva-interacting protein gene from *Streptococcus mutans*. *J. Bact.* **169,** 5167–5173.

Verma, I. M., Temple, G. F., Fan, H. and Baltimore, D. (1972) In vitro synthesis of DNA complementary to rabbit reticulocyte 10S RNA. *Nature New Biology* **235,** 163–167.

Walter, P., Green, S., Greene, G., Krust, A., Bornert, J. M., Jeltsch, J. M., Staub, A., Jensen, E., Scrace, G., Waterfield, M. and Chambon, P. (1985). Cloning of the human estrogen receptor cDNA. *Proc. Natl. Acad. Sci. USA* **82,** 7889–7893.

Weiler, I. J., Lew, D. and Shapiro, D. J. (1987) The *Xenopus laevis* estrogen receptor: sequence homology with human and avian receptors and identification of multiple estrogen receptor messenger ribonucleic acids. *Molecular Endocrinology* **1,** 355–362.

Weinberger, C., Hollenberg, S. M., Ong, E. S., Harmon, J. M., Brower, S. T.,

Cidlowski, J., Thompson, E. B., Rosenfeld, M. and Evans, R. M. (1985) Identification of human glucocorticoid receptor complementary DNA clones by epitope selection. *Science* **228,** 740–742.

Weinberger, C., Thompson, C. C., Ong, E. S., Lebo, R., Grudl, D. J. and Evans, R. M. (1986) Tne c-erbA gene encodes a thyroid hormone receptor. *Nature* **324,** 641–646.

Williams, B. G. and Blattner, F. R. (1980) Bacteriophage lambda vectors for DNA cloning. In: Setlow, J. K. and Hollaender, A. (eds.) *Genetic Engineering*, Vol. 2. Plenum, New York, pp. 201–281.

Young, R. A. and Davis, R. W. (1983) Yeast RNA polymerase II genes: isolation with antibody probes. *Science* **222,** 778–782.

Young, R. A., Mehra, V., Sweester, D., Buchanan, T., Clark-Curtiss, J., Davis, R. W. and Bloom, B. R. (1985) Genes for the major protein antigens of the leprosy parasite mycobacterium lepae. *Nature* **316,** 450–452.

15

The nuclear receptor family: cloning, structure and function

H. Gronemeyer, S. Green, J. M. Jeltsch, V. Kumar, A. Krust and P. Chambon
Institut de Chimie Biologique, Faculté de Médecine, Strasbourg, France

I. INTRODUCTION

The study of biological mechanisms which regulate patterns of gene expression both during the development of eukaryotic organisms and in terminally differentiated cells is one of the major subjects of molecular biology. At the level of transcription this regulation appears to be achieved through positive and negative effects mediated by trans-acting proteins interacting with cis-acting DNA promoter elements (Ptashne, 1986). Steroid hormones trigger complex development and physiological processes. The structural information provided by a given steroid hormone is converted into regulatory transcriptional signals by interaction with specific protein receptors. On binding of the hormone, the receptor gains the ability to alter specifically gene transcription — a process which appears to result from the binding of the hormone–receptor complex to 'enhancer' elements of target gene promoters (Yamamoto, 1985). Thus understanding (1) how the receptor protein is able to specifically recognize its cognate hormone, (2) how this binding results in the interaction of the receptor with target gene promoter elements, (3) how this interaction leads to initiation of specific gene transcription, and (4) what the role of the hormone is in these processes, are prerequisites to understanding the molecular mechanism of steroid-hormone-controlled gene activity. Such an analysis is now possible as a result of the cloning of cDNAs corresponding to the estrogen (Walter *et al.*, 1985; Krust *et al.*, 1986; Koike *et al.*, 1987; Weiler *et al.*, 1987; White *et al.*, 1987), glucocorticoid (Miesfeld *et al.*, 1984; Weinberger *et al.*, 1985a,b; Govindan *et al.*, 1985; Danielson *et al.*, 1986), progesterone (Jeltsch *et al.*, 1986; Conneely *et al.*, 1986; Loosfelt *et al.*, 1986; Misrahi *et al.*, 1987), and vitamin D (McDonnell *et al.*, 1987) receptors. We review here the results obtained recently, with a particular emphasis on the work performed in our laboratory.

II. cDNA CLONING

A. Human estrogen receptor

Two independent approaches were used to isolate cDNAs corresponding to the estrogen receptor (Walter *et al.*, 1985) from the human breast cancer cell line MCF-7. PolyA^+ RNA was fractionated on sucrose gradients and assayed for the presence of receptor mRNA by *in vitro* translation and immunoprecipitation with monoclonal antibodies directed against the human estrogen receptor. Fractions enriched in receptor mRNA were pooled and used for randomly primed cDNA synthesis and cloning into λgt vectors. The λgt11–cDNA expression library was then screened with a panel of monoclonal anti-estrogen receptor antibodies. Out of more than 1 000 000 phages screened, five individual positive clones were identified. In parallel, synthetic oligonucleotide probes were constructed on the basis of sequence information obtained from cyanogen bromide cleaved peptides of homogeneous receptor preparations. Two strategies were used for the design of oligonucleotide probes. They were synthesized either as a mixture of 24 14-mers covering all possible codon choices of the least ambiguous region or as two single 41-mer oligonucleotides, choosing the most common codon for each amino acid after taking into consideration the frequency of human codon usage. A screening of 300 000 λgt10 cDNA clones resulted in the identification of five positive clones. The clone with the longest cDNA insert (2.1 kb), designated λOR8, crossreacted with the insert of all other clones, irrespective of the strategy of isolation.

B. Chicken progesterone receptor

Two routes were chosen to clone the cDNA corresponding to the chicken progesterone receptor. On the basis of the longest amino acid sequence determined for the tryptic peptide T7 (Simpson *et al.*, 1987; see Chapter 10) derived from digestion of the form B receptor, a synthetic oligonucleotide probe shown in Fig. 1 was

A SYNTHETIC OLIGONUCLEOTIDE DESIGNED

TO DETECT THE cPR -cDNA

OLIGO 84-mer	GAC	GCC	GGC	CCC	GGC	GAG	GGC	GGC	CTG	GCC
CPR-cDNA	...	..G	..G	...	..A	..A	...	..A	...	..G
AA-SEQUENCE	D	A	G	P	G	E	G	G	L	A

CCC	GCC	GCC	GCC	GCC	TCC	CCC	GCC	GCC	GTG
...	...	..G	..T	..T	...	..G	..G	...	...
P	A	A	A	A	S	P	A	A	V

GAG	CCC	GGC	GCC	GGC	CAG	GAC	TAC
...	..G	...	..G	..G	...	...	...
E	P	G	A	G	Q	D	Y

Fig. 1 — A synthetic 84-mer oligonucleotide ('oligo 84-mer') designed on the basis of amino acid sequence information ('AA-sequence') obtained from tryptic digests of homogeneous chicken progesterone receptor form B (Simpson *et al.*, 1987) is compared with the correct cDNA sequence ('cPR-cDNA') of the corresponding region. Dots indicate identity. The overall homology of the designed probe is 83.3%.

synthesized. This sequence, complementary to the mRNA, was designed such that the codon frequency of all known chicken genes was considered and, in ambiguous cases, G was preferred to A so as to allow for G–T mismatches. In this case, a long probe was preferred to a mixture of short oligonucleotides. Further details for the design of oligonucleotide probes deduced from amino acid sequence data are given in a review by Lathe (1985) on this subject. The progesterone receptor probe ('84-mer') was then used to check Northern blots of polyA^+ RNA derived from chicken oviducts in different hormonal milieu (DES stimulated, withdrawn, secondary stimulated, laying hen) as well as Southern blots. The results obtained were depressing, since the probe hybridized only to ribosomal RNA, even under stringent conditions. Southern blots showed a smear of hybridization signals. We know now that the sequence chosen was 84% correct (see Fig. 1) but because of the high GC content (86.5%) only ribosomal RNA or the corresponding gene family was detected on Northern or Southern blots. In fact, even when using the correct sequence, we had severe problems with ribosomal RNA crosshybridization on Northern blots.

In parallel, a λgt11 cDNA expression library was prepared from size selected (>4 kb) laying hen oviduct poly(A^+) RNA and screened with monospecific anti-progesterone receptor form A or form B antibodies. For this the antisera were obtained as described in Chapters 3 and 10 and passed over affinity columns to which homogeneous form A or form B receptor had been covalently coupled. Additionally, the eluted monospecific antibodies were cleared of anti-*Escherichia coli* protein antibodies either by affinity chromatography using *E. coli* lysate or by absorption to *E. coli* proteins bound to nitrocellulose filters. Screening of 10^6 colonies with the antibodies gave two clones that were positive after amplification and rescreening. The β-galactosidase fusion proteins of the two clones, after induction with isopropyl-β-D-thiogalactopyranoside (IPTG), had molecular weights of 125 and 130 kDa (Fig. 2A, lanes 1 and 3). Inserts of 250 and 390 bp were found in the corresponding cDNA clones. Immunoblots of the IPTG-induced bacterial lysates confirmed that the fusion proteins crossreacted with the monospecific antiserum (Fig. 2B, middle). Interestingly, both the anti-form A and anti-form B antibodies recognized both fusion proteins. As a control, β-galactosidase did not react with any of the antibodies (Fig. 2B, right).

To demonstrate that the epitopes expressed by the two clones, λcPR1 and λcPR2, correspond to epitopes effectively present on progesterone receptor forms A and B, we selected the corresponding antibodies from anti-progesterone receptor form B antiserum by affinity chromatography on resins containing either one of the two covalently bound purified fusion proteins. These antibodies were used as immunoprobes on immunoblots of crude and partially purified progesterone receptor. Fig. 2C shows the fluorograph of partially purified preparations of progesterone receptor form A (79 kDa) (lane 1) and a mixture of progesterone receptor forms A and B (109 kDa) (lane 2) crosslinked with tritiated progestin R5020. Fig. 2D demonstrates the crossreactivity of anti-progesterone receptor form B antiserum with progesterone receptor form A. A crude preparation of a mixture of progesterone receptor forms A and B (Fig. 2D, lane 1), a partially purified progesterone receptor form A (lane 2), and a partially purified mixture of progesterone receptor forms A and B (lane 3) were separated on 7.5% SDS–polyacrylamide gels and analyzed by immunoblotting. The major bands that were visualized with a crude

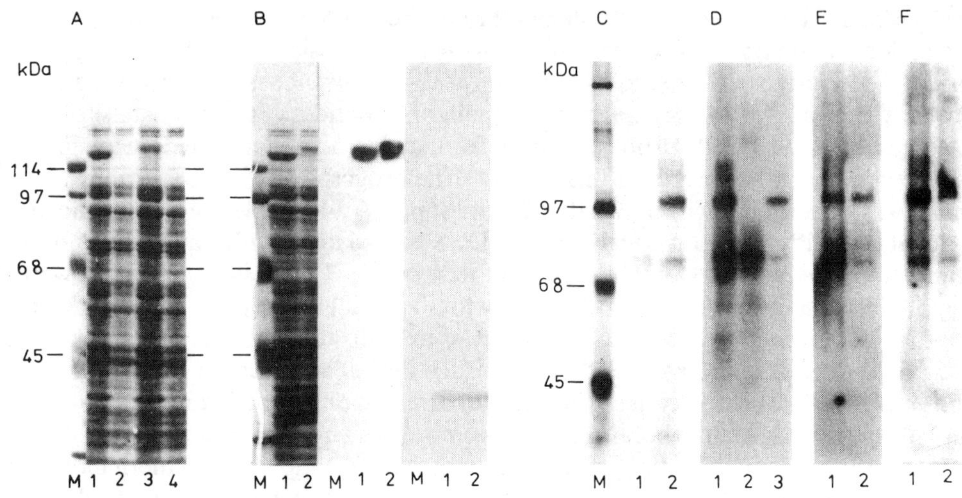

Fig. 2 — Two cDNA clones contain epitopes present in both progesterone receptor forms A and B. (A) Coomassie blue-stained SDS lysate of non-induced (lanes 2 and 4) and induced (1 mM isopropyl-β-D-thiogalactopyranoside (IPTG), lanes 1 and 3), recombinant phage lysogens in *E. coli* Y1089. (B) Three immunoblots of crude lysates from IPTG lysogens were either stained with Coomassie blue (left) or probed with monospecific anti-progesterone receptor form B (middle) or preimmune serum (right). All sera were stripped of *E. coli* and β-galactosidase antibodies by filtration through corresponding affinity resins. Each immunoblot contained purified *E. coli* β-galactosidase in the marker lane (M) to control for crossreaction with β-galactosidase. (C) Fluorograph of SDS–7.5% polyacrylamide gel of photoaffinity labeled partially purified progesterone receptor form A (lane 1) and partially purified mixture of progesterone receptor forms A and B (lane 2). (D) Immunoblot of a crude progesterone receptor preparation (lane 1) and of the same partially purified progesterone receptor form A (lane 2) and mixture of progesterone receptor forms A and B (lane 3) as in (C), probed with crude anti-progesterone receptor form B antiserum. (E), (F) Immunoblots of the same progesterone receptor preparations as shown in C and D, probed with fusion protein-selected antibodies derived from crude anti-progesterone receptor form B antiserum. Lanes 1, crude mixture of progesterone receptor forms A and B; lanes 2, partially purified preparation of forms A and B. The immunocomplexes in (B)–(F) were visualized by autoradiography after incubating the blots with ^{125}I-labeled *Staphylococcus aureus* protein A, and (D)–(F) SDS–7.5% polyacrylamide gels. The position of each marker (lanes M) is indicated.

anti-progesterone receptor form B antiserum have clearly the same apparent molecular masses as those revealed by fluorography in Fig. 2C. The same preparations of crude and partially purified mixtures of progesterone receptor forms A and B as those used in Fig. 2D were then analyzed with the two populations of fusion protein-selected antibodies. Fig. 2E and 2F correspond to immunoblots with λcPR2 and λcPR1 fusion protein-selected antibodies respectively. Again, two bands were revealed with the same apparent molecular masses as the progesterone receptor forms A and B visualized in Fig. 2C. Thus, we concluded that the λcPR1 and λcPR2 fusion proteins contain epitopes that are present in both progesterone receptor form A and form B.

To demonstrate that the cDNA inserts present in λcPR1 and λcPR2 were derived from the same mRNA, both inserts were used to screen a randomly primed λgt10 library prepared from fractionated laying-hen oviduct poly(A)$^+$ RNA. A 1.1 kb cDNA clone, λcPR3, containing the sequences found in the inserts of both λcPR1 and λcPR2, was isolated. Sequencing of the λcPR3 cDNA revealed that in three regions the deduced amino acid sequences are identical to the sequences of three peptides which were isolated from tryptic digests of apparently homogeneous progesterone receptor form B (Simpson *et al.*, 1987; see also Chapter 10). By isolating various overlapping cDNAs as well as the corresponding genomic sequences, the chicken progesterone receptor mRNA was determined as being 4472 nucleotides long.

It is important to mention that during the entire cloning work we did not find any evidence for the existence of two different genes, encoding forms A and B. Furthermore, we have not found any indication that form A could be generated from a different RNA by an alternative termination of transcription or alternative splicing.

The cloning of a so-called 'chicken progesterone receptor B antigen' was reported (Zarucki-Schulz *et al.*, 1984). However, it is now known that this cDNA corresponds to a heat-shock protein of a similar molecular weight to the progesterone receptor form B (Kulomaa *et al.*, 1986; Sargan *et al.*, 1986; Kleinsek *et al.*, 1988). The same group reported subsequently the cloning of a chicken progesterone receptor cDNA which is apparently identical to the one obtained in this laboratory (Conneely *et al.*, 1986).

C. Human glucocorticoid receptor

The strategy chosen in our laboratory for the cloning of the cDNA corresponding to the human glucocorticoid receptor was based on the crossreactivity of antibodies which had been raised against the rat glucocorticoid receptor with the human glucocorticoid receptor from MCF-7 breast cancer cells (Govindan *et al.*, 1985). Initially, the MCF-7 receptor was characterized by triamcinolone acetonide mediated photoaffinity labeling and immunoblotting. Monospecific antisera were prepared and used to screen an MCF-7 λgt11 cDNA expression library. Several clones were isolated and, using epitope selection techniques similar to those described above for the progesterone receptor, the identity of the clones with the human glucocorticoid receptor gene derived cDNA was demonstrated. Using polysome immunoadsorption techniques (Miesfeld *et al.*, 1984) or approaches basically similar to those described above (Weinberger *et al.*, 1985a), two other groups succeeded in cloning the rat and the human glucocorticoid receptor. Further details may be found in the original literature as well as in the previous chapter by J. M. Jeltsch on cloning strategies.

III. STRUCTURE

A. Human estrogen receptor mRNA

The sequencing of several overlapping MCF-7 estrogen receptor cDNA clones showed that the corresponding mRNA is 6322 nucleotides in length (Green *et al.*, 1986; Fig. 3). The major open reading frame of 1785 nucleotides encodes a 66 182 Da protein. The sequence of the 5′ end of the estrogen receptor mRNA was determined from a genomic clone isolated from a human library, and the transcription start site

```
-128 GAGTTGTGCCTGGAGTGATGTTTAAGCCAATGTCAGGGCAAGGCAACAGTCCCTGGCCGTCCTCCAGCACCTTTGTAATGCATATGAGCTCGGGAGACCAGTACTTAAAGTTGGAGGCCC
         ↓+1
  -8 GGGAGCCCAGGAGCTGGCGGAGGGCGTTCGTCCTGGGAGCTGCACTTGCTCCGTCGGGTCGCCGGCTTCACCGGACCGCAGGCTCCCGGGGCAGGGCCGGGGCCAGAGCTCGCGTGTCGG
                         ORF
 113 CGGGACATGCGCTGCGTCGCCTCTAACCTCGGGCTGTGCTCTTTTTCCAGGTGGCCCGCCGGTTTCTGAGCCTTCTGCCCTGCGGGGACACGGTCTGCACCCTGCCCGCGGCCACGGACC

 233 ATG ACC ATG ACC CTC CAC ACC AAA GCA TCT GGG ATG GCC CTA CTG CAT CAG ATC CAA GGG AAC GAG CTG GAG CCC CTG AAC CGT CCG CAG
     Met Thr Met Thr Leu His Thr Lys Ala Ser Gly Met Ala Leu Leu His Gln Ile Gln Gly Asn Glu Leu Glu Pro Leu Asn Arg Pro Gln  30

 323 CTC AAG ATC CCC CTG GAG CGG CCC CTG GGC GAG GTG TAC CTG GAC AGC AGC AAG CCC GCC GTG TAC AAC TAC CCC GAG GGC GCC GCC TAC
     Leu Lys Ile Pro Leu Glu Arg Pro Leu Gly Glu Val Tyr Leu Asp Ser Ser Lys Pro Ala Val Tyr Asn Tyr Pro Glu Gly Ala Ala Tyr  60

 413 GAG TTC AAC GCC GCG GCC GCC GCC AAC GCG CAG GTC TAC GGT CAG ACC GGC CTC CCC TAC GGC CCC GGG TCT GAG GCT GCG GCG TTC GGC
     Glu Phe Asn Ala Ala Ala Ala Ala Asn Ala Gln Val Tyr Gly Gln Thr Gly Leu Pro Tyr Gly Pro Gly Ser Glu Ala Ala Ala Phe Gly  90

 503 TCC AAC GGC CTG GGG GGT TTC CCC CCA CTC AAC AGC GTG TCT CCG AGC CCG CTG ATG CTA CTG CAC CCG CCG CCG CAG CTG TCG CCT TTC
     Ser Asn Gly Leu Gly Gly Phe Pro Pro Leu Asn Ser Val Ser Pro Ser Pro Leu Met Leu Leu His Pro Pro Pro Gln Leu Ser Pro Phe 120

 593 CTG CAG CCC CAC GGC CAG CAG GTG CCC TAC TAC CTG GAG AAC GAG CCC AGC GGC TAC ACG GTG CGC GAG GCC GGC CCG CCG GCA TTC TAC
     Leu Gln Pro His Gly Gln Gln Val Pro Tyr Tyr Leu Glu Asn Glu Pro Ser Gly Tyr Thr Val Arg Glu Ala Gly Pro Pro Ala Phe Tyr 150

 683 AGG CCA AAT TCA GAT AAT CGA CGC CAG GGT GGC AGA GAA AGA TTG GCC AGT ACC AAT GAC AAG GGA AGT ATG GCT ATG GAA TCT GCC AAG
     Arg Pro Asn Ser Asp Asn Arg Arg Gln Gly Gly Arg Glu Arg Leu Ala Ser Thr Asn Asp Lys Gly Ser Met Ala Met Glu Ser Ala Lys 180

 773 GAG ACT CGC TAC TGT GCA GTG TGC AAT GAC TAT GCT TCA GGC TAC CAT TAT GGA GTC TGG TCC TGT GAG GGC TGC AAG GCC TTC TTC AAG
     Glu Thr Arg Tyr Cys Ala Val Cys Asn Asp Tyr Ala Ser Gly Tyr His Tyr Gly Val Trp Ser Cys Glu Gly Cys Lys Ala Phe Phe Lys 210

 863 AGA AGT ATT CAA GGA CAT AAC GAC TAT ATG TGT CCA GCC ACC AAC CAG TGC ACC ATT GAT AAA AAC AGG AGG AAG AGC TGC CAG GCC TGC
     Arg Ser Ile Gln Gly His Asn Asp Tyr Met Cys Pro Ala Thr Asn Gln Cys Thr Ile Asp Lys Asn Arg Arg Lys Ser Cys Gln Ala Cys 240

 953 CGG CTC CGC AAA TGC TAC GAA GTG GGA ATG ATG AAA GGT GGG ATA CGA AAA GAC CGA AGA GGA GGG AGA ATG TTG AAA CAC AAG CGC CAG
     Arg Leu Arg Lys Cys Tyr Glu Val Gly Met Met Lys Gly Gly Ile Arg Lys Asp Arg Arg Gly Gly Arg Met Leu Lys His Lys Arg Gln 270

1043 AGA GAT GAT GGG GAG GGC AGG GGT GAA GTG GGG TCT GCT GGA GAC ATG AGA GCT GCC AAC CTT TGG CCA AGC CCG CTC ATG ATC AAA CGC
     Arg Asp Asp Gly Glu Gly Arg Gly Glu Val Gly Ser Ala Gly Asp Met Arg Ala Ala Asn Leu Trp Pro Ser Pro Leu Met Ile Lys Arg 300

1133 TCT AAG AAG AAC AGC CTG GCC TTG TCC CTG ACG GCC GAC CAG ATG GTC AGT GCC TTG TTG GAT GCT GAG CCC CCC ATA CTC TAT TCC GAG
     Ser Lys Lys Asn Ser Leu Ala Leu Ser Leu Thr Ala Asp Gln Met Val Ser Ala Leu Leu Asp Ala Glu Pro Pro Ile Leu Tyr Ser Glu 330

1223 TAT GAT CCT ACC AGA CCC TTC AGT GAA GCT TCG ATG ATG GGC TTA CTG ACC AAC CTG GCA GAC AGG GAG CTG GTT CAC ATG ATC AAC TGG
     Tyr Asp Pro Thr Arg Pro Phe Ser Glu Ala Ser Met Met Gly Leu Leu Thr Asn Leu Ala Asp Arg Glu Leu Val His Met Ile Asn Trp 360

1313 GCG AAG AGG GTG CCA GGC TTT GTG GAT TTG ACC CTC CAT GAT CAG GTC CAC CTT CTA GAA TGT GCC TGG CTA GAG ATC CTG ATG ATT GGT
     Ala Lys Arg Val Pro Gly Phe Val Asp Leu Thr Leu His Asp Gln Val His Leu Leu Glu Cys Ala Trp Leu Glu Ile Leu Met Ile Gly 390

1403 CTC GTC TGG CGC TCC ATG GAG CAC CCA GTG AAG CTA CTG TTT GCT CCT AAC TTG CTC TTG GAC AGG AAC CAG GGA AAA TGT GTA GAG GGC
     Leu Val Trp Arg Ser Met Glu His Pro Val Lys Leu Leu Phe Ala Pro Asn Leu Leu Leu Asp Arg Asn Gln Gly Lys Cys Val Glu Gly 420

1493 ATG GTG GAG ATC TTC GAC ATG CTG CTG GCT ACA TCA TCT CGG TTC CGC ATG ATG AAT CTG CAG GGA GAG GAG TTT GTG TGC CTC AAA TCT
     Met Val Glu Ile Phe Asp Met Leu Leu Ala Thr Ser Ser Arg Phe Arg Met Met Asn Leu Gln Gly Glu Glu Phe Val Cys Leu Lys Ser 450

1583 ATT ATT TTG CTT AAT TCT GGA GTG TAC ACA TTT CTG TCC AGC ACC CTG AAG TCT CTG GAA GAG AAG GAC CAT ATC CAC CGA GTC CTG GAC
     Ile Ile Leu Leu Asn Ser Gly Val Tyr Thr Phe Leu Ser Ser Thr Leu Lys Ser Leu Glu Glu Lys Asp His Ile His Arg Val Leu Asp 480

1673 AAG ATC ACA GAC ACT TTG ATC CAC CTG ATG GCC AAG GCA GGC CTG ACC CTG CAG CAG CAG CAC CAG CGG CTG GCC CAG CTC CTC CTC ATC
     Lys Ile Thr Asp Thr Leu Ile His Leu Met Ala Lys Ala Gly Leu Thr Leu Gln Gln Gln His Gln Arg Leu Ala Gln Leu Leu Leu Ile 510

1763 CTC TCC CAC ATC AGG CAC ATG AGT AAC AAA GGC ATG GAG CAT CTG TAC AGC ATG AAG TGC AAG AAC GTG GTG CCC CTC TAT GAC CTG CTG
     Leu Ser His Ile Arg His Met Ser Asn Lys Gly Met Glu His Leu Tyr Ser Met Lys Cys Lys Asn Val Val Pro Leu Tyr Asp Leu Leu 540

1853 CTG GAG ATG CTG GAC GCC CAC CGC CTA CAT GCG CCC ACT AGC CGT GGA GGG GCA TCC GTG GAG GAG ACG GAC CAA AGC CAC TTG GCC ACT
     Leu Glu Met Leu Asp Ala His Arg Leu His Ala Pro Thr Ser Arg Gly Gly Ala Ser Val Glu Glu Thr Asp Gln Ser His Leu Ala Thr 570

1943 GCG GGC TCT ACT TCA TCG CAT TCC TTG CAA AAG TAT TAC ATC ACG GGG GAG GCA GAG GGT TTC CCT GCC ACA GTC TGA GAG CTC CCT GGC
     Ala Gly Ser Thr Ser Ser His Ser Leu Gln Lys Tyr Tyr Ile Thr Gly Glu Ala Glu Gly Phe Pro Ala Thr Val ***                 595

2033 TCCCACACGGTTCAGATAATCCCTGCTGCATTTTACCCTCATCATGCACCACTTTAGCCAAATTCTGTCTCCTGCATACACTCCGGCATGCATCCAACACCAATGGCTTTCTAGATGAGT
2153 GGCCATTCATTTGCTTGCTCAGTTCTTAGTGGCACATCTTCTGTCTTCTGTTGGGAACAGCCAAAGGGATTCCAAGGCTAAATCTTTGTAACAGCTCTCTTTCCCCCTTGCTATGTTACT
2273 AAGCGTGAGGATTCCCGTAGCTCTTCACAGCTGAACTCAGTCTATGGGTTGGGGCTCAGATAACTCTGTGCATTTAAGCTACTTGTAGAGACCCAGGCCTGGAGAGTAGACATTTTGCCT
2393 CTGATAAGCACTTTTTAAATGGCTCTAAGAATAAGCCACAGCAAAGAATTTAAAGTGGCTCCTTTAATTGGTGACTTGGAGAAAGCTAGGTCAAGGGTTTATTATAGCACCCTCTTGTAT
2513 TCCTATGGCAATGCATCCTTTTATGAAAGTGGTACACCTTAAAGCTTTTATATGACTGTAGCAGAGTATCTGGTGATTGTCAATTCACTTCCCCCTATAGGAATACAAGGGGCCACACAG
2633 GGAAGGCAGATCCCCTAGTTGGCCAAGACTTATTTTAACTTGATACACTGCAGATTCAGAGTGTCCTGAAGCTCTGCCTCTGGCTTTCCGGTCATGGGTTCCAGTTAATTCATGCCTCCC
2753 ATGGACCTATGGAGAGCAACAAGTTGATCTTAGTTAAGTCTCCCTATATGAGGGATAAGTTCCTGATTTTTGTTTTTATTTTTGTGTTACAAAAGAAAGCCCTCCCTCCCTGAACTTGCA
2873 GTAAGGTCAGCTTCAGGACCTGTTCCAGTGGGCACTGTACTTGGATCTTCCCGGCGTGTGTGTGCCTTACACAGGGGTGAACTGTTCACTGTGGTGATGCATGATGAGGGTAAATGGTAG
2993 TTGAAAGGAGCAGGGGCCCTGGTGTTGCATTTAGCCCTGGGGCATGGAGCTGAACAGTACTTGTGCAGGATTGTTGTGGCTACTAGAGAACAAGAGGGAAAGTAGGGCAGAAACTGGATA
3113 CAGTTCTGAGCACAGCCAGACTTGCTCAGGTGGCCCTGCACAGGCTGCAGCTACCTAGGAACATTCCTTGCAGACCCCGCATTGCCTTTGGGGGTGCCCTGGGATCCCTGGGGTAGTCCA
3233 GCTCTTATTCATTTCCCAGCGTGGCCCTGGTTGGAAGAAGCAGCTGTCAAGTTGTAGACAGCTGTGTTCCTACAATTGGCCCAGCACCCTGGGGCACGGGAGAAGGGTGGGGACCGTTGC
3353 TGTCACTACTCAGGCTGACTGGGGCCTGGTCAGATTACGTATGCCCTTGGTGGTTTAGAGATAATCCAAAATCAGGGTTTGGTTTGGGGAAGAAAATCCTCCCCCTTCCTCCCCCGCCCC
3473 GTTCCCTACCGCCTCCACTCCTGCCAGCTCATTTCCTTCAATTTCCTTTGACCTATAGGCTAAAAAAGAAAGGCTCATTCCAGCCACAGGGCAGCCTTCCCTGGGCCTTTGCTTCTCTAG
3593 CACAATTATGGGTTACTTCCTTTTTCTTAACAAAAAAGAATGTTTGATTTCCTCTGGGTGACCTTATTGTCTGTAATTGAAACCCTATTGAGAGGTGATGTCTGTGTTAGCCAATGACCC
3713 AGGTAGCTGCTCGGGCTTCTCTTGGTATGTCTTGTTTGGAAAAGTGGATTTCATTCATTTCTGATTGTCCAGTTAAGTGATCACCAAAGGACTGAGAATCTGGGAGGGCAAAAAAAAAA
3833 AAAAAGTTTTTATGTGCACTTAAATTTGGGGACAATTTTATGTATCTGTGTTAAGGATATGCTTAAGAACATAATTCTTTTGTTGCTGTTTGTTTAAGAAGCACCTTAGTTTGTTTAAGA
3953 AGCACCTTATATAGTATAATATATATTTTTTTGAAATTACATTGCTTGTTTATCAGACAATTGAATGTAGTAATTCTGTTCTGGATTTAATTTGACTGGGTTAACATGCAAAAACCAAGG
4073 AAAAATATTTAGTTTTTTTTTTTTTTTTTTGTATACTTTTCAAGCTACCTTGTCATGTATACAGTCATTTATGCCTAAAGCCTGGTGATTATTCATTTAAATGAAGATCACATTTCATATC
4193 AACTTTTGTATCCACAGTAGACAAAATAGCACTAATCCAGATGCCTATTGTTGGATATTGAATGACAGACAATCTTATGTAGCAAAGATTATGCCTGAAAAGGAAAATTATTCAGGGCAG
4313 CTAATTTTGCTTTTACCAAAATATCAGTAGTAATATTTTTGGACAGTAGCTAATGGGTCAGTGGGTTCTTTTTAATGTTTATACTTAGATTTTCTTTTAAAAAAATTAAAATAAAACAAA
4433 AAAAATTTCTAGGACTAGACGATGTAATACCAGCTAAAGCCAAACAATTATACAGTGGAAGGTTTTACATTATTCATCCAATGTGTTTCTATTCATGTTAAGATACTACTACATTTGAAG
4553 TGGGCAGAGAACATCAGATGATTGAAATGTTCGCCCAGGGGTCTCCAGCAACTTTGGAAATCTCTTTGTATTTTTACTTGAAGTGCCACTAATGGACAGCAGATATTTTCTGGCTGATGT
4673 TGGTATTGGGTGTAGGAACATGATTTAAAAAAAAAACTCTTGCCTCTGCTTTCCCCCACTCTGAGGCAAGTTAAAATGTAAAAGATGTGATTTATCTGGGGGGCTCAGGTATGGTGGGGA
4793 AGTGGATTCAGGAATCTGGGGAATGGCAAATATATTAAGAAGAGTATTGAAAGTATTTGGAGGAAAATGGTTAATTCTGGGTGTGCACCAAGGTTCAGTAGAGTCCACTTCTGCCCTGGA
4913 GACCACAAATCAACTAGCTCCATTTACAGCCATTTCTAAAATGGCAGCTTCAGTTCTAGAGAAGAAAGAACAACATCAGCAGTAAAGTCCATGGAATAGCTAGTGGTCTGTGTTTCTTTT
5033 CGCCATTGCCTAGCTTGCCGTAATGATTCTATAATGCCATCATGCAGCAATTATGAGAGGCTAGGTCATCCAAAGAGAAGACCCTATCAATGTAGGTTGCAAAATCTAACCCCTAAGGAA
5153 GTGCAGTCTTTGATTTGATTTCCCTAGTAACCTTGCAGATATGTTTAACCAAGCCATAGCCCATGCCTTTTGAGGGCTGAACAAATAAGGGACTTACTGATAATTTACTTTTGATCACAT
5273 TAAGGTGTTCTCACCTTGAAATCTTATACACTGAAATGGCCATTGATTTAGGCCACTGGCTTAGAGTACTCCTTCCCCTGCATGACACTGATTACAAATACTTTCCTATTCATACTTTCC
5393 AATTATGAGATGGACTGTGGGTACTGGGAGTGATCACTAACACCATAGTAATGTCTAATATTCACAGGCAGATCTGCTTGGGGAAGCTAGTTATGTGAAAGGCAAATAAAGTCATACAGT
5513 AGCTCAAAAGGCAACCATAATTCTCTTTGGTGCAAGTCTTGGGAGCGTGATCTAGATTACACTGCACCATTCCCAAGTTAATCCCCTGAAAACTTACTCTCAACTGGAGCAAATGAACTT
5633 TGGTCCCAAATATCCATCTTTTCAGTAGCGTTAATTATGCTCTGTTTCCAACTGCATTTCCTTTCCAATTGAATTAAAGTGTGGCCTCGTTTTTAGTCATTTAAAATTGTTTTCTAAGTA
5753 ATTGCTGCCTCTATTATGGCACTTCAATTTTGCACTGTCTTTTGAGATTCAAGAAAAATTTCTATTCATTTTTTTGCATCCAATTGTGCCTGAACTTTTAAAATATGTAAATGCTGCCAT
5873 GTTCCAAACCCATCGTCAGTGTGTGTGTTTAGAGCTGTGCACCCTAGAAACAACATACTTGTCCCATGAGCAGGTGCCTGAGACACAGACCCCTTTGCATTCACAGAGAGGTCATTGGTT
5993 ATAGAGACTTGAATTAATAAGTGACATTATGCCAGTTTCTGTTCTCTCACAGGTGATAAACAATGCTTTTTGTGCACTACATACTCTTCAGTGTAGAGCTCTTGTTTTATGGGAAAAGGC
6113 TCAAATGCCAAATTGTGTTTGATGGATTAATATGCCCTTTTGCCGATGCATACTATTACTGATGTGACTCGGTTTTGTCGCAGCTTTGCTTTGTTTAATGAAACACACTTGTAAACCTCT
6233 TTTGCACTTTGAAAAAGAATCCAGCGGGATGCTCGAGCACCTGTAAACAATTTTCTCAACCTATTTGATGTTCAAATAAAGAATTAAACTAAAAAAAAAAAAAAA
```

was determined both by S1 nuclease mapping and primer extension (Green *et al.*, 1986). The region immediately upstream of the start site contains a TATA-like box at position −27 and a CAAT-like box at position −103.

The 5′-untranslated region of 232 nucleotides is unusually long, inviting the speculation that it may participate in a post-transcriptional control (Kozak, 1984a) of receptor expression. Most notably, this region contains a short, 20-codon open reading frame (Green *et al.*, 1986; boxed in Fig. 3) while that of the chicken estrogen receptor contains two short open reading frames (Krust *et al.*, 1986). Similarly, the rat (Koike *et al.*, 1987) and *Xenopus* (Weiler *et al.*, 1987) estrogen receptor cDNAs contain similar short 5′ open reading frames of 18 and 7 codons respectively (see Fig. 6A). It is unknown whether these upstream open reading frames are translated *in vivo*, but it is unlikely that the predicted peptides could play any conserved function as there is a complete lack of sequence homology between the four species (Fig. 6A; see also Krust *et al.*, 1986). Interestingly, the mRNAs of both the human (Weinberger *et al.*, 1986) and the chicken (Sap *et al.*, 1986) c-erb-A genes, which belong to the same gene superfamily as steroid hormone receptors (see below), contain at least two open reading frames in their 5′-untranslated regions. In the case of human c-erb-A (Weinberger *et al.*, 1986), this region contains two such open reading frames; one of them is unusually long with the potential to code for a 70 amino acid long peptide.

It has been demonstrated that the insertion of ATG codons in the 5′-untranslated region of eukaryotic genes usually results in a reduction of translation from the downstream open reading frame and that this is only partially restored if translation termination codons are introduced inframe with the upstream ATGs (Liu *et al.*, 1984; Kozak *et al.*, 1984b). Although the effect on translation efficiency of short open reading frames preceding the major coding sequence remains to be experimentally elucidated, their appearance has been noted in a number of tightly regulated genes, e.g. several oncogenes, and it has been suggested that they may serve regulatory purposes (Kozak, 1986). Strong evidence for such a role for short open reading frames comes from the study of the yeast transcription factor GCN4, which is involved in the general regulation of many amino acid biosynthetic genes. The GCN4 mRNA has a 577 bp leader sequence containing four small upstream open reading frames. By constructing deletion mutants (Tzamarias *et al.*, 1986) or point mutations in the upstream initiation codons (Mueller and Hinnebusch, 1986), it has been demonstrated that this region is responsible for a complex pattern of translational control. The presence of a single intact upstream AUG (any of the four) is sufficient to repress downstream translation, but the two 3′ proximal of these four AUGs are more inhibitory than the most 5′-proximal AUG codon. However, under different conditions of amino acid availability, e.g. starvation, it is the 5′-AUG which is

◄ Fig. 3 — Full-length nucleotide sequence and deduced amino acid sequence of the MCF-7 estrogen receptor mRNA. The 5′-flanking sequence of the human estrogen receptor gene is shown from −128 to the cap site at nucleotide +1 (arrow). The estrogen receptor mRNA sequence begins at nucleotide +1 and ends at the start of the poly(A) tail at nucleotide +6322. The numbers on the left refer to the position of the nucleotides and those on the right to that of the amino acids. The CAAT (GCCAATG) and TATA (TACTTAAA) box homologies in the 5′-flanking sequence and the three putative polyadenylation signal sites (AATAAA) are all underlined. The upstream 20-codon open reading frame (ORF) is boxed.

required for efficient GCN4 expression. In addition, it has been shown that at least two trans-acting genes are acting on the upstream open reading frames to modulate the level of GCN4 translation (Mueller and Hinnebusch, 1986; Tzamarias *et al.*, 1986). Thus it is perhaps relevant to the expression of the v-erb-A oncogenic protein that, because of its fusion with the viral gag gene, it lacks this potential translational control region in addition to 12 *N*-terminal amino acids.

The 3′-untranslated region of the human estrogen receptor is more than 4 kb long, a feature in common with the human (Hollenberg *et al.*, 1985) and rat (Miesfeld *et al.*, 1986) glucocorticoid receptors, the *Xenopus* estrogen receptor (Weiler *et al.*, 1987), and probably also the chicken estrogen receptor (Krust *et al.*, 1986). Three putative polyadenylation signal sites (AATAAA) are present in the human estrogen receptor mRNA (see Fig. 3), but only the most 3′ signal is used efficiently, since only one RNA species of 6.3 kb is observed on RNA blots (Walter *et al.*, 1985). In contrast, RNA blots of human, rat, and mouse glucocorticoid receptors reveal at least two messenger populations, apparently because different polyadenylation signals are used (Danielson *et al.*, 1986; Hollenberg *et al.*, 1985; Miesfeld *et al.*, 1986). Multiple species are also observed for the chicken (Sap *et al.*, 1986) and human (Weinberger *et al.*, 1986) c-erb-A mRNAs, the chicken, rabbit, and human progesterone receptors (Conneely *et al.*, 1986, 1987; Loosfelt *et al.* 1986; Misrahi *et al.*, 1987; Gronemeyer *et al.*, 1987), and the *Xenopus* estrogen receptor (Weiler *et al.*, 1987). The physiological significance of long 3′-untranslated regions is unknown, but it has been demonstrated that the purified glucocorticoid receptor recognizes specific DNA-binding sites within this region of the rat glucocorticoid receptor gene (Okret *et al.*, 1986). Such a finding may have important implications for the autoregulation of glucocorticoid receptor expression which may possibly involve post-transcriptional regulation.

B. Chicken progesterone receptor mRNA

Similar to the estrogen receptor mRNA, the 4472 nucleotide long chicken progesterone receptor mRNA contains a 366 nucleotide long 5′- and a 1745 nucleotide long 3′-untranslated region (Fig. 4). However, apart from an AUG codon, directly followed by a UGA termination codon, which is present in the 5′-untranslated region (position −290), there are no upstream open reading frames. The AUG codon at position 367 which is preceded by an in-frame terminator at position 208 initiates an open reading frame encoding 786 amino acids. The unusually long 3′-untranslated region has four potential polyadenylation signal sites.

Fig. 4 — The chicken progesterone receptor: full-length cDNA nucleotide sequence and reduced amino acid sequence. The 5′-flanking sequence of the chicken progesterone receptor gene is shown from −354 to the cap site (+1, arrow). The progesterone receptor mRNA sequence begins at nucleotide +1 and ends at nucleotide +4472. The numbers on the left refer to the position of the nucleotides and those on the right to that of the amino acids. The three oligonucleotides (oligo 1, 2 and 3) used for probe synthesis are given below the corresponding sequence, the most 5′ end of the cloned cDNAs (close to the BglI site, underlined) is indicated by a filled triangle at +29. Asterisks indicate the in-frame termination codon upstream and downstrean of methionine 1. The two only methionines present in the 5′ half of the open reading frame are boxed. An ATGTGA motif, present in the 5′ non-translated region is marked by dashes, a polyglutamic acid region is indicated by dots. Peptide sequences which were found in tryptic digests of homogeneous chicken progesterone receptor form B (Simpson *et al.*, 1987) are underlined. The central cysteine-rich DNA binding domain is boxed. In the 3′-untranslated region four potential polyadenylation signals sites are underlined. ➡

```
-354 ACTGGAAAATGGAAGATGTATTCATGCTAACAGGGCGTACAGGAACTTAAACCTGAAAAACACAGCAAACGTCTGAAAGAGAGGGTGAAGAACCAGGAAGGAGAGGACATATCTTCCTCC

-234 CCGCGCTCCGTAACAGCGATTTGCCCAGCCTGAGTGCAGCCAGTATTGCACTCCTCGACCAGGACGGACACCTGCAAGAGCCGGGAGACGGGGTGAGATGGGGCTGCGAAGGCAGGAGGC

-114 GGCCGGGCGGCTAAGAGCGGAGCCGAGAGGGAACGGGAGCGGTGGGAGGGTCGGAGCAGGCGGAGCGAAGGAGGGGCCAGGCCGTGCCGGGAGCTGCCGGAGGGAGTCGCCGGTGCCCGA
                                 TCGGCTCTCCCTTGCCCT oligo 1                                                           +1
        BglI
   7 GAGCCGAGCGCCGAAGGGGCGAGAAGGCGGCTGCCTGCCGGGGGCAGCGGGGAAGCGGGAGGGGCGAACCCGGGCGCCTCTCAGCAGTACGGGCGGCGGCGGACGGGGAGCGTCCAGGG
                                                                                 GAGAGTCGTCATGCCCGCCG oligo 2

 127 GAGCGCGGCGAGGAGCAGGGCGGGCAGCGGGGACGAGGCGCCGTCCCCGCCCGCCCTCTCGCTGCCCCTCGGAAGACGGCTGACGCTCAGATCGCTCTCCCCTCCGCAGGACCGTCCGG
                                                                              ***           AGCGAGAGGGGAGGCGT oligo 3

 247 GCGGCGATCGGCGCACGGTCGGGGCTCCTCGGAACGCAGCGCCATGTGAACGGGAGCCCCGTCCGGGCCGCGAGGAGCCGCGCCGGAGGTTTGGCGGCGTCCAGCGCTGCTCCGGGCAGC

 367 ATG ACC GAG GTG AAG AGC AAG GAG ACC CGA GCG CCT TCG TCC GCC CGA GAC GGG GCG GTC CTG CTG CAG GCT CCC CCC AGC CGC GGG GAA
     Met Thr Glu Val Lys Ser Lys Glu Thr Arg Ala Pro Ser Ser Ala Arg Asp Gly Ala Val Leu Leu Gln Ala Pro Pro Ser Arg Gly Glu   30

 457 GCG GAG GGC ATC GAC GTC GCC CTG GAC GGG CTG CTC TAC CCC AGA AGC AGC GAC GAG GAG GAG GAG GAG GAG GAG AAC GAG GAG GAG GAG
     Ala Glu Gly Ile Asp Val Ala Leu Asp Gly Leu Leu Tyr Pro Arg Ser Ser Asp Glu Glu Glu Glu Glu Glu Glu Asn Glu Glu Glu Glu   60

 547 GAG GAG GAA GAA CCG CAG CAG CGG GAG GAG GAG GAA GAG GAG GAG GAG GAG GAT CGG GAC TGC CCC TCG TAC CGG CCG GGC GGC GGC TCG
     Glu Glu Glu Glu Pro Gln Gln Arg Glu Glu Glu Glu Glu Glu Glu Glu Glu Asp Arg Asp Cys Pro Ser Tyr Arg Pro Gly Gly Gly Ser   90

 637 CTC TCC AAG GAC TGT CTG GAC AGC GTC CTG GAC ACC TTC CTG GCG CCC GCA GCC CAT GCC GCG CCC TGG TCC CTA TTC GGG CCG GAG GTG
     Leu Ser Lys Asp Cys Leu Asp Ser Val Leu Asp Thr Phe Leu Ala Pro Ala Ala His Ala Ala Pro Trp Ser Leu Phe Gly Pro Glu Val  120

 727 CCG GAA GTG CCC GTC GCC CCG ATG AGC CGC GGC CCC GAG CAG AAG GCT GTG GAC GCC GGC CCG GGG GCT CCC GGT CCC TCG CAG CCG CGA
     Pro Glu Val Pro Val Ala Pro Met Ser Arg Gly Pro Glu Gln Lys Ala Val Asp Ala Gly Pro Gly Ala Pro Gly Pro Ser Gln Pro Arg  150

 817 CCC GGG GCT CCG CTG TGG CCG GGC GCC GAC TCC CTG AAC GTC GCC GTG AAA GCG CGT CCG GGA CCC GAG GAC GCC AGC GAG AAC AGA GCG
     Pro Gly Ala Pro Leu Trp Pro Gly Ala Asp Ser Leu Asn Val Ala Val Lys Ala Arg Pro Gly Pro Glu Asp Ala Ser Glu Asn Arg Ala  180

 907 CCG GGG CTG CCC GGC GCC GAG GAG CGC GGC TTC CCG GAG CGG GAC GCG GGG CCC GGA GAA GGC GGA CTG GCG CCC GCC GCG GCT GCT TCC
     Pro Gly Leu Pro Gly Ala Glu Glu Arg Gly Phe Pro Glu Arg Asp Ala Gly Pro Gly Glu Gly Gly Leu Ala Pro Ala Ala Ala Ala Ser  210

 997 CCG GCG GCC GTG GAG CCG GGC GCG GGG CAG GAC TAC CTG CAC GTG CCC ATC CTG CCG CTC AAC TCT GCC TTC CTG GCC TCG CGC ACG CGG
     Pro Ala Ala Val Glu Pro Gly Ala Gly Gln Asp Tyr Leu His Val Pro Ile Leu Pro Leu Asn Ser Ala Phe Leu Ala Ser Arg Thr Arg  240

1087 CAG CTG CTG GAC GTG GAG GCG GCG TAC GAC GGC AGC GCC TTC GGG CCG CGC TCC TCG CCT TCC GTC CCC GCC GCG GAC TTG GCG GAG TAC
     Gln Leu Leu Asp Val Glu Ala Ala Tyr Asp Gly Ser Ala Phe Gly Pro Arg Ser Ser Pro Ser Val Pro Ala Ala Asp Leu Ala Glu Tyr  270

1177 GGC TAC CCG CCG CCC GAC GGC AAG GAG GGC CCC TTC GCT TAC GGC GAG TTC CAG AGC GCG CTG AAG ATC AAG GAG GAG GGC GTC GGT CTG
     Gly Tyr Pro Pro Pro Asp Gly Lys Glu Gly Pro Phe Ala Tyr Gly Glu Phe Gln Ser Ala Leu Lys Ile Lys Glu Glu Gly Val Gly Leu  300

1267 CCC GCC GCT CCC CCC CCG TTC CTG GGC GCC AAG GCC GCC CCC GCC GAC TTC GCG CAG CCC CCG CGG GCC GGG CAG GAG CCC TCG CTG GAG
     Pro Ala Ala Pro Pro Pro Phe Leu Gly Ala Lys Ala Ala Pro Ala Asp Phe Ala Gln Pro Pro Arg Ala Gly Gln Glu Pro Ser Leu Glu  330

1357 TGC GTC CTC TAC AAG GCC GAG CCG CCC CTC CTG CCC GGC GCA TAC GGG CCG CCC GCG GCC CCC GAC AGC CTG CCG TCC ACC TCG GCC GCG
     Cys Val Leu Tyr Lys Ala Glu Pro Pro Leu Leu Pro Gly Ala Tyr Gly Pro Pro Ala Ala Pro Asp Ser Leu Pro Ser Thr Ser Ala Ala  360

1447 CCG CCC GGC CTC TAC TCG CCG CTG GGC CTT AAC GGG CAC CAC CAG GCG CTG GGC TTC CCC GCG GCC GTG CTC AAG GAG GGC CTG CCC CAG
     Pro Pro Gly Leu Tyr Ser Pro Leu Gly Leu Asn Gly His His Gln Ala Leu Gly Phe Pro Ala Ala Val Leu Lys Glu Gly Leu Pro Gln  390

1537 CTG TGC CCG CCC TAC CTC GGC TAC GTG CGG CCA GAC ACA GAA ACC AGC CAG AGC TCC CAG TAC AGC TTT GAA TCG CTA CCC CAG AAG ATT
     Leu Cys Pro Pro Tyr Leu Gly Tyr Val Arg Pro Asp Thr Glu Thr Ser Gln Ser Ser Gln Tyr Ser Phe Glu Ser Leu Pro Gln Lys Ile  420

1627 TGT CTC ATC TGT GGT GAT GAG GCT TCT GGT TGC CAC TAC GGA GTA CTC ACC TGT GGA AGC TGT AAA GTC TTC TTT AAA AGG GCA ATG GAA
     Cys Leu Ile Cys Gly Asp Glu Ala Ser Gly Cys His Tyr Gly Val Leu Thr Cys Gly Ser Cys Lys Val Phe Phe Lys Arg Ala Met Glu  450

1717 GGG CAG CAC AAC TAT TTA TGT GCT GGA AGA AAT GAC TGC ATA GTT GAT AAA ATT CGT AGG AAG AAC TGT CCA GCG TGT CGC TTG AGG AAG
     Gly Gln His Asn Tyr Leu Cys Ala Gly Arg Asn Asp Cys Ile Val Asp Lys Ile Arg Arg Lys Asn Cys Pro Ala Cys Arg Leu Arg Lys  480

1807 TGC TGT CAA GCT GGA ATG GTC CTG GGA GGT CGA AAA TTT AAA AAG CTT AAC AAA ATG AAG GTT GTG AGG ACA TTA GAT GTT GCA CTT CAG
     Cys Cys Gln Ala Gly Met Val Leu Gly Gly Arg Lys Phe Lys Lys Leu Asn Lys Met Lys Val Val Arg Thr Leu Asp Val Ala Leu Gln  510

1897 CAG CCA GCC GTG CTT CAG GAT GAA ACC CAA TCT CTA ACG CAA AGG CTG TCC TTT TCT CCA AAT CAA GAA ATA CCG TTT GTT CCC CCA ATG
     Gln Pro Ala Val Leu Gln Asp Glu Thr Gln Ser Leu Thr Gln Arg Leu Ser Phe Ser Pro Asn Gln Glu Ile Pro Phe Val Pro Pro Met  540

1987 ATA AGT GTT TTG CGA GGC ATT GAG CCC GAA GTT GTC TAT GCT GGT TAT GAC AAT ACA AAA CCT GAA ACA CCA AGT TCC TTG CTG ACC AGT
     Ile Ser Val Leu Arg Gly Ile Glu Pro Glu Val Val Tyr Ala Gly Tyr Asp Asn Thr Lys Pro Glu Thr Pro Ser Ser Leu Leu Thr Ser  570

2077 CTA AAT CAT CTT TGT GAG AGG CAA CTG CTT TGT GTA GTC AAG TGG TCA AAA TTG CTA CCA GGA TTT CGG AAT TTA CAT ATT GAT GAC CAG
     Leu Asn His Leu Cys Glu Arg Gln Leu Leu Cys Val Val Lys Trp Ser Lys Leu Leu Pro Gly Phe Arg Asn Leu His Ile Asp Asp Gln  600

2167 ATA ACT CTC ATC CAG TAT TCT TGG ATG AGT CTA ATG GTT TTT GCA ATG GGA TGG AGA TCA TAC AAA CAT GTC AGT GGT CAG ATG TTG TAT
     Ile Thr Leu Ile Gln Tyr Ser Trp Met Ser Leu Met Val Phe Ala Met Gly Trp Arg Ser Tyr Lys His Val Ser Gly Gln Met Leu Tyr  630

2257 TTT GCA CCG GAT CTG ATT CTG AAT GAA CAG AGG ATG AAA GAA TCA TCG TTC TAT TCA CTG TGT CTC TCC ATG TGG CAG CTC CCA CAG GAG
     Phe Ala Pro Asp Leu Ile Leu Asn Glu Gln Arg Met Lys Glu Ser Ser Phe Tyr Ser Leu Cys Leu Ser Met Trp Gln Leu Pro Gln Glu  660

2347 TTT GTC AGA CTC CAA GTT AGT CAA GAG GAG TTT CTA TGT ATG AAA GCA CTA CTG CTT CTC AAT ACA ATT CCT TTG GAA GGT CTA AGA AGT
     Phe Val Arg Leu Gln Val Ser Gln Glu Glu Phe Leu Cys Met Lys Ala Leu Leu Leu Leu Asn Thr Ile Pro Leu Glu Gly Leu Arg Ser  690

2437 CAA AGC CAA TTT GAT GAG ATG AGA ACG AGT TAC ATT AGA GAA CTG GTG AAG GCC ATT GGT TTG CGC CAG AAA GGT GTG GTG GCT AAC TCC
     Gln Ser Gln Phe Asp Glu Met Arg Thr Ser Tyr Ile Arg Glu Leu Val Lys Ala Ile Gly Leu Arg Gln Lys Gly Val Val Ala Asn Ser  720

2527 CAG CGC TTC TAT CAG CTT ACA AAA CTG ATG GAT TCC ATG CAT GAT CTA GTG AAA CAG CTC CAT CTC TTC TGT CTG AAC ACA TTT CTC CAG
     Gln Arg Phe Tyr Gln Leu Thr Lys Leu Met Asp Ser Met His Asp Leu Val Lys Gln Leu His Leu Phe Cys Leu Asn Thr Phe Leu Gln  750

2617 TCC CGT GCA CTG AGC GTT GAA TTC CCT GAG ATG ATG TCA GAG GTG ATT GCT GCA CAG CTG CCC AAG ATT CTG GCA GGG ATG GTG AAA CCT
     Ser Arg Ala Leu Ser Val Glu Phe Pro Glu Met Met Ser Glu Val Ile Ala Ala Gln Leu Pro Lys Ile Leu Ala Gly Met Val Lys Pro  780

2707 CTT CTC TTT CAC AAG AAG TGA AAT GTC TTT TCT CTT ATC GTA GAG ATG ATT TCT GGG TCG GTT TTT GTT TGT TTA TTT TGG TCA AGA TTT
     Leu Leu Phe His Lys Lys ***                                                                                          786

2797 TGCAATGTCAGGACTTCAGTGTATTTGTCATACCCTGCCTACTGCTTTATACTTGTAACATACACAAGCCATTTAAGCTTGATGTACATATTGTGAGTTCCTGATTCCTTAACTGCAAAA
2917 ATCACAACAGTCGTAAGAAGTGTTTTGTAAACTTGGGTAGATTGCTTTTTAGATATGCGTAAGAATGGCCGAATGAATAGAGTTTTCAGATTTCTAAGAACAGTTATCTTATACATTTTT
3037 CTTACATATTACACACTTTCAAATTTCTGATATGACATGGTGTAAACCTTTTAATACACTGTGTGCATGGGTACACGTGGGCATGTGCATGGTCCTAAACATTTTGAGGGAATTAAAATG
3157 TATATTGCCCACACTTTCTCACGGATCTTTGACTTTTTATTTACTAATCTCCTTCACAGAAATATCCTTTCAAAGTCAGCTGTTAAGTTTAAGTAGATATCACTGCAGTACCTTCATGGG
3277 TAACTGAACTGCTCTCTGCAAAATATGTAAATATGGATTCATGTTCTTCTCTGTGCTCATGCTGTATTGAAGTTCTGATATCAGTGTGCTTCTTTCTGCATCATCAGTGCAGTTCTCCT
3397 TCAAAAATCAGAAAGCAGAAGAAGATTTCAGTATATGTTATACCAGAAGTAGGATGTATGTTAACCCATACAGCTTCAGTTTGTGTTATTTTCCTTCAAACTTTGGCACTTAGTAAAGCT
3517 AAAAAACATTGTTGTAATCTTAAATTTATCAACTTAAAAGTTATGTATTTCAGAAGTTTTAAATAGAAATGTTTTTCCTGACTACCACTCTCAATAATCTTGATGGTTTGAATTCCCGTG
3637 CTGAATACCCAACCAGCCACTACCATCAAATTTTAGAGAGGTTAGAAACAAAAAGAAGTGGAAAATATGACCACAAACAGTTTGATAAACATAACATTCATCTTAATACCATTTAGTGAA
3757 GCTACATATTTTTTACTTTAAAGGCATGCTGTTATACTTGGAGTAACTGTATGAGGAGTTGGGCTAAACAAGCATTGTACATTCAAGAAGAATGTCATGCTTTTCATGGAAATGAAATGT
3877 TTGACTAAAAAATGATAGTGTGATGGTTGTTGTTCCTTTGGTATGGACATGAAGCTACTCATCTTTGAGGGAGTAAACTGCTGATATATTCAGACAAACAGGAATGGATAAAGCCCAAGA
3997 AGAATTTTGGATTATTAAAATCTGATAAATTTTTACCAGCTCACTAGAAATGATACCCAGCTTCTCTAGCACTAATCAACGTAGGAAACTGCAAGTCTTGAATAACTGCTGACCATACGC
4117 TATTGCTTTTCCCCATTAGACACTGGATGCATCCATTGTGGTAGCATTCTACTTATGGTTTGTAACATGGGATGTATGGACTTGTCTTTGTCAGGTAAGCTAAGGCCCATCAAGAGGAAC
4237 CTGTGTAACAGTTAGGAATAAAAGCATCCTACATAGTACAAAACCCACTGGAGTTGTGGCAAAGAACTGACCCAAGGTAAAATGCTCTAAGCAGACTTTTGTGTTTGTTCACTAGGTTCA
4357 TAACTTTTAACTGCCCCAAAACAAGCTATTTCTTACTTATTCAATACAGTCTGACTTTCATCTGTGATCCTTAAAGCAATTGTGTTTTAATAAATAACACAAGATTCCGTTTCTAC-(A)21
```

The region containing the 5′ end of the chicken progesterone receptor mRNA was first localized by Northern blotting of poly(A)$^+$ RNA from laying-hen oviduct using the three single-stranded probes D, E and F described in Fig. 5C. Two clear bands of ~4.5 and 3.3 kb and a fainter band of 6.5 kb were revealed using probes D

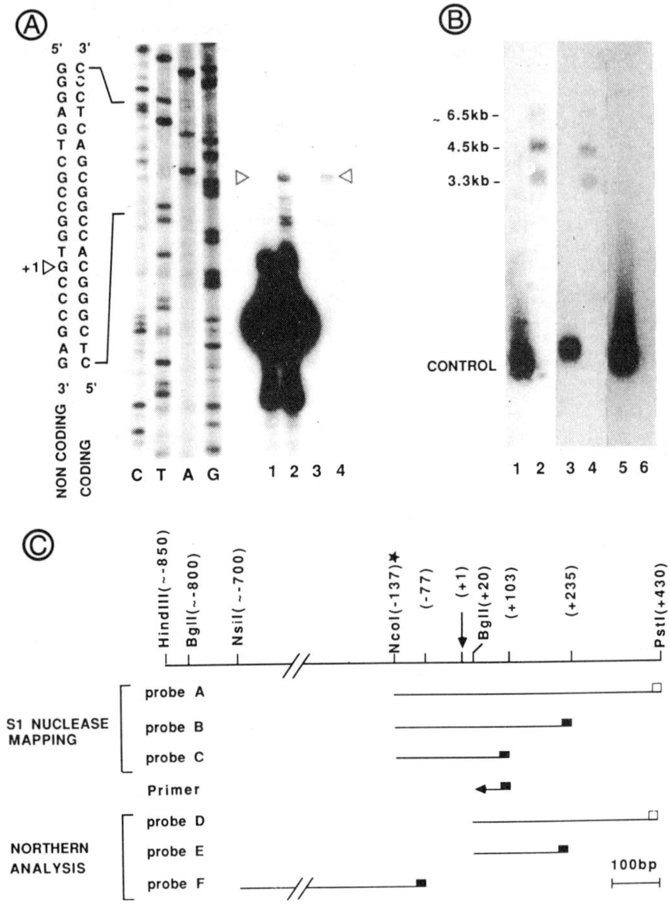

Fig. 5 — Mapping of the chicken progesterone receptor mRNA 5′ end. (A) A 6% sequencing gel showing S1 nuclease (lanes 3 and 4) and primer extension (lanes 1 and 2) mapping using poly(A)$^+$ RNA from laying-hen oviduct (lanes 2 and 4) or from HeLa cells (lanes 1 and 3). The lanes CTAG show the sequence ladder obtained using the dideoxy method and oligo 2 as primer elongated in the presence of [α-^{35}S]dATP. Open triangles indicate the position of the major transcription start site. (B) Northern blot analysis using 15 μg of RNA per slot. Lanes 1, 3 and 5 consist of total HeLa cell RNA mixed with 2 ng of the HindIII–PstI fragment (see (C)) as a positive control; lanes 2, 4 and 6 are poly(A)$^+$ RNA from oviduct. The probes used for hybridization were D (lanes 1, 2), E (lanes 3, 4) and F (lanes 5, 6). (C) Map of the 5′ region of the chicken progesterone receptor gene showing the probes and the Primer used in A and B. The M13 universal primer (open squares) was used to synthesize probes A and D, oligonucleotide 3 was used for the synthesis of probes B and E, oligonucleotide 2 for the synthesis of probe C and the Primer, and oligonucleotide 1 for probe F. The sequences of oligoneucleotides (1–3) indicated by the filled squares are depicted in Fig. 4. In all cases probe synthesis was carried out in the presence of [α-^{32}P]dATP, followed by digestion with NcoI (probes A, B and C), BglI (probes D, E and the Primer) and NsiI (probe F) and purification. The NcoI site (marked with an asterisk) at −137 was introduced by site-directed mutagenesis. (From Gronemeyer *et al.* (1987).)

and E, whereas no RNA hybridizing with probe F could be detected (Fig. 5B, lanes 2, 4 and 6 respectively). A mixture of HeLa cell total RNA and HindIII–PstI genomic DNA fragment (Control, see Fig. 5B) was run in parallel in lanes 1, 3 and 5. As expected, only the control DNA fragment was revealed in these lanes. These results indicated that the 5′ end of the chicken progesterone receptor mRNA was located within a few hundred bases upstream of the BglI (+20) site (Fig. 5C).

The primer extension technique was then used to map more precisely the position of the cap site. A single-stranded ^{32}P-labeled primer cut at the BglI (+20) site (Primer, see Fig. 5C) was synthesized by elongation of the synthetic oligonucleotide 2 (Fig. 4), using the HindIII–PstI genomic DNA fragment (Fig. 5C) inserted in M13 as a template. When elongated in the presence of oviduct poly(A)$^+$ RNA with reverse transcriptase and cold dNTPs, this labeled primer yielded a major product (Fig. 5A, lane 2) which mapped at the position of a C in the mRNA coding strand of the DNA (see the sequence ladder of the DNA obtained using the dideoxy method with [^{35}S] dATP and the oligonucleotide 2 (Fig. 4) as a primer). Some additional bands were visible both upstream and downstream of the main band, suggesting the existence of multiple initiation sites. The 5′ end of the mRNA was also mapped with S1 nuclease using a ^{32}P-labeled probe C (Fig. 5C). After hybridization with oviduct poly(A)$^+$ mRNA and S1 nuclease digestion, a major S1 nuclease-resistant DNA fragment (Fig. 5A, lane 4) migrated at the same position as the reverse transcriptase extended primer (compare lanes 2 and 4). In this case also, several minor bands, possibly corresponding to multiple initiation sites, were visible on the original autoradiogram. Under identical conditions, no primer extension or S1 nuclease-resistant products were seen using HeLa cell poly(A)$^+$ RNA (lanes 1 and 3). Thus the 5′ end of the chicken progesterone receptor mRNA is most probably localized in a region centered on the G indicated as +1 in Figs. 1 and 2A and there may be multiple initiation sites. That the 5′ end of the chicken progesterone receptor mRNA is located in this region was also suported by S1 nuclease mapping experiments (not shown) performed with ^{32}P-labeled probes A and B (Fig. 2C), which resulted in protected fragments of the expected lengths.

All the above results lead us to conclude that the length of chicken progesterone receptor mRNA is very close to 4.5 kb, which is in good agreement with the estimated size of the major poly(A)$^+$ RNA species revealed on Northern blots with chicken progesterone receptor cDNA probes (see above). The 3.3 kb poly(A)$^+$ mRNA corresponds to a truncated form of the chicken progesterone receptor mRNA which has a 5′ end similar to that of the 4.5 kb species but is interrupted in the coding region after amino acid 451. The origin of the 6.5 kb species is unknown but it may correspond to a splicing intermediate. It is important to note that, with the exception of the cDNA clones which correspond to the 3.3 kb mRNA species, we did not find any cDNA clones with a sequence that is not collinear with that of the 4.5 kb mRNA in any of the chicken oviduct cDNA libraries which have been used. In addition, the study of genomic clones did not provide any evidence that the chicken genome may contain different progesterone receptor genes (unpublished results).

The progesterone receptor mRNA open reading frame encodes a protein with a deduced molecular weight of 85 743 Da. This is a marked discrepancy with the molecular weight determined from SDS gels. As described below, an unusual poly-GAG stretch, encoding a poly-glutamic acid region in the form B receptor, is responsible for an aberrant migration of this protein on SDS–polyacrylamide gels.

C. Sequence comparison between estrogen receptors for four different species

Important functional domains of the estrogen receptor should be conserved during evolution. An initial comparison performed in our laboratory between the primary amino acid sequence of the chicken and human estrogen receptors, which diverged approximately 100 000 000 years ago, revealed six regions (A–F) of differing homology (Krust *et al.*, 1986; see Fig. 6).

The optimal alignment of the chicken (589 amino acids) and human (595 amino acids) estrogen receptor amino acid sequences is shown in Fig. 6B. The two proteins show 80% overall identity. There are three regions of high homology, A (87%), C (100%), and E (94%), and three regions of relatively low homology, B (56%), D (38%), and F (41%). The 83 amino acid long region C is unusual in its high content of cysteine, lysine, and arginine. It is completely conserved between the human, rat, and chicken estrogen receptors, and there is only one amino acid change in the *Xenopus* homologue in this region (Fig. 6B), suggesting that it plays a crucial role in receptor function. It is this region that is responsible for the recognition of target gene sequences by the estrogen receptor (see below) and therefore corresponds to the DNA-binding domain of the receptor. A hydropathicity plot (Fig. 7) further illustrates the segmented character of the human estrogen receptor (Green *et al.*, 1986). This plot demonstrates that the DNA-binding region C is flanked by short hydrophilic regions that may play hinge functions, whereas the region E is mainly hydrophobic. Because of its high degree of conservation, region E is also likely to be functionally important, and it was suggested that, owing to its hydrophobic character, this region contains the steroid hormone binding domain (Krust *et al.*, 1986; Green *et al.*, 1986).

The first 37 amino acids (region A) of the human and chicken estrogen receptors are relatively highly conserved (87%), whereas this region is not conserved in the *Xenopus* receptor (Krust *et al.*, 1986; see Fig. 6B). The significance of this conservation, which is also observed for the rat receptor, is unknown and there appears to be little or no sequence homology of this region with any of the other members of the receptor family (see below). Surprisingly, regions C and E of the estrogen receptors of mammals and birds show a striking homology with the corresponding regions of the amphibian *Xenopus* from which they diverged about 350 000 000 years ago (Fig. 6B). Apparently these two regions are under intensive selective pressure to maintain a specific amino acid sequence. Also, the comparison of all four sequences identifies two short regions within region B which are also perfectly conserved, corresponding to positions 53–57 and 127–134 of the human estrogen receptor. The significance of these sequences remains to be established.

Fig. 6 — Comparison of the chicken and human estrogen receptor amino acid sequences. (A) Comparison of the small amino acid open reading frames of the human (HER), rat (RER), chicken (CER) and *Xenopus* (XER) estrogen receptors. (B) Comparison of the human, rat, chicken, and *Xenopus* estrogen receptor sequences. Only non-identical amino acids are shown for the rat, chicken, and the *Xenopus* sequences. Gaps (—) were introduced into the sequence in order to obtain maximal alignment of identical amino acids. The numbers on the left-hand side refer to the amino acid sequence. The sequence was divided into six regions (A–F) based on sequence homology between the four receptors. The three most highly conserved regions (A, amino acids 1–37; C, 180–262; E, 302–552 (amino acid numbers with respect to human estrogen receptor)) are boxed. The CER sequence, differing in one amino acid, has also been determined by Maxwell *et al.* (1987). ►

```
A.

hER   1  MRCVASNLGLCSFSRWPAGF 20
rER   1  MEHFWKDAFEPAGWLIRC   18
cER   1  MYCCGNMISAEPQNRGWCFF 20
cER   1  MFLAHCH 7
xER   1  MSHYHCQ 7

B.

hER   1  MTMTLHTKASGMALLHQIQGNELEPLNRPQLKIPLERPLGEVYLDSSKPA
rER   1                                  M M A     V N               A
cER   1                VT         T   T S    S SDM VE N TG
xER   1     P PN TT VTF       SS    T T  P    S     MTVENNRTG

hER  51  VYNYPEGAAYEFNAAAA-----ANAQVYGQTGLPYGPGSEAAAFGSNGLG
rER  51   F                AAAAG S P    SSIT           A S
cER  51   F       T D ---------GTT P   S T S A T  --S   SS A
xER  51  IF      TT D A     --------P  SSAS S AAS  --T   SS T

hER  96  GFPPLNSVSPSPLMLLHPPPQLSPFLQPHGQQVPYYLENEPSGYTVREAG
rER 101  A  Q                HV    H   H           A A  DT
cER  90    HS  N P   VVF QTA       IHH S          QGSFGM   A       B
xER  91   LHT  N P   VVF AKL       IHH          S QGTFA     A

hER 146  PPAFYRPNSDNRRQGGRERLASTNDKGSMAMESAKETRYCAVCNDYASGY
rER 151        S       N     S SSE  N I
cER 140         S     HSI   MS   E   LS   T
xER 141    T   SSS     S    MS A    PPSM  T         S

hER 196  HYGVWSCEGCKAFFKRSIQGHNDYMCPATNQCTIDKNRRKSCQACRLRKC   C
rER 201
cER 190
xER 191

hER 246  YEVGMMKGGIRKDRRGGRMLKHKRQRDDGEGRGEVGSAGDMRAANLWPSP
rER 251                              L   N M TS
cER 240                   E M Q    EEQDS NGEASSTEL  PT  T     D
xER 241                    L      KEEQ QKND DPSEIRT SITVN-

hER 296  LMIKRSKKNSLALSLTADQMVSALLDAEPPILYSEYDPTRPFSEASMMGL
rER 301   V  HT    P                 LI     S
cER 290   VV HN    P      E       E     V      N    N      T
xER 291  SV--K MKL PV      E LI    ME  A  V   H S K L       T

hER 346  LTNLADRELVHMINWAKRVPGFVDLTLHDQVHLLECAWLEILMIGLVWRS
rER 351                        G  N
cER 340
xER 338                                                V  I

hER 396  MEHPVKLLFAPNLLLDRNQGKCVEGMVEIFDMLLATSSRFRMMNLQGEEF
rER 401       G
cER 390       G                                AA                   E
xER 388  V    G  S           R    L      VT AT     R R

hER 446  VCLKSIILLNSGVYTFLSSTLKSLEEKDHIHRVLDKITDTLIHLMAKAGL
rER 451                                      N
cER 440                           R Y                   S
xER 438  I                    E    DT L  II     I   V F   S

hER 496  TLQQQHQRLAQLLLILSHIRHMSNKGMEHLYSMKCKNVVPLYDLLLEMLD
rER 501        R                         N
cER 490  S     R                         N
xER 488  S    QR

hER 546  AHRLHAPTSRGGASVEETDQSHLATAGSTSSHSLQKYYI-TGEAEGFPATV 595
rER 551         A  M VPP  PS  Q T TS   A     T    -PP       N I 600   F
cER 540         AA SA PM  ENRNQ T  -PA        SF   NSK E SMQN I 589
xER 538      I T KDKTTT-Q  DSRSPPT TVNGA PC   P  T-DT EVSLQS    586
```

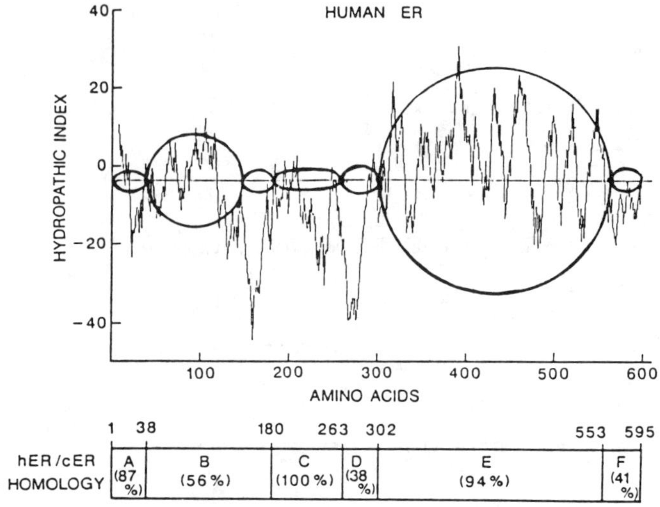

Fig. 7 — Hydropathicity plot of the human estrogen receptor amino acid sequence. A computer program, using the algorithm described by Kyte and Doolittle (1982), was used to display the hydropathic nature of the human estrogen receptor using a window of 15 amino acids. Negative values represent hydrophilic regions and positive values hydrophobic regions. The numbers on the horizontal axis correspond to the amino acid sequence beginning at the N terminus. The ellipses drawn on the plot serve to show the correlation between the hydropathicity of the receptor and its division into six regions (A–F) based on the comparison of the chicken and human estrogen receptors shown below the plot.

D. Homology between steroid hormone receptors and erb-A sequences

Comparison of the estrogen receptor amino acid sequence with those in the NBRF protein database revealed that the product of the v-erb-A gene of the avian erythroblastosis virus (AEV) was significantly homologous (Krust *et al.*, 1986; Green *et al.*, 1986). There is a high degree of homology in region C (55%), and less in region E (17%), suggesting that the cellular counterpart of the viral gene, c-erb-A, should also belong to the steroid hormone receptor gene family. A similar degree of homology was also observed between the human glucocorticoid receptor and v-erb-A (Weinberger *et al.*, 1985a,b; see below). Recently, cDNAs corresponding to the chicken (Sap *et al.*, 1986) and human (Weinberger *et al.*, 1986) c-erb-A genes have been cloned, and the protein expressed from these cDNAs has biochemical properties very similar to those of the receptor for the thyroid hormones tri-iodothyronine (T3) and thyroxine (T4). Both c-erb-A proteins bind thyroid hormones with high affinity and, as is known for the thyroid hormone receptor, the chicken v-erb-A product is localized in the nucleus (Sap *et al.*, 1986). The homology between the erb-A proteins and the steroid hormone receptors supports the previous suggestion that both receptor systems modulate gene expression by a similar mechanism (Oppenheimer and Samuels, 1983).

E. A superfamily of nuclear receptor genes

Comparison of the primary amino acid sequence of all steroid and thyroid hormone receptors available to date (human (Green *et al.*, 1986), chicken (Krust *et al.*, 1986), rat (Koike *et al.*, 1987), mouse (White *et al.*, 1987) and *Xenopus* (Weiler *et al.*, 1987) estrogen receptors; human (Hollenberg *et al.*, 1985), rat (Miesfeld *et al.*, 1986) and mouse (Danielson *et al.*, 1986) glucocorticoid receptors; chicken (Gronemeyer *et al.*, 1987; Conneely *et al.*, 1987), rabbit (Loosfelt *et al.*, 1986) and human (Misrahi *et al.*, 1987) progesterone receptors, chicken vitamin D receptor (McDonnell *et al.*, 1987); chicken (Sap *et al.*, 1986), human (Weinberger *et al.*, 1986; Benbrook and Pfahl, 1987) and rat (Thompson *et al.*, 1987) c-erb-A–thyroid hormone receptors; human mineralocortoid receptor (Arriza *et al.*, 1987)) reveals a strikingly similar structural organization (Fig. 8). Regions which are homologous to the two highly conserved regions C and E of the estrogen receptor are found in the progesterone, glucocorticoid, and vitamin D receptors and also in the c-erb-A gene products

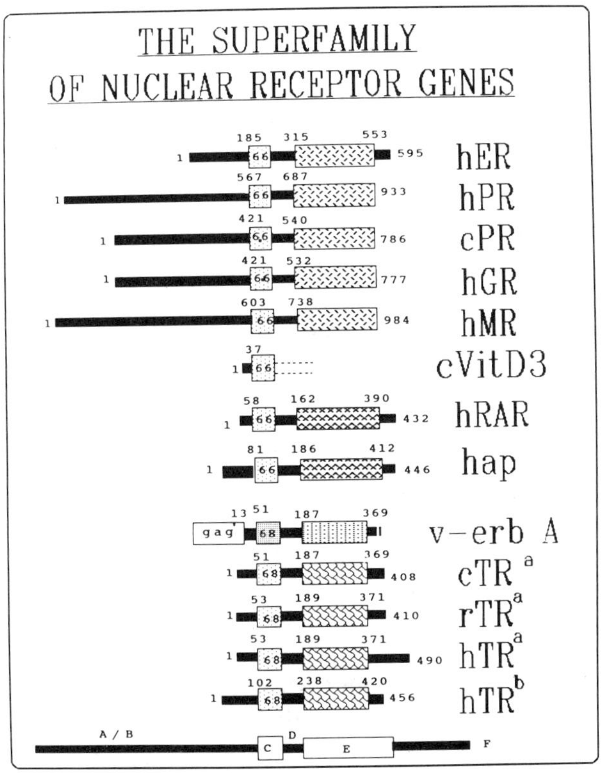

Fig. 8 — Schematic alignment of nuclear receptor sequences. The amino acid sequences of the human estrogen receptor (hER), human (hPR) and chicken progesterone receptor (cPR), human glucocorticoid receptor (hGR), human mineralocorticoid receptor (hMR), chicken vitamin D receptor (cVitD3), human retinoic acid receptor (hRAR) and the related hap protein, and the chicken (cTR[a]), human (hTR[a] and hTR[b]) and rat (rTR[a]) thyroid hormone receptor–c-erb-A proteins as well as the v-erb-A–gap fusion protein were deduced from their cDNA sequences. For details and references see text. Alignment is based on amino acid similarity, and the numbers given refer to the position of amino acid residues. The division of the receptors into six regions, based on the crosshomologies of their sequences, is shown at the bottom of the figure. The two highly conserved regions, representing the DNA–binding (region C) and the hormone–binding (region E) domains, are shown as blocks. Little or no significant homology is observed when comparing other regions of the receptors (solid black lines). A different pattern of regions E indicates the different families of ligands. v-erb-A does not bind thyroids. The number of amino acids that constitute region C is indicated within the corresponding symbol.

(Fig. 8). Most importantly, the retinoic acid receptor subfamily, which has been cloned recently (Petkovich *et al.*, 1987; Giguère *et al.*, 1987; de Thé *et al.*, 1987), and which binds a ligand that is not clearly related to the steroid or thyroid family, exhibits a similar organization (the ligand for hap remains to be identified but should belong to the retinoids (de Thé *et al.*, 1987)). Region C, containing the DNA-binding domain (see below), is the region of highest homology between all these receptors, with homologies varying between 45% and 91%. For the estrogen receptor, this region has been 99% conserved during 350 000 000 years of evolution. The ligand-binding domain, region E, exhibits a lower overall homology ranging from 15% to 55%. Region D, which may function as a hinge between the DNA- and hormone-binding domains, is not conserved, either in sequence or length, between any of the receptors. It is of interest that the difference in size between the receptors is due primarily to the difference in length of region A–B (steroid receptors) and F (thyroid receptors) (see Fig. 8). Region A–B is encoded by a single exon in both the progesterone and the estrogen receptors (unpublished results from our laboratory), and no sequence homologies between the various receptors are apparent in this region, indicating an early divergence in evolution.

It appears, therefore, that a family of proteins, whose ability to activate gene transcription is regulated by the binding of the hormone, has evolved from a common ancestral receptor gene which already contained the structural elements for both DNA and ligand binding. In this respect it is interesting to note that the invertebrate *Drosophila*, whose evolutionary line diverged from that of mammals approximately 600 000 000 years ago, contains tight nuclear-binding proteins for the steroid 20-hydroxy-ecdysone (Gronemeyer *et al.*, 1983). As yeast has been shown to contain 17 β-estradiol (Feldman *et al.*, 1984a,b) and since yeast contains estrogen-binding proteins (Feldman *et al.*, 1982; Burshell *et al.*, 1984), it is tempting to speculate that the combination of the ancestral motifs for DNA and hormone binding might have appeared very early in eukaryote evolution. In fact, it has been shown recently in our laboratory that chimeras between the yeast transcriptional activator GAL4 and the human estrogen receptor are functionally active in transactivating transcription from estrogen target genes (Webster *et al.*, 1988).

In the case of the glucocorticoid and estrogen receptors, the responsive elements of some of their target genes belong to the class of cis-acting DNA elements termed 'enhancers' (see below). Thus it is likely that other members of the nuclear receptor family are also transcription factors that regulate gene expression through interaction with specific enhancer elements as shown recently, for example, for the c-erb-A–thyroid hormone receptor and the growth hormone gene (Glass *et al.*, 1987).

There has been considerable progress in the chromosomal localization of human receptors. The human retinoic acid receptor (Petkovich *et al.*, 1987) as well as a human c-erb-A gene have been located on chromosome 17 (Spurr *et al.*, 1984). The latter probably corresponds to a recently cloned thyroid hormone receptor (Benbrook and Pfahl, 1987; Thompson *et al.*, 1987). The human estrogen (Walter *et al.*, 1985) and glucocorticoid (Gehring *et al.*, 1985; Weinberger *et al.*, 1985) receptors are located on chromosomes 6 and 5 respectively. The human progesterone receptor has been located on chromosome 11 (Law *et al.*, 1987; Mattei *et al.*, 1987). Weinberger *et al.* (1986) have shown that their cloned c-erb-A–thyroid hormone receptor gene is located on human chromosome 3 and suggested the existence of other related genes present on the same chromosome. Furthermore, in a human hepatocellular carcinoma, the hepatitis B virus genome integrated into a gene, now called hap (de Thé *et al.*, 1987), located on chromosome 3, which has sequence homology with region C of steroid hormone receptors and erb-A (Dejean *et al.*, 1986). Thus, it is likely that there are additional members of this superfamily of transcriptional regulatory factors still to be isolated (see Fig. 10).

1. The DNA-binding domain

Although the different steroid and thyroid hormone receptors trigger very different physiological and developmental programs, their putative DNA-binding domains (region C, see below) are highly homologous. At first sight, this may seem surprising, because it is this domain which interacts with specific promoter sequences of target genes, thus activating gene transcription (see below and Green and Chambon, 1987). Sequence analysis of the DNA-binding region of the steroid and thyroid hormone receptors shows, however, that at least part of this homology is apparently due to the requirement of a specific structural organization. The most striking feature of region C is the conserved arrangement of nine cysteines, which is common to all receptors as shown in Fig. 9. From this alignment, it is already apparent that this domain consists of two halves, each with a specific arrangement of cysteines. In fact, these halves are encoded in different exons in both the progesterone and the estrogen receptors (Jeltsch *et al.*, 1986; unpublished work of our laboratory).

The N-terminal half of domain C is characterized by the motif Cys–X_2–Cys–X_{13}–Cys–X_2–Cys. This structure is reminiscent of the DNA-binding region of a class

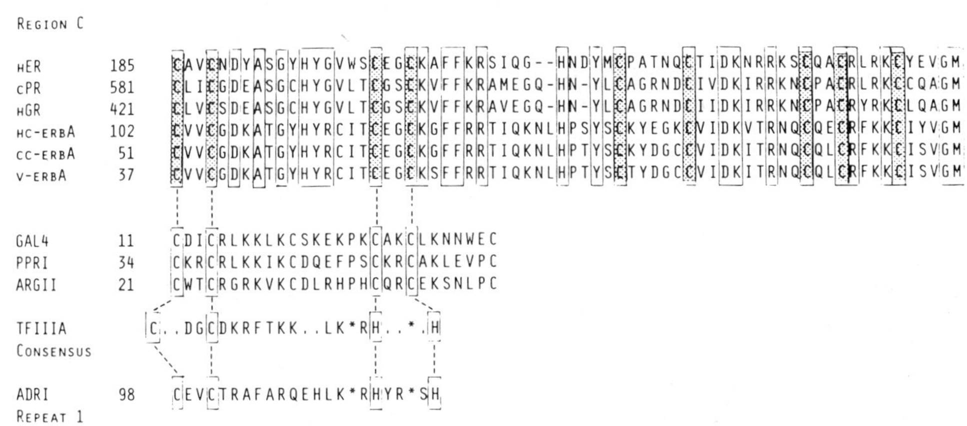

Fig. 9 — Alignment of receptor DNA-binding domains. Shown are the human estrogen (hER), the chicken progesterone (cPR), and the human glucocorticoid (hGR) receptor together with human c-erb-A (hc-erbA), the chicken c-erb-A (cc-erb-A) and the v-erb-A protein. Boxed areas indicate complete amino acid identity between all the sequences, and dashes (—) indicate amino acid gaps for optimal alignment. Conserved cysteins are shaded. The Cys–X_2–Cys–X_{13}–Cys–X_2–Cys motif present in the putative DNA-binding domain of the GAL4, PPR1 and ArgII yeast regulatory proteins is aligned with the corresponding motif of the steroid hormone receptors and erbA gene product. The consensus motif of the *Xenopus* transcription factor TFIIIA and the corresponding repeat 1 of the yeast regulatory proteins ADR1 are also shown. The pairs of cysteine and histidine residues dictating the postulated 'finger' structures are boxed. Asterisks indicate positions where amino acid insertions may occur, and dots in the consensus sequence indicate variable residues. Numbers indicate the position of amino acid residues in each sequence.

	Chromosome	Reference
Steroid receptors		
Estrogen	6	Walter *et al.*, 1985
Glucocorticoid	5	Gehring *et al.*, 1985
Progestin	11 q22	Mattei *et al.*, 1987
Mineralocorticoid	4	Arriza *et al.*, 1987
Vitamin D	?	McDonell *et al.*, 1987
Thyroid receptors		
Alpha (c-erb A1)	17q11.12–q12	Thompson *et al.*, 1987
Beta	3	Weinberger *et al.*, 1986
erb A2 crosshybridizing	17q21.3	Gosden *et al.*, 1986
	17q25	Gosden *et al.*, 1986
Retinoic acid receptors		
Retinoic acid	17q21.1	Petkovitch *et al.*, 1987
hap	3	de Thé *et al.*, 1987
Retinoic acid receptor crosshybridizing	several?	Giguére *et al.*, 1987

Fig. 10 — Chromosomal localization of steroid, thyroid–c-erb-A and retinoic acid receptors. For the erb-A2 and retinoic acid receptor crosshybridizing genes and for hap the ligand is not yet definitely defined.

of eukaryotic transcription factors that contain similar cysteine and histidine arrangements (see Miller *et al.*, 1985; Berg, 1986a,b; for a recent review see Klug and Rhodes, 1987; Fig. 9). In the case of prokaryotic transcription regulatory factors, a helix–turn–helix secondary structure is responsible for the recognition of specific DNA sequences (Pabo and Sauer, 1984); no such structure is predicted for region C of the hormone receptors (Krust *et al.*, 1986). In the case of the *Xenopus* 5S rRNA transcription factor TFIIIA (Miller *et al.*, 1985), a cysteine- and histidine-containing motif is repeated nine times (see Fig. 9). Each repeat is believed to fold into DNA-binding loops or 'fingers' with the invariant cysteine and histidine pairs chelating Zn^{2+} ions with specific DNA contacts being made by some of the other amino acids. Recently, from their DNA sequences, more proteins have been proposed to contain regions with the potential to fold into similar structures. These include the *Drosophila* developmental genes KRÜPPEL (Rosenberg *et al.*, 1986; Schuh *et al.*, 1986), HUNCHBACK (Tautz *et al.*, 1987), SERENDIPITY (Vincent *et al.*, 1985; Vincent, 1986) and SNAIL (Boulay *et al.*, 1987) and the yeast ADR1 protein (Hartshorne *et al.*, 1986). Mouse genes with sequence homologies to Krüppel and the potential to form 'DNA-binding fingers' have been described recently (Chowdhury *et al.*, 1987).

Assuming that cysteine and histidine are functionally equivalent with respect to the chelation of Zn^{2+} or other metal ions (Miller *et al.*, 1985), the DNA-binding domain C of steroid–thyroid hormone receptors might fold into similar structures. Interestingly, other transcriptional regulatory proteins, such as the yeast GAL4, PPRI, and ARGII proteins (Laughon and Gesteland, 1984; Kammerer *et al.*, 1984; Messenguy *et al.*, 1986; see Fig. 9) contain amino acid sequences with cysteine motifs identical to those of the steroid–thyroid hormone receptors. In fact, in the case of GAL4 strong evidence has been presented for the requirement of Zn^{2+} in the structural organization of the DNA-binding domain (Johnston, 1987). A GAL^- mutant due to a Pro→Leu change (as a consequence of a point mutation two amino acids from one of the Cys residues), could be rescued by high concentrations of Zn^{2+}. Note, however, that protein kinase C, which is not known to be a DNA-binding protein, also contains two copies of this cysteine motif in its catalytic domain (Ohno *et al.*, 1987). Evidence in favor of the presence of zinc or other divalent metals in steroid hormone receptors has been presented, and it has been suggested that metals may play a role in nuclear binding (Shyamala and Yeh, 1975; Lohmar and Toft, 1975; Schmidt *et al.*, 1981; Colvard and Wilson, 1984; Sabbah *et al.*, 1987). If DNA-binding 'fingers' exist, then it is likely that the DNA-binding specificity of the receptors is determined by the non-conserved amino acids of the 'finger', while the highly conserved residues provide the scaffolding of this domain. It is therefore of particular interest that there are only six amino acid residue differences between the chicken progesterone and human glucocorticoid receptors in region C, two of which are conservative (Jeltsch *et al.*, 1986; Conneely *et al.*, 1986; see Fig. 9). One would expect that this difference is sufficient for the receptors either to recognize different binding sites or to bind to the same sites with different affinities. Data obtained by footprinting analysis of the 5′ regulatory region of the chicken lysozyme gene (von der Ahe *et al.*, 1986) suggest that the two receptors recognize, with different affinities, overlapping, but not identical, binding motifs in a promoter distal region, whereas a more downstream binding site for the rat glucocorticoid receptor is not recognized by the chicken progesterone receptor. Despite these apparent differences in the DNA protection by the progesterone and glucocorticoid receptor, a 15 bp long DNA sequence has been demonstrated to be sufficient to confer both glucocorticoid and progesterone inducibility on reporter genes (Strähle *et al.*, 1987). Moreover, point mutations in the recognition sequence affected inducibility by both glucocorticoids and progesterone. Thus it appears that the differences observed in the footprinting studies for the binding of both receptors reflect differences in the way each receptor contracts the DNA but that both use a common hormone responsive element as recognition sequence. It should be stressed, however, that usually more than one hormone responsive element is located upstream of target genes and that there is good evidence for synergism (Jantzen *et al.*, 1987). This synergism might be the result of cooperative binding as has been demonstrated recently for the yeast transcription factor GAL4 (Gininger and Ptashne, 1988). Synergism may be an important component of hormonal regulation and differences in target gene inducibility may reflect differences in cooperativity of the same receptors (Jantzen *et al.*, 1987) or even different receptors. With respect to single response elements, it will be challenging to define the specific interaction of a glucocorticoid and estrogen receptor with their responsive elements which apparently differ by only a single base pair (Klock *et al.*, 1987).

2. The hormone-binding domain

Comparison of the hydrophobic hormone-binding domains, region E, of the steroid and thyroid hormone receptors is shown in Fig. 11. No specific structural features can be deduced from the primary amino acid sequence, though some residues are conserved at equivalent positions in all receptors. The highest overall homology between the E domains of the various receptors is found when comparing the glucocorticoid, mineralocorticoid and progesterone receptor (Fig. 11). This is not surprising, since crossreactivity of their ligands is well known.

One part of region E, which is well conserved between all the receptors, is located between amino acids 358 and 395 of the human estorogen receptor (see Fig. 11). The possible role of this region in ligand binding or in the mechanism by which receptors

THE LIGAND BINDING DOMAIN E

```
HER         315  MVSALLDAEPPILYSEYDPTRPFSEASMMGLLTNLADREL
HGR         532  LVSLLEVIEPEVLYAGYDSSVPDSTWRIMTTLNMLGGRQV
HMR         738  PVMVLENIEPEVLYAGYDSSKPDTAENLLSTLNRLAGKQM
HPR         687  LINLLMSIEPDVIYAGHDNTKPDTSSSLLTSLNQLGERQL
CPR         540  MISVLRGIEPEVVYAGYDNTKPETPSSLLTSLNHLCERQL
HRAR        170  PALCQLGKYTTNNSSEQRVSLDIDLWDKFSE---LSTKCI
HTR   T     189  RKFLPEDIGQSPIVSMPDGDKVDLEA--FSEFTKIITPAI
HTR   beta  238  PKFLPEDIGQAPIVNAPEGGKVDLEA--FSHFTKIITPAI

HER              VHMINWAKRVPGFVDLTLHDQVHLLECAWLEILMIGLVWR
HGR              IAAVKWAKAIPGFRNLHLDDQMTLLQYSWMFLMAFALGWR
HMR              IQVVKWAKVLPGFKNLPLEDQITLIQYSWMCLSSFALSWR
HPR              LSVVKWSKSLPGFRNLHIDDQITLIQYSWMSLMVFGLGWR
CPR              LCVVKWSKLLPGFRNLHIDDQITLIQYSWMSLMVFAMGWR
HRAR             IKTVEFAKQLPGFTTLTIADQITLLKAACLDILILRICTR
HTR   T          TRVVDFAKKLPMFSELPCEDQIILLKGCCMEIMSLRAAVR
HTR   beta       TRVVDFAKKLPMFCELPCEDQIILLKGCCMEIMSLRAAVR

HER              SMEHPV---KLLFAPNLLLDRNQGK--CVEGMVE-IFDML
HGR              SYRQ-SSANLLCFAPDLIINE-QRMTLPCMYDQCKHMLYV
HMR              SYKH-TNSQFLYFAPDLVFNE-EKMHQSAMYELCQGMHQI
HPR              SYKH-VSGQMLYFAPDLIILNE-QRMKESSFYSLCLTMWQI
CPR              SYKH-VSGQMLYFAPDLIILNE-QRMKESSFYSLCLSMWQL
HRAR             -YTPEQDTMT--FSDGLTLNRTQMHNAGFGPLTDLVFAFA
HTR   T          -YDPESDTLTLSGEMAVK--REQLKNGGLGVVSDAIFELG
HTR   beta       -YDPESETLTLNGEMAVI--RGQLKNGGLGVVSDAIFDLG

HER              LATSSRFRMMNLQGEEFVCLKSIILLNSGVYTFLSSTLKS
HGR              SSELHRL---QVSYEEYLCMKTLLLLSSVPKDGLKSQELF
HMR              SLQFVRL---QLTFEEYTIMKVLLLLSTIPKDGLKSQAAF
HPR              PQEFVKL---QVSQEEFLCMKVLLLLNTIPLEGLRSQTQF
CPR              PQEFVRL---QVSQEEFLCMKALLLLNTIPIEGLRSQSQF
HRAR             NQLLPLEMDDA---ETGL-ISAICLICGD-RQDLEQPDR-
HTR   T          KSLSAFNLDDT---EVAL-LQAVLLMSTD-RSGLLCVDK-
HTR   beta       MSLSSFNLDDT---EVAL-LQAVLLMSSD-RPGLACVER-

HER              LEEK--DHIHRVLDKITDTLIHLMAKAGLTLQQQHQRLAQ
HGR              DEIR--MTYIKELGK------AIVKREGNSSQ-NWQRFYQ
HMR              EEMR--TNYIKELRK------MVTKCPNNSGQ-SWQRRYQ
HPR              EEMR--SSYIRELIK------AIGLRQKGVVS-SSQRFYQ
CPR              DEMR--TSYIRELVK------AIGLRQKGVVA-NSQRFYQ
HRAR             VDM-LQEPLI-EALK-----VYVRKRRPSRPHMF-PKM--
HTR   T          IE-KSQEAYLLAFE------HYVNHRKHNIPH-FWPKL--
HTR   beta       IE-KYQDSFLLAFE------HYINYRKHHVTH-FWPKL--

HER              LLLIL-SHIRHMSNKGMEHLYSMKCKNVV-------PLYD
HGR              LTKLLDS--MHEVVENL-LNYCFQTFLD-K-TMSIE-FPE
HMR              LTKLLDS--MHDLVSDL-LEFCFYTFRESH-ALKVE-FPA
HPR              LTKLLDN--LHDLVKQL-HLYCLNTEIQSR-ALSVE-FPE
CPR              LTKLMDS--MHDLVKQL-HLFCLKTFLQSR-ALSVE-FPE
HRAR             LMKITDLR-SI-------SAKGAERVITLKMEIPGSM-PP
HTR   T          LMK------120 RESIDUES-------------------
HTR   beta       LMKVTDLR-MI-------GACHASRFLHMKVECPTELLPP

HER              LLLEMLDAH--RLHAP----------          595*
HGR              MLAEIITNQIPKYSNGNIKKLLFHKK*         777
HMR              MLVEIIS-------------------          984*
HPR              MMSEVIAAQLPKILAGMVKPLLFHKK*         933
CPR              MMSEVIAAQLPKILAGMVKPLLFHKK*         786
HRAR             LIQEMLENSEG---------------          432*
HTR   T          --------------------------          490*
HTR   beta       LFLEVFED*                           456
```

Fig. 11 — Alignment of the ligand binding domains. Shown are the human estrogen (hER), glucocorticoid (hGR), mineralocorticoid (hMR), human (hPR) and chicken (cPR) progesterone, human retinoic acid (hRAR) receptors together with the human thyroid hormone α, cloned from a testis library (hTR T), and β (hTR β) receptors. Numbers refer to amino acid positions. An asterisk indicates the carboxy terminus.

activate gene transcription is unknown. It is interesting to note that the C terminus of the human glucocorticoid receptor, which has no apparent sequence homology with either the estrogen receptor or c-erb-A, is essential for binding glucocorticoids (Giguère *et al.*, 1986), whereas the corresponding region of the estrogen receptor is not required for hormone binding (Kumar *et al.*, 1986). In this respect, it is noteworthy that v-erb-A, which is unable to bind thyroid hormones (Sap *et al.*, 1986), is missing nine amino acids from its C terminus when compared with the chicken c-erb-A.

Thus, it appears that region E may form a hydrophobic pocket with some of the non-conserved amino acids providing ligand specificity and invariant amino acids maintaining the structure of the pocket.

IV. FUNCTION

A. The hormone-binding domain

1. Expression of the human estrogen receptor cDNA in transfected cells

The part of the human estrogen receptor cDNA containing the entire coding region (see Fig. 3) was cloned into the eukaryotic expression vector pKCR2 (Breathnach and Harris, 1983) downstream of the SV40 early promoter, producing the human estrogen receptor expression vector HE0 (Green *et al.*, 1986; Kumar *et al.*, 1986). HE0 was expressed transiently in HeLa cells, and its ability to direct the synthesis of estrogen receptor was tested using both estrogen receptor monoclonal antibodies and a hormone-binding assay. Fig. 12A shows that the expression of HE0 resulted in the appearance of a protein (lane 3, arrowhead) which had the same molecular weight as the estrogen receptor present in MCF-7 cells (lane 4). This protein was not detected in HeLa cells (lane 1) or in HeLa cells transfected with the parent pKCR2 vector (lane 2). Cytosol extracts prepared from HeLa cells transfected with HE0 were tested for their ability to bind estradiol (Fig. 12B). Whereas only background binding was detected when using pKCR2 (open squares), significant binding was seen when using HE0 (triangles). The protein exhibited an affinity for estradiol (K_d= 0.4 nM) that was very similar to that of the MCF-7 (circles) estrogen receptor (K_d= 0.5 nM (Fig. 12B, inset)). Thus, the expression of the estrogen receptor cDNA in HeLa cells yields a protein which, using these criteria, is indistinguishable from the MCF-7 estrogen receptor.

The cDNA insert of HE0 was also inserted downstream of the T7 promoter of the plasmid vector Bluescribe (Stratagene), allowing large amounts of estrogen receptor mRNA to be synthesized *in vitro*. Translation of this mRNA in a rabbit reticulocyte lysate produced a 66 kDa protein which bound estradiol with high affinity ($K_d \sim$ 1 nM) (Kumar *et al.*, 1986; data not shown).

2. The hormone-binding domain is contained in region E

A series of contiguous human estrogen receptor deletion mutants (HE1 to HE9; Kumar *et al.*, 1986; see Fig. 13) were created using site-directed mutagenesis in order to study the structure–function relationship of the human estrogen receptor.

Two approaches were used to localize the estradiol-binding domain of the human estrogen receptor. The first approach involved the transient expression of the human estrogen receptor mutants in HeLa cells and the second used the coupled *in vitro*

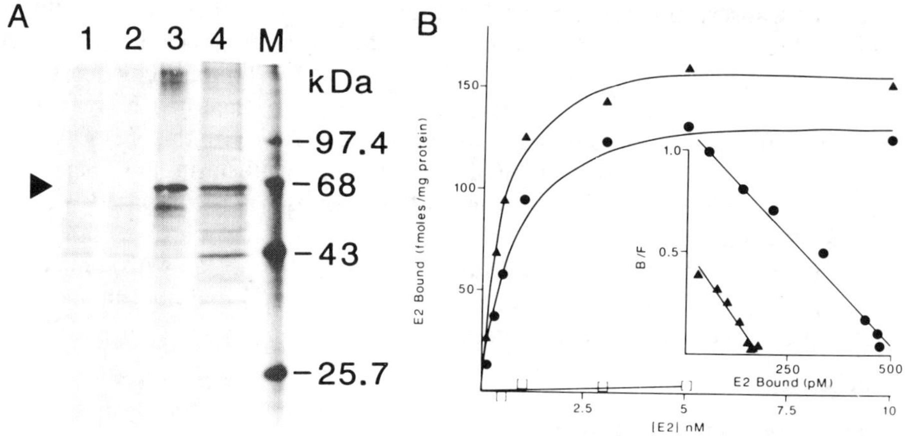

Fig. 12 — Expression of the human estrogen receptor (ER) cDNA in HeLa cells. (A) The size of the protein encoded by the cloned estrogen receptor cDNA is identical to that of the MCF-7 estrogen receptor. HE0 was transfected into HeLa cells and the proteins were labeled with [^{35}S]methionine. Cytoplasmic proteins were immunoprecipitated with estrogen receptor antibodies, separated by SDS–PAGE (7%), and revealed by fluorography. Lane 1, wild-type HeLa cells; lane 2, HeLa cells transfected with pKCR2; lane 3, HeLa cells transfected with HE0; lane 4, wild-type MCF-7 cells. The arrow shows the position of the 66 kDa protein. Lane M shows ^{14}C-labeled protein markers: phosphorylase-b (97.4 kDa), bovine serum albumin (68 kDa), ovalbumin (43 kDa), α-chymotrypsinogen (25.7 kDa). (B) The estrogen receptor cDNA-coded protein binds estradiol with the same affinity as the MCF-7 estrogen receptor. Cytosol extracts, prepared from MCF-7 cells (●) or HeLa cells transfected with pKCR2 (□) or HE0 (▲) were incubated overnight at 0°C in the presence of [^{3}H]17β-estradiol (E2) with or without a 100-fold excess of unlabeled 17β-estradiol. The inset shows a Scatchard plot of the same data.

transcription–translation system (see above). In both cases it was found that mutants HE1 to HE4 bound estradiol normally, whereas mutants HE5 to HE9 did not. That mutants HE5 to HE9 were expressed in HeLa cells was verified by immunoprecipitation and electrophoresis. In each case, with the exception of HE7, a protein of ~60 kDa was expressed with about the same efficiency as HE0 (Kumar *et al.*, 1986). This size corresponds to that expected following the deletion of approximately 65 amino acids in these mutants. These results suggested that region E contained the hormone-binding domain and therefore two additional deletion mutants (HE12 and HE13; Fig. 13) were created to delineate further the region responsible. The deletion of HE12 removes a portion of region D up to the boundary of region E, which begins at amino acid 302. The deletion in HE13 removes all of region F from amino acids 553 to 595. Both HE12 and HE13 bound estradiol with an affinity similar to HE0 (data not shown).

An estrogen receptor mutant (HE14) consisting of only the C-terminal half of the estrogen receptor, which contains half of region D and all of regions E and F, was expressed in HeLa cells or synthesized *in vitro*. HE14 was able to bind estradiol with an affinity very similar to that of HE0 (K_d~1 nM), indicating that hormone binding is independent of the N-terminal half of the receptor (Kumar *et al.*, 1986).

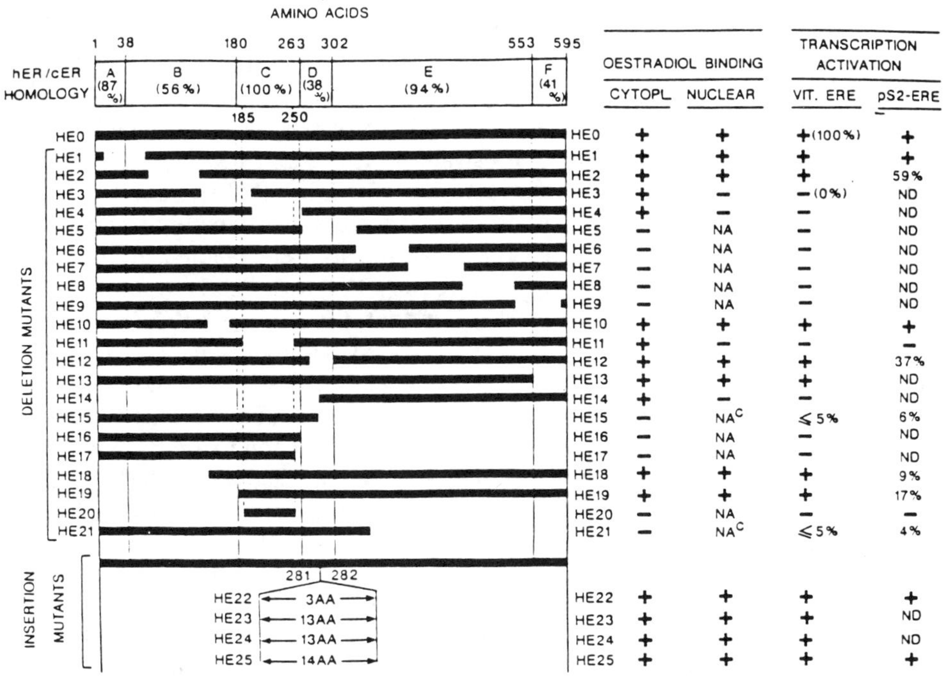

Fig. 13 — Analysis of human estrogen receptor mutants. The structure of each of the human estrogen receptor mutants which were analyzed for estradiol binding, in both the cytoplasm and nucleus, and for their abilities to activate transcription of the two reporter genes vit-tk-CAT (VIT·ERE) and pS2-CAT (pS2·ERE) is shown schematically. The division of the human estrogen receptor into six regions (A–F) is shown at the top of the figure together with the percentage homologies between the human and chicken estrogen receptors in these regions. Other numbers refer to amino acid positions. HE0 represents the wild-type receptor as indicated by the solid line from amino acid 1 to 595. A gap indicates a deletion, and the arrows for mutant HE22 to HE25 indicate insertions between amino acids 281 and 282. The sequence of the insertions is as follows: HE22: SRA; HE23: SRALALAIDISRA; HE24: SRGMGMGIDISRA; HE25: SRPIDSLLSLLSRA (single-letter amino acid code). The results of the analysis of the mutants are shown along the right-hand side of the figure. A positive sign (+) indicates wild-type activity and a negative sign (−) no activity. Values which fell between these two extremes are indicated as percentages of wild-type (100%). The suffix 'c' for mutants HE15 and HE21 in the column for nuclear estradiol binding indicate that these two mutants compete with HE0 in transcription competition assays. ND, not determined. NA, not applicable.

The estradiol-binding domain is therefore located in the hydrophobic region between amino acids 301 and 552. As noted above, this region may form a hydrophobic pocket, the structure of which is destroyed by deletions in this region.

B. The tight nuclear-binding domain

The unoccupied human estrogen receptor is only loosely associated with the cell's nuclear compartment, since the majority of the receptor is found in the cytosolic extract when cells grown in the absence of estrogens are homogenized in low-salt buffers. In contrast, the hormone–receptor complex is tightly associated with the nuclear compartment from which it can be extracted using a high concentration of salt (e.g. 0.4 M KCl). Thus, if MCF-7 cells are labeled *in vivo* with [^{3}H]estradiol, 78% of the receptor-bound estradiol is tightly associated with the nucleus (Kumar *et al.*, 1986). When cytosol and nuclear extracts were prepared from HeLa cells transfected with HE0 in the absence of estradiol, at least 80% of the unoccupied receptor was found in the cytosolic extract, demonstrating that the hormone-free receptor was not tightly associated with the HeLa cell nuclear compartment (our unpublished results). However, when these same cells were labeled *in vivo* with [^{3}H]estradiol, >90% of the receptor-bound estradiol was tightly bound to the nucleus (Kumar *et al.*, 1986), demonstrating that the cloned MCF-7 estrogen receptor expressed in HeLa cells becomes tightly associated with nuclear sites when bound to estradiol. This same assay was used to analyze the ability of those mutants which bound estradiol to associate tightly with the nucleus (Fig. 13). Of the initial four mutants tested (HE1 to HE4), only HE1 and HE2 were found tightly associated with the nucleus. Thus, any deletion which altered region C abolished tight nuclear association. Deletion mutants HE10, HE11 and HE12 were constructed to delineate further the region important for tight nuclear association (Fig. 13). While HE10 behaved like HE0 for nuclear binding, only about 20% of HE12 and none of HE11 bound tightly (Kumar *et al.*, 1986; data not shown; see Fig. 13). Measurement of the cellular distribution of HE11, which lacks the cysteine and basic amino acid rich domain of region C, showed that >80% of the receptor is found in the cytosolic fraction, compared with only 3% for the wild-type receptor HE0 (Kumar *et al.*, 1986). We conclude that region A–B is dispensable for tight nuclear binding, whereas at least the highly conserved region C and to some extent the less conserved region D (HE12) are required.

C. Target gene specificity

1. Assay for activation of target gene transcription

Using co-transfection experiments, it was shown that HE0 stimulated the transcription of an estrogen-inducible reporter gene and that this stimulation was hormone dependent (Druege *et al.*, 1986). Two different estrogen-regulated reporter genes were used to analyze the effect of human estrogen receptor mutations on activation of estrogen-responsive elements (ERE). One of these, vit-tk-CAT, consists of the *Xenopus* vitellogenin A2 upstream region (−331 to −87) fused to the Herpes simplex virus thymidine kinase (tk) promoter (−105 to +51) downstream of which lies the *E. coli* chloramphenicol acetyl transferase (CAT) gene (Klein-Hitpass *et al.*, 1986). The second reporter gene was constructed from the human estrogen-responsive gene, pS2, which was isolated, as was the human estrogen receptor sequence used in this study, from MCF-7 cells (Masiakowski *et al.*, 1982; Brown *et al.*, 1984; Jeltsch *et al.*, 1987). The 5′-flanking promoter region of pS2 (~−1100 to +10) was fused upstream of the CAT indicator gene. Half-maximal stimulation of vit-tk-CAT required approximately 200 pM estradiol (data not shown), and all assays were performed with 10 nM estradiol. Titration experiments performed using 1 μg of vit-tk-CAT or 10 μg of pS2-CAT showed that maximal activity was obtained using approximately 100 ng and 250 ng of the human estrogen receptor expression vector

HE0 respectively (data not shown). Experiments with vit-tk-CAT were performed using between 10 ng and 5 μg of the receptor mutant expression vectors, while those with pS2-CAT were performed using 1 and 5 μg.

2. *Region C is responsible for the specific recognition of the hormone responsive element*

Assuming that steroid receptors activate gene transcription by a mechanism which involves the specific recognition of promoter sequences of the target gene, we reasoned that the replacement of the human estrogen receptor DNA-binding domain with that of the human glucocorticoid receptor may produce a chimeric receptor capable of activating glucocorticoid-responsive genes in the presence of estradiol. Since region C, which is responsible for tight nuclear binding, was likely to contain the DNA-binding domain (see above), we decided to prepare such a chimeric receptor by replacing the 66 highly conserved amino acids of region C of the human estrogen receptor by those of the human glucocorticoid receptor (Green and Chambon, 1987; Fig. 14A).

To facilitate the exchange of region C between the two receptors, unique restriction enzyme recognition sites were created at the 5′-(KpnI) and 3′-(XhoI) ends of this region in the cDNAs of both the human estrogen receptor and human glucocorticoid receptor (HE28 and GR·CAS; Fig. 14A). The estrogen receptor cysteine-rich cassette (ER·CAS) was replaced by that of the glucocorticoid receptor (GR·CAS) to produce the novel human estrogen–glucocorticoid receptor chimera ER–GR·CAS (Fig. 14A). Both HE28 and ER–GR·CAS bound estradiol as effectively as HE0 when expressed in HeLa cells (data not shown). These receptor expression vectors, together with HE0, were tested in the presence and absence of estradiol for their ability to activate the estrogen-inducible reporter gene vit-tk-CAT in HeLa cells. In the presence of estradiol, CAT activity was stimulated by both HE0 and HE28, whereas ER–GR·CAS and the parent expression vector, pKCR2 (which does not contain receptor sequences) had no effect (Fig. 14B). These receptors were also tested using the glucocorticoid-inducible reporter gene, MMTV-CAT. This gene contains the mouse mammary tumor virus (MMTV) long terminal repeat (−631 to +125) upstream of the CAT gene (Cato *et al.*, 1986). Whereas HE0, HE28, and pKCR2 were not able to stimulate MMTV-CAT, stimulation with ER–GR·CAS was seen only in the presence of estradiol (Fig. 14B), and the concentration of estradiol required for activation was similar to that required for HE0 to stimulate vit-tk-CAT (our unpublished results), suggesting that the mechanism by which estradiol activates these receptors is similar. Thus region C of steroid receptors determines the specificity of target gene activation by the hormone.

The reverse 'finger-swapping' experiment has been performed which gives a GR–ER·CAS chimera which is able to transactivate transcription from the ERE of estrogen receptor target genes under the influence of glucocorticoids (Kumar *et al.*, 1987). A very important application of this technique is the identification of ligands of unknown receptor genes belonging to the same superfamily (see Figs. 8 and 10) as demonstrated in our laboratory by Petkovich *et al.* (1987). In their case a cDNA was isolated on the basis of crosshybridization with a synthetic oligonucleotide probe (see also Chapter 14) which contained an open reading frame with the structural characteristics of a 'finger' receptor (Fig. 8). No ligand or target gene of this receptor

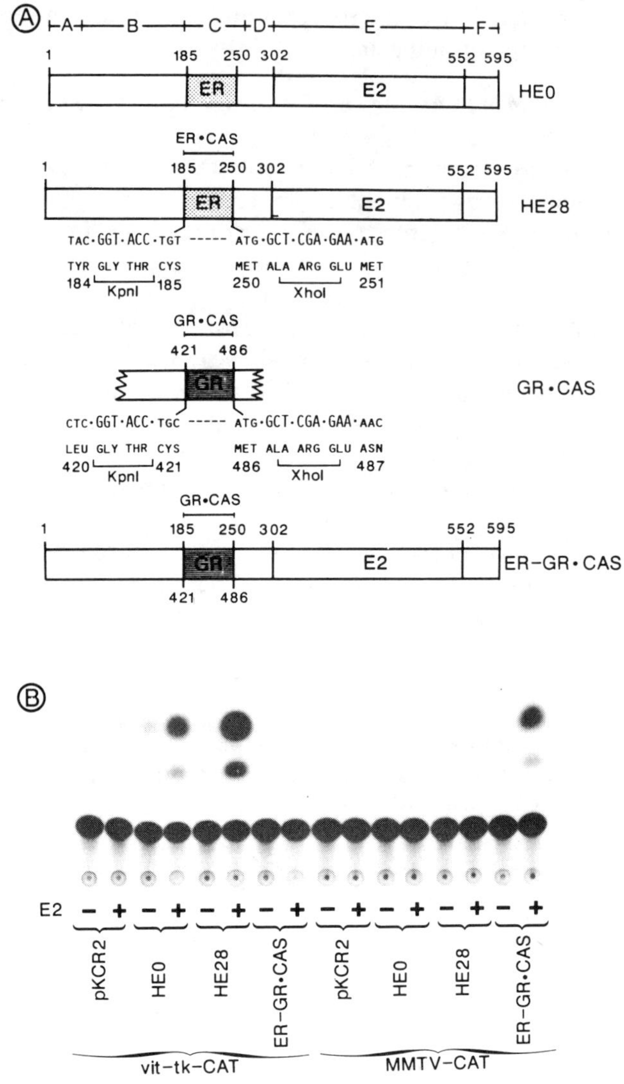

Fig. 14 — (A) Construction of the chimeric receptor ER–GR·CAS. The structures of HE0, HE28 and GR·CAS are shown schematically; the numbers refer to amino acid positions. The division of wild-type hER (HE0) into regions A–F is shown at the top of the figure. The putative DNA-binding domains within region C of the human estrogen (ER) and glucocorticoid (GR) receptors together with the estrogen receptor estrogen-binding domain (E2) are shaded. The KpnI and XhoI sites, created within region C of the estrogen and glucocorticoid receptors to allow the exchange of this region as a cassette (ER·CAS and GR·CAS), are indicated with the additional amino acids introduced into the HE28 protein. (B) The receptor chimera, ER–GR·CAS, activates the glucocorticoid-responsive reporter gene (MMTV-CAT) in the presence of estradiol. Each of the expression vectors (pKCR2, HE0, HE28, and ER–GR·CAS) was co-transfected into HeLa cells together with either the estrogen-responsive (vit-tk-CAT) or the glucocorticoid-responsive (MMTV-CAT) reporter genes in the absence (–) or presence (+) of 10^{-8} M estradiol.

was known. Therefore, Petkovich *et al.* constructed the chimera XR–ER·CAS in which the finger region of the unknown receptor (XR) was replaced with that of the estrogen receptor (illustrated in Fig. 15) and cloned into the eukaryotic expression

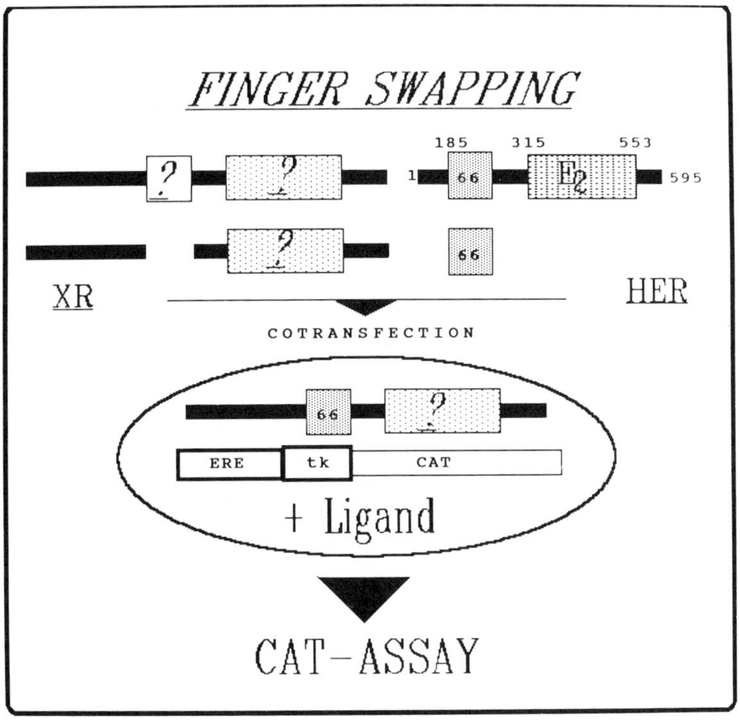

Fig. 15 — Schematic illustration of the 'Finger-swapping' technique. An unknown member of the nuclear receptor family (XR) is identified by its homology of region C (empty box) and E (dotted box). The region C of XR is replaced by the corresponding region of the human estrogen receptor (HER). In co-transfection assays, using for example vit-tk-CAT (containing the ERE), putative ligands are checked for transactivation of transcription from the ERE by CAT assay.

vector pKCR2. Subsequently, a variety of potential ligands were tested for their influence on vit-tk-CAT expression in transient transfection assays. In this way XR was identified as the human retinoic acid receptor (RAR). Other members of the nuclear receptor superfamily which are not yet cloned (Fig. 10) may be identified by a similar approach.

D. Activation of transcription

1. Region C is essential for activation of transcription

Each of the initial series of human estrogen receptor deletion mutants (HE1 to HE9; see Fig. 13) was tested for its ability to activate gene transcription using the estrogen-inducible reporter genes vit-tk-CAT and pS2-CAT. None of the deletion mutants which removed part of region C (HE3, HE4) was able to activate gene transcription (Fig. 13). Three other deletion mutants (HE10, HE11 and HE12) were used to localize further the important region. Only HE11 failed to stimulate transcription, indicating that the highly conserved cysteine-rich region C is important for the activation of transcription.

2. Region E is essential for efficient activation of transcription

Using the same series of human estrogen receptor deletion mutants described above (HE1 to HE13; see Fig. 13), it was found that receptors containing large deletions within the hormone-binding domain (HE5 to HE9) fail to activate vit-tk-CAT transcription, whereas HE13 had wild-type activity. These results parallel those obtained with the hormone-binding assay (Kumar *et al.*, 1986; see above and Fig. 13). They suggested that the hormone-binding domain was indispensable for gene activation but did not exclude the possibility that the only function of the unoccupied hormone-binding domain is to prevent the receptor from either binding to the ERE and/or activating transcription. We therefore examined whether the N-terminal half of the human estrogen receptor, containing the DNA-binding domain but lacking all the hormone-binding domain, could constitutively stimulate transcription. Mutants HE15, HE16, and HE17, which lack regions E and F, and HE21, which lacks a large portion of region E and all of region F, were prepared for this purpose (see Fig. 13). In the absence of estradiol, HE15 and HE21 stimulated vit-tk-CAT to a level which was ≤5% of that observed under identical conditions using HE0. No stimulation was seen when using HE16 and HE17. It is important to note, however, that due to the presence of the thymidine kinase promoter, the basal CAT activity of the vit-tk-CAT reporter gene in the absence of either human estrogen receptor or estradiol is in the same range, thus making exact estimations of the receptor-induced stimulation difficult. Similar experiments were performed using HE15 and HE21 with the other reporter gene, pS2-CAT. A stimulation of approximately 5% of that seen with HE0 was observed. Although the activity of HE15 and HE21 is very low, it is nevertheless clear that they do possess some constitutive activity when compared with HE11, which lacks the DNA-binding domain and is totally unable to stimulate CAT activity (see Fig. 13).

We then created a severely deleted human estrogen receptor mutant (HE20; see Fig. 13) containing the highly conserved 66 amino acids of region C flanked by a Kozak translation initiation consensus sequence (Kozak, 1984a) (N terminal) and a stop codon (C terminal). This domain is responsible for defining the receptor's target gene specificity (see above). No stimulation of either vit-tk-CAT or pS2-CAT could be detected when using HE20 (Fig. 13).

Explanations for why little or no stimulation is observed when using these C-terminal truncated receptors (HE15, HE16, HE17, HE20, and HE21) include their possible instability or their inability to enter the cell nucleus and therefore to interact with the ERE. However, immunoprecipitation studies using human estrogen receptor monoclonal antibodies showed that these receptor mutants (except for HE20, which does not contain the epitopes for these antibodies) are expressed and can be detected in the nucleus (our unpublished results).

More interestingly, these mutants may poorly activate gene expression because they lack a domain located in regions D or E, which is important for activation of transcription, and/or because they have a weak affinity for the ERE. Competition experiments performed in HeLa cells by co-transfecting these human estrogen receptor mutants (HE15, HE16, HE17, HE20, or HE21) with HE0 and vit-tk-CAT showed that two of these C-terminal deletion mutants (HE15 and HE21) could efficiently 'repress' the ability of HE0 to stimulate the expression of the reporter gene (Kumar *et al.*, 1987). Furthermore, competition with HE0 was eliminated by deleting the cysteine-rich region C from the competing receptor expression vector, suggesting that DNA binding is important for efficient competition. We interpret these results as showing that HE15 and HE21 can bind efficiently to the ERE but, since they are poor transcriptional activators, they reduce the stimulation seen with HE0.

3. *Differential effect of region A–B on activation of the vitellogenin and pS2 EREs*

It was previously speculated that region A–B may be important for activating gene transcription (Krust *et al.*, 1986; Kumar *et al.*, 1986). Therefore human estrogen receptors with N-terminal deletions were analyzed for their ability to stimulate expression of the two estrogen reporter genes. Co-transfections of HE1, HE2, and HE10 with vit-tk-CAT demonstrated that none of these receptor mutants was defective in its ability to activate gene transcription and that in each case stimulation was estrogen dependent (Kumar *et al.*, 1987; see Fig. 13). A similar analysis using the reporter gene pS2-CAT showed the same to be true in this system, except that HE2 had only approximately 60% of the stimulatory activity of HE0 (see Fig. 13).

Together the deletions represented by HE1, HE2, and HE10 cover almost all of region A–B. Two deletion mutants, which removed almost (HE18) or all (HE19) of region A–B (Fig. 13), were constructed to eliminate the possibility of there being several functionally redundant elements in this region. Both HE18 and HE19, when expressed in HeLa cells, bound hormone (Fig. 13) with the same affinity as HE0. Measurement of the ability of these hormone–receptor complexes to bind tightly to the nucleus showed that HE18 had nuclear levels approximately 40% of those of HE0, while the value for HE19 was about 70%. Experiments performed using vit-tk-CAT indicated that each of these mutants stimulated CAT activity to the same extent as HE0 (Fig. 13) and that deletion of region A–B had no effect on transcription initiated at the thymidine kinase promoter start site (Kumar *et al.*, 1987). Surprisingly, however, pS2–CAT activity was stimulated by only 9% with HE18 and 17% with HE19 (Fig. 13), indicating that the two receptor mutants have different effects on the two types of estrogen-responsive reporter genes. The same relative decrease was observed using various amounts of the expression vectors suggesting that this effect cannot simply be ascribed to a lower affinity of the receptor mutants for the pS2 ERE.

Thus it appears that while the presence of region A–B is not absolutely required for stimulation of transcription it may be necessary to obtain maximal stimulation with some estrogen-responsive genes.

4. *Does region D act as a hinge?*

Since region D was postulated to act as a hinge between the DNA and hormone binding domains (see above), we investigated the effect of varying the length of region D on the stimulation of transcription. One deletion mutant (HE12) and four insertion mutants (HE22–HE25) were analyzed (Fig. 13). HE12 contains a deletion of 30 amino acids which are relatively poorly conserved between the chicken and human estrogen receptors (Krust *et al.*, 1986). This mutant receptor bound estradiol with the same affinity as HE0 (data not shown), but the amount of the receptor–hormone complex which bound tightly to the nucleus was only about 20% that observed with HE0 (Kumar *et al.*, 1986, 1987). Examination of the ability of HE12 to activate vit-tk-CAT demonstrated that, despite the reduction in tight nuclear binding, HE12 stimulated transcription equally as well as HE0 (Fig. 13). In contrast, HE12 showed only 37% of the stimulatory activity of HE0 when used with the pS2–CAT reporter gene (Fig. 13), again indicating that the requirements for efficient activation of transcription of the two estrogen-responsive reporter genes by the human estrogen receptor are different.

Insertions of between 3 and 14 amino acids were made in the middle of region D (HE22–HE25; see Fig. 13). These mutants bound estradiol as effectively as HE0 and the receptor–hormone complexes bound tightly to the nucleus. The ability of these insertion mutants to stimulate vit-tk-CAT and pS2-CAT was not significantly different from that of HE0.

Thus the length of region D (39 amino acids) may be altered significantly (between the deletion of 30 amino acids and the insertion of 14 amino acids) without a loss of receptor function. Clearly, when compared with HE0, HE12 is somewhat impaired in its ability to bind tightly to the nucleus and to activate pS2 gene transcription, but the fact that this receptor works at all indicates that the exact distance between the DNA and hormone-binding domains is not critical when using the present assays.

E. Functional domains of the glucocorticoid and progesterone receptors

1. *Glucocorticoid receptor*

As described above, regions C and E of the estrogen receptor share a large degree of homology with the glucocorticoid receptor. Therefore it is not surprising that mutational analysis of the rat (Rusconi *et al.*, 1986; Godowski *et al.*, 1987; Rusconi and Yamamoto, 1987), mouse (Danielson *et al.*, 1986) and human (Giguère *et al.*, 1986) glucocorticoid receptors have shown that these two regions play the same roles in both receptors.

Giguère *et al.* (1986) prepared a series of 27 human glucocorticoid receptor mutants, each with a single three or four amino acid insertion at specific positions of the 777 amino acid sequence. These mutants were tested for their ability to bind hormone or to activate gene transcription from the MMTV-CAT reporter gene after transient expression in monkey kidney cells. Their results show that the hormone-binding domain is localized in the C-terminal region of the glucocorticoid receptor (from amino acid 550 to 777). Similar conclusions were made from an analysis of deletion mutants of the rat glucocorticoid receptor (Rusconi and Yamamoto, 1987). Furthermore, analysis of mouse glucocorticoid receptors has shown that two independent point mutations in this C-terminal domain impair hormone binding (Danielson *et al.*, 1986).

Steroid hormone receptors normally activate gene transcription only after binding their cognate hormone. Certain deletion mutants of both the rat and human glucocorticoid receptor, which lack all or part of the hormone binding domain, are able to constitutively activate MMTV-CAT (Godowski *et al.*, 1987; Hollenberg *et al.*, 1987). In the case of one human glucocorticoid receptor deletion mutant this constitutive activity is at least as great as that of the occupied wild type receptor (Hollenberg *et al.*, 1987). One possible interpretation of these results is that the unoccupied hormone binding domain exerts some negative influence on the activity of the receptor which is relieved when the hormone binds.

The results with insertion mutants of the human glucocorticoid receptor (Giguère *et al.*, 1987) also confirm the importance of region C for activation of gene transcription, since insertions which alter the spacing of the cysteines in this region either inactivate or drastically reduce the ability of the receptor to stimulate MMTV-CAT transcription. A point mutation in this region of a mouse nt^- glucocorticoid receptor, which changes an arginine to a histidine, reduces the ability of the receptor to bind tightly to the nucleus (Danielson *et al.*, 1986). The integrity of this region is also essential for specific DNA binding *in vitro*. In the case of both the rat (Rusconi *et al.*, 1986) and the human (Hollenberg *et al.*, 1987) glucocorticoid receptors any deletions which remove part of this highly conserved region result in the loss of specific DNA binding using an *in vitro* DNA-binding assay.

As discussed above, there is no apparent sequence homology within region A–B of the estrogen receptor between any of the members of the steroid and thyroid hormone receptor family. Insertion and deletion mutants within the human glucocorticoid receptor have defined a region known as Tau1 between amino acids 77 to 262 which is essential in order to activate MMTV-CAT transcription fully (Giguère *et al.*, 1986; Hollenberg *et al.*, 1987). Deletion mutants of the rat glucocorticoid receptor, within the region equivalent to the region A–B of the estrogen receptor, also appear to reduce the receptor's ability to activate MMTV-CAT transcription in transient assays and endogenous MMTV and α-1-acid glycoprotein (AGP) transcription in stably transformed cell lines (Miesfeld *et al.*, 1987). In the case of both the rat and the human receptors the removal of N-terminal sequences extending close to the cysteine-rich region results in receptors which have approximately 10% of wild type activity. That some activity remains when this region is deleted suggests that whilst this region may not be absolutely essential for receptor function, it is clearly important for full activity. Therefore these results can be interpreted as indicating either that this region plays a modulatory role (Hollenberg *et al.*, 1987) or that the deletion of this region results in a loss of structural constraints provided by sequences located outside of the essential functional domains (Miesfeld *et al.*, 1987).

2. *Progesterone receptor*

An amino acid sequence comparison between the chick progesterone receptor and its human (Misrahi *et al.*, 1987; A. Krust, U. Stropp, P. Kastner and P. Chambon, unpublished results) and rabbit (Loosfelt *et al.*, 1986) counterparts reveals two highly conserved regions. The cysteine-rich region C spanning from amino acid number 410 to 495 (see Fig. 4 and top of Fig. 16) exhibits a 100% conservation, whereas region E from amino acid 540 to the C terminus is 87% conserved, when compared with the progesterone receptor of mammals. These two regions are separated by a region D of lower homology (~60%). In contrast, the N-terminal region (termed A–B by analogy with the corresponding region of the estrogen

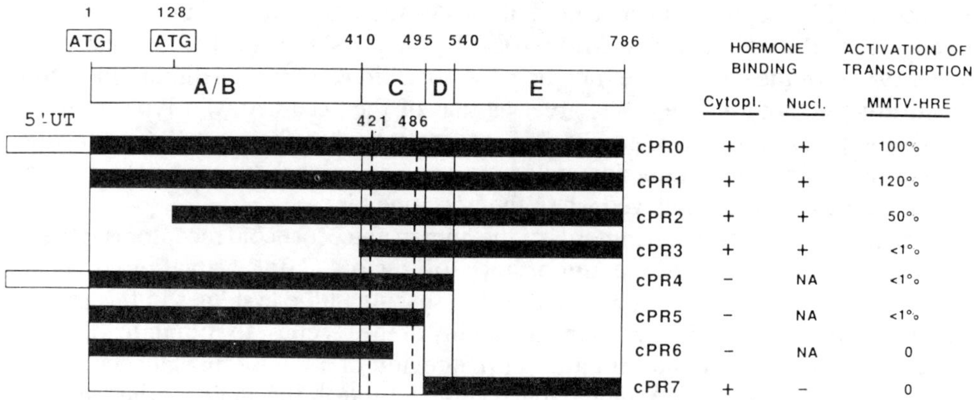

Fig. 16 — Domain structure and functional properties of the chicken progesterone receptor and its deletion mutants. On the basis of an amino acid sequence comparison with its human and rabbit counterparts, the chicken progesterone receptor has been subdivided into two highly conserved regions C (AA 410–495) and E (AA 540–786), exhibiting 100% and 87% homology with the mammalian progesterone receptors respectively, separated by a region D of lower homology (~60%) and an N-terminal region A–B, differing considerably in both length and amino acid sequences from the mammalian progesterone receptors. The nomenclature of these regions is in analogy with that of Krust *et al.* (1986); the amino acid position of their boundaries is given at the top, as well as those of the two only ATGs present in region A–B. The central cysteine-rich core of region C (amino acids 421–486, see text) is indicated by dashed lines. A solid black bar represents the chicken progesterone receptor sequences present in the deletion mutants (cPR2–cPR7) with the deleted portion as a gap. The wild-type chicken progesterone receptor sequence, cPR0, contains all 786 amino acids as well as the 5′-untranslated sequence (5′-UT) downstream of position +29 (see Fig. 4), indicated as an open box. The same 5′ region is also present in cPR4 whereas it is deleted in the other recombinants. The mutants cPR1–cPR3 and cPR5–cPR7 all contain 'consensus' Kozak sequences at their translational start sites. On the right side under 'HORMONE BINDING' the ability of the various mutants to bind either progestins in cytoplasmic extracts ('Cytopl.') or the ability of the progestin–receptor complex to bind tightly to HeLa cell nuclei ('Nucl.') is shown. +, wild-type progesterone receptor characteristics; −, the absence of specific progestin binding. The ability of chicken progesterone receptor wild-type and mutant constructs to transactivate transcription by interacting with the MMTV-HRE is shown on the very right ('ACTIVATION OF TRANSCRIPTION'). The data, which correspond to the average of at least three independent quantitative S1 nuclease analyses with different plasmid preparations, were obtained from densitometric scans of the autoradiograms and were corrected for transcription from the cotransfected internal control recombinant pG1B. The values are relative to cPRO taken as 100%. NA, not applicable.

receptor (Krust *et al.*, 1986)) of these progesterone receptors diverges very significantly in both sequence and length (556, 556 and 409 amino acids for the human, rabbit and chicken progesterone receptors respectively). It is in fact this region which is responsible for the size difference between the chicken progesterone receptor and mammalian progesterone receptors. Functional studies of the estrogen and glucocorticoid receptors have shown that the 66 amino acid long core of region C, which is characterized by a specific cysteine-repeat motif (amino acids 421–486 in the chicken progesterone receptor; see the boxed sequence in Fig. 4 and also top of Fig. 16) contains the domain responsible for 'tight' nuclear binding and specific recognition of the hormone responsive elements of the target genes (Green and Chambon, 1987;

Kumar *et al.*, 1987, 1988). It has also been shown that region E of these receptors contains the hormone-binding domain and that this domain can function independently (Kumar *et al.*, 1986).

On the basis of these results, we constructed the expression vector cPR3 which contains the C, D and E regions (see Fig. 16). When transfected in HeLa cells grown in an hormone-free medium, cPR3 resulted in the appearance of a truncated chicken progesterone receptor protein which bound progesterone with 'wild type' characteristics indistinguishable from those of the protein encoded in cPR0 or cPR1 (see above; Fig. 3, 'cytopl.', and data not shown). 'Tight' nuclear binding was also observed with the truncated receptor encoded in cPR3 (Fig. 3, 'nucl.', and data not shown). These results suggest very strongly that regions C and E of the chicken progesterone receptor have the same functions as those previously established for the corresponding domains of the estrogen and glucocorticoid receptors (see above). In particular it appears that the highly conserved region E can function as an independent progestin-binding domain because expression of the isolated region E resulted in the appearance of a truncated protein capable of binding progestins (Fig. 16). As expected, no hormone binding could be detected in cells transfected with recombinants in which region E was deleted (cPR4, cPR5 or cPR6, see below and Fig. 16).

It has been shown that not only the glucocorticoid receptor, but also the progesterone receptor, can activate initiation of transcription *in vivo* from the MMTV LTR by interaction with its HRE (Cato *et al.*, 1986, 1987). This observation prompted us to investigate whether the 'cloned' chicken progesterone receptor transactivates transcription from the MMTV LTR. cPR0 and cPR1 were transfected in HeLa cells grown in the presence or absence of R5020, together with the MMTV-CAT reporter gene (Cato *et al.*, 1987). A clear hormone-dependent stimulation of CAT expression was observed with cPR0, cPR1 or cPR2 (Fig. 17A): compare lanes 3, 5 and 7 with lanes 4, 6 and 8; see also lanes 1 and 2 which correspond to transfections with the parent expression vector pKCR2. The results shown in Fig. 17A indicate also that the truncated chicken progesterone receptors encoded in cPR3 (lanes 9 and 10) and cPR5 (lanes 11 and 12) had a drastically decreased stimulatory activity which, as expected, was independent of the presence of the hormone in the case of cPR5. No stimulation could be detected when cPR6 (lanes 13 and 14) was used.

In order to quantify more rigorously the transcriptional stimulatory activity of the various deletion mutants, RNA was determined by quantitative S1 nuclease analyses using an MMTV LTR based reported recombinant in which the CAT sequence was replaced by a promoterless rabbit β-globin gene (MMTV-globin or MG). In addition, in order to be able to correct for variations in transfection efficiencies, an internal control recombinant (pG1B) containing the rabbit β-globin gene and its promoter was co-transfected. Using this system, the RNAs initiated at the MMTV LTR start site (+1 MG in Fig. 17B) and at the globin start site of pG1B (+1 pG1B) could be quantitatively determined by scanning the autoradiograms of polyacrylamide–urea gels, following hybridization with a single stranded ^{32}P-labeled DNA probe and nuclease S1 digestion. The results of a representative transfection experiment in HeLa cells are displayed in Fig. 17B, and the average values obtained by scanning at least three similar autoradiograms are given in Fig. 16, relative to

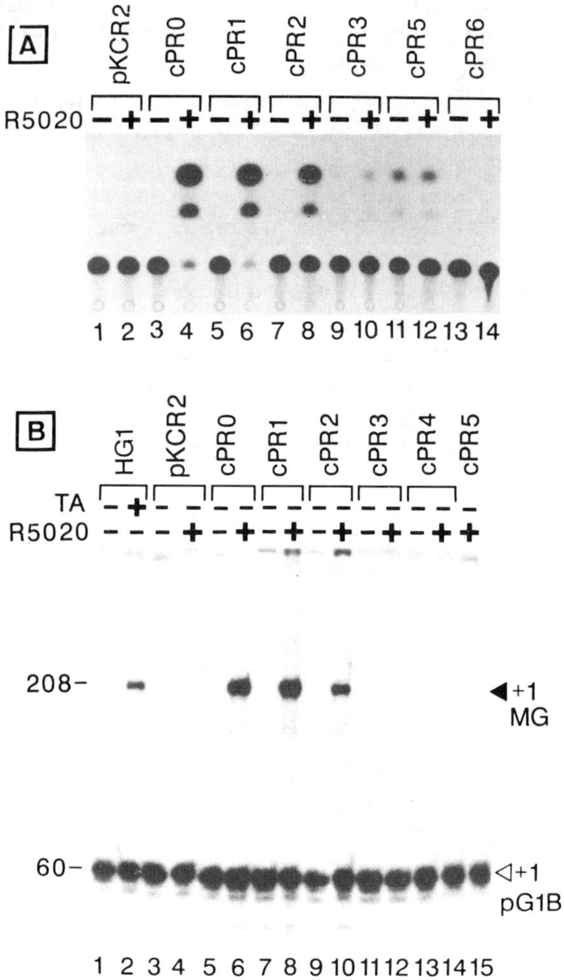

Fig. 17 — Activation of transcription by the chicken progesterone receptor and its deletion mutants. (A) CAT assay using the cytosol of HeLa cells transiently transfected with 1 μg of the chicken progesterone receptor expression vectors (lanes 3–14, as indicated) or with the parent vector pKCR2 (lanes 1 and 2) and grown in the presence (+) or absence (−) of 5 nM R5020. The reporter plasmid used was MMTV-CAT (Cato *et al.*, 1987). The amounts of extracts for the CAT assay were normalized by measuring the β-galactosidase activity generated from the co-transfected vector pCH110 (Pharmacia) which encodes the *E. coli* β-galactosidase. (B) Quantitative S1 nuclease analysis. The human glucocorticoid expression vector (HG1, lanes 1 and 2), the parent vector pKCR2 (lanes 3 and 4) or the chicken progesterone receptor expression vectors as indicated (lanes 5–15) were co-transfected in HeLa cells together with an MMTV-globin reporter recombinant (MG) and the internal reference vector pG1B (for description of these recombinants, see Part I of this book). The cells were grown in the presence (+) or absence (−) of 5 nM progestin (R5020), or triamcinolone acetonide (TA) as indicated. After isolation of the RNA, hybridization and S1 nuclease treatment were resolved on denaturing polyacrylamide gels. The fragments seen on the autoradiogram correspond to RNA initiated at the start site of transcription in the MMTV LTR (208 nucleotides in length, indicated by +1 MG) and in the internal control recombinant (60 nucleotides in length indicated by +1 pG1B).

those obtained with cPR0 (taken as 100%, after correction for transcription from the internal control pG1B). A very strong stimulation (>100-fold) was observed with cPR0 in the presence of the progestin R5020. In fact, this stimulation was ~4-fold higher than that obtained under identical conditions (but in the presence of triamcinolone acetonide) with HG1, a pKCR2-based vector expressing the glucocorticoid receptor (Kumar *et al.*, 1987).

The stimulatory activity of cPR1 was very close to that of cPR0 whereas that of cPR2 which encodes a truncated receptor which may correspond to the 'natural' chicken progesterone receptor form A (see below) was decreased by ~2-fold. In agreement with the results of the CAT assay, deletion of the A–B region to amino acid 405 resulted in a decrease in stimulation of transcription by at least two orders of magnitude (cPR3 in Fig. 17B, lanes 11 and 12). This is particularly striking because the truncated protein encoded in cPR3 bound progesterone with 'wild-type' characteristics and was 'tightly' bound to the nucleus in the presence of progestins (see above). Finally, very little (cPR4 and cPR5, Fig. 17B, lanes 13–15) or no (cPR6) stimulation of transcription was observed with the other deletion mutants, as expected from the results with MMTV-CAT. Recently we have constructed mutants in which 35 amino acids were added N terminally to cPR3, as well as further mutants with different C-terminal truncations in region E. The results obtained support the view that both A–B and E are required for efficient activation of the MMTV-HRE by the progesterone receptor.

F. Origin of two chicken progesterone receptors

On the basis of the results obtained during the cloning of the chicken progesterone receptor cDNA and gene, several possibilities for the origin of two receptor forms with extensive structural and immunological similarities could be excluded. For example, we did not find any evidence for the existence of a second receptor gene or for a cDNA encoding the hormone binding domain which was not collinear with the one described in Fig. 2. Since both forms bind progestins with identical characteristics and since the hormone-binding domain was localized in the carboxy-terminal region, two possibilities were considered as the most likely ones: (i) internal initiation from the only AUG present in the 5′ half of the RNA, or (ii) proteolysis in this region. If internal initiation actually occurred in the chick oviduct, an expression vector encoding only the C-terminal region of AUG 128 (see Fig. 2) should produce a protein of the same length as the natural form A. Fig. 18 shows that *in vitro* expression of cPRG2 (containing the sequences for amino acid 128 to the carboxy terminus) in fact produces a protein with a molecular weight similar to that of the photoaffinity labeled form A. Additionally, extracts of Cos-1 cells transiently transfected with expression vectors encoding the same progesterone receptor region exhibited a 79 kDa band similar to the chick oviduct receptor form A on immunoblots (Fig. 19). Although *in vitro* internal initiation accounts, at least in part, for the production of a 79 kDa protein (Gronemeyer *et al.*, 1987), following transient transfection of Cos-1 cells only traces of form A could be detected. These results were obtained even when expressing a cDNA containing most of the 5′-untranslated region, the ATGTGA motif, and the natural Lozak's sequences which should most accurately reproduce normal translational regulation (Fig. 19, lane 2; see also Gronemeyer *et al.*, 1987).

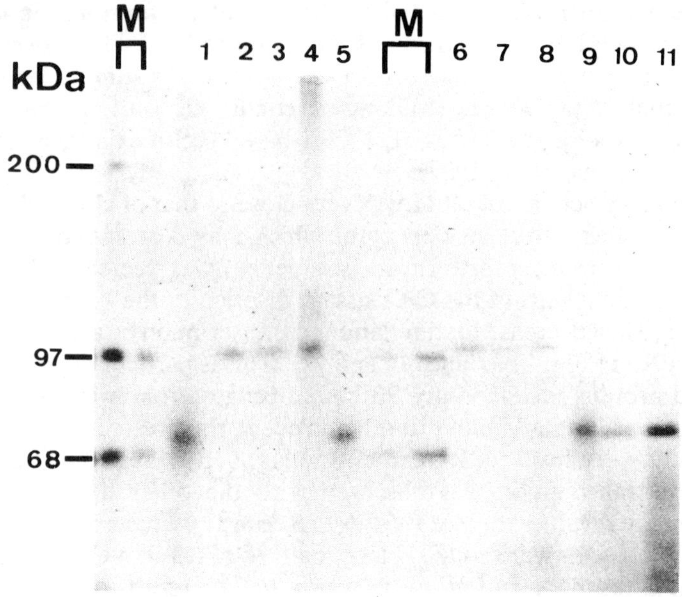

Fig. 18 — Comparison of *in vitro* and *in vivo* expressed chick oviduct progesterone receptor with the natural progesterone receptor forms from oviduct. Partially purified and [^{3}H]R5020-crosslinked chick oviduct progesterone receptors form A (lanes 1, 5 and 9) and form B (lanes 2, 4 and 6) and crosslinked cytosol of Cos-1 cells, transiently transfected with cPR0 (lane 3) are separated on a 7.5% SDS-polyacrylamide gel together with the *in vitro* synthesized form B (lanes 7 and 8) or form A (expressing from amino acid 128, lanes 10 and 11) (see also Gronemeyer *et al.*, 1987). M, marker lanes; the molecular weights of the nucleus are given on the left.

Therefore, we conclude that form B is the full-length chicken receptor whereas form A corresponds to an N terminally truncated receptor generated either by internal initiation of translation from AUG 128 or by proteolysis in the vicinity of this AUG. Both processes appear to be tissue specific as we could not mimic them in our transient transfection systems.

G. How do steroid hormone receptors activate specific gene transcription?

In conclusion, the results described above indicate that the steroid hormone receptors are composed of at least two physically independent domains, one which binds DNA, and therefore determines target gene specificity, and one which binds hormone. The mechanisms by which hormone binding 'activates' the receptor and the hormone–receptor complex stimulates gene transcription still remain unsolved. The presence of the hormone is clearly essential for gene activation *in vivo*, but recent reports suggest that in the case of the glucocorticoid receptor (Willmann and Beato, 1986) and progesterone receptor (Bailly *et al.*, 1986) the hormone plays no role in determining the specificity of the DNA binding of the receptor *in vitro* (reviewed in Green and Chambon, 1986). This, together with the fact that the hormone-free glucocorticoid receptor appears to be unable to recognize specific

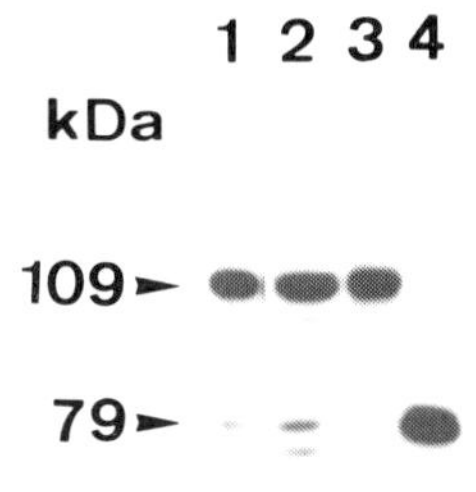

Fig. 19 — Western blot of cytoplasmic extracts of Cos-1 cells transiently transfected with cPR0 (lane 2), cPR2 (lane 3) or cPR2 (lane 4). A Western blot of a partially purified chick oviduct progesterone receptor form B preparation, run on the same gel, is shown for comparison (lane 1). The positions of form B (109 kDa) and form A (79 kDa), still present in the form B preparation, are indicated.

responsive elements of a target gene *in vivo* (Becker *et al.*, 1986), has led to the speculation that the role of the hormone *in vivo* is to 'unmask' a pre-formed DNA-binding domain hidden in the unoccupied receptor by an as-yet unknown structure. This might be, for example, a part of the receptor itself or the 90 kDa heat shock protein (Joab *et al.*, 1984; reviewed in Green and Chambon, 1986). If this assumption is correct, then our own results indicate that the function of the hormone-binding domain cannot simply be to mediate the masking of the DNA binding domain, since mutants where the hormone-binding domain, contained in the C-terminal half of the estradiol receptor, is deleted cannot efficiently activate gene transcription whereas they still recognize efficiently the ERE (see above). Thus, the occupied hormone-binding domain may be necessary for activation, directly or indirectly, of the transcription machinery. This assumption is further supported by experiments in which the transcriptional activating domain of GAL4, a yeast transcription factor, was replaced with the hormone-binding domain of the estrogen receptor, yielding a chimera which is fully active in the transactivation of transcription from an estrogen-responsive element (Webster *et al.*, 1988). The same may not be true for the glucocorticoid receptor where gene transcription can be stimulated efficiently when using some, but not all, truncated glucocorticoid receptors in which the hormone-binding domain has been removed (Godowski *et al.*, 1987; Hollenberg *et al.*, 1987). It cannot be excluded that in these cases the truncation might generate a transcriptional activating domain pre-existing in the receptor in a different conformation. It is of interest that the estrogen receptor mutants HE15 and HE21 are still able to stimulate gene transcription, even if stimulation is only 5% of that seen with HE0. This suggests that region C together with a small amount of region D (i.e. the difference between HE19 and HE15) contain all that is necessary for DNA binding and at least an element of that which is required for gene activation, but that the integrity of the hormone binding domain is required for efficient activation of transcription.

A structrual and functional analysis of a mutant glucocorticoid receptor protein (nt^i) found in a glucocorticoid-resistant mouse lymphoma cell line has demonstrated that it contains both a hormone-binding domain and a DNA-binding domain which has an increased affinity for non-specific DNA (Payvar and Wrange, 1983). Since these cells are resistant to the normal killing action of glucocorticoids, the receptor is presumed to be non-functional. Structural comparison of the nt^i mutant (~40 kDa) with the wild type glucocorticoid receptor (Hollenberg *et al.*, 1985; Danielson *et al.*, 1986; Miesfeld *et al.*, 1986) suggests that this mutant has lost the region which corresponds to region A–B of the estrogen receptor. It has been proposed that the domain which is missing in the nt^i mutant is important for modulating the activity of the glucocorticoid receptor (Westphal *et al.*, 1984). A portion of the human glucocorticoid receptor region A–B (referred to as Tau1) is important for activation of transcription Giguère *et al.*, 1986) and a chicken progesterone receptor devoid of region A–B is unable to stimulate transcription (see above). In contrast, mutants in which region A–B of the rat glucocorticoid (Miesfeld *et al.*, 1987) or human estrogen (see above) receptors are deleted can, in some cases (see differences between pS2 and vit-tk-CAT expression above), still fully stimulate gene transcription. It appears therefore that the A–B region contains a domain of steroid receptors important for activating the transcription machinery but that its requirement may not be universal.

At the present time, there are clearly insufficient data to present a satisfactory model of steroid hormone receptor action. It is evident that it is the highly conserved region C which determines, presumably through DNA binding, the specificity of the receptor–target gene interaction. The role of the hormone-binding domain in regulating this interaction, the role of region A–B, and the location of the domain(s) activating the transcription machinery still need to be determined.

Finally, it should be stressed that the present studies have been performed using 'transient assays', in which the activated hormone-responsive elements are not integrated in the genome and are probably not in their 'normal' chromatin structure. The role of the various regions of the receptor, and in particular that of region A–B, may become more apparent in future studies when using responsive genes in their 'normal' environment.

ACKNOWLEDGMENTS

We are grateful to our numerous colleagues who have participated in various aspects of the work; their names are apparent from the list of references. We gratefully acknowledge the help of the secretarial staff and of B. Boulay and C. Werlé in the preparation of the figures. This work was supported by grants from the CNRS (A.I.V.), the INSERM (grants CNAMTS), the Ministères de la Recherche et de la Technologie, the Fondation pour la Recherche Médicale and the Association pour la Recherche sur le Cancer.

REFERENCES

Arriza, J. L., Weinberger, C., Cerelli, G., Glaser, T. M., Handelin, B. L., Housman, D. E. and Evans, R. M. (1978) Cloning of human mineralocorticoid receptor complementary DNA: structural and functional kinship with the glucocorticoid receptor. *Science* **237**, 268–275.

Bailly, A., LePage, C., Rauch, M. and Milgrom, E. (1986) Sequence-specific DNA binding of the progesterone receptor to the uteroglobin gene: effects of hormone, antihormone and receptor phosphorylation. *EMBO J.* **5**, 3235–3241.

Becker, P. B., Gloss, B., Schmid, W., Strähle, U. and Schutz, G. (1986) *In vivo* protein–DNA interactions in a glucocorticoid response element require the presence of the hormone. *Nature* **324**, 686–688.

Benbrook, D. and Pfahl, M. (1987) A novel thyroid hormone receptor encoded by a cDNA clone from a human testis library. *Science* **238**, 788–791.

Berg, J. M. (1986a) Potential metal-binding domains in nucleic acid binding proteins. *Science* **232**, 485–487.

Berg, J. M. (1986b) More metal-binding fingers. *Nature* **319**, 264–265.

Boulay J. L., Dennefeld, C. and Alberga, A. (1987) The *Drosophila* developmental gene snail encodes a protein with nucleic acid binding fingers. *Nature* **330**, 395–398.

Breathnach, R. and Harris, B. A. (1983) Plasmids for the cloning and expression of full-length double-stranded cDNAs under control of the SV40 early or late gene promoter. *Nucleic Acids Res.* **11**, 7119–7136.

Brown, A. M. C., Jeltsch, J. M., Roberts, M. and Chambon, P. (1984) Activation of pS2 gene transcription is a primary response to estrogen in the human breast cancer cell line MCF-7. *Proc. Natl. Acad. Sci. USA* **81**, 6344–6348.

Burshell, A., Stathis, P., Do, Y., Miller, S. and Feldman, D. (1984) Characterization of an estrogen-binding protein in the yeast *Saccharomyces cerevisiae. J. Biol. Chem.* **259**, 3450–3456.

Cato, A. C. B., Miksicek, R., Schütz, G., Arnemann, J. and Beato, M. (1986) The hormone regulatory element of mouse mammary tumour virus mediates progesterone induction. *EMBO J.* **5**, 2237–2240.

Cato, A. C. B., Henderson, D. and Ponta, H. (1987) The hormone response element of the mouse mammary tumour virus DNA mediates the progestin and androgen induction of transcription in the proviral long terminal repeat region. *EMBO J.* **6**, 363–368.

Chowdhury, K., Deutsch, U. and Gruss, P. (1987) A multigene family encoding several "finger" structures is present and differentially active in mammalian genomes. *Cell* **48**, 771–778.

Colvard, D. S. and Wilson, E. M. (1984) Zinc potentiation of androgen receptor binding to nuclei *in vitro. Biochemistry* **23**, 3471.

Conneely, O. M., Sullivan, W. P., Toft, D. O., Birnbaumer, M., Cook, R. G., Maxwell, B. L., Zarucki-Schulz, T., Greene, G. L., Schrader, W. T. and O'Malley, B. W. (1986) Molecular cloning of the chicken progesterone receptor. *Science* **233**, 767–770.

Conneely, O. M., Dobson, A. D. W., Tsai, M. J., Beattie, W. G., Toft, D. O., Huckaby, C. S., Zarucki, T., Schrader, W. T. and O'Malley, B. W. (1987) Sequence and expression of a functional chicken progesterone receptor. *Mol. Endocrinol.* **1**, 517–525.

Danielson, M., Northrop, J. P. and Ringold, G. M. (1986) The mouse glucocorticoid receptor: mapping of functional domains by cloning, sequencing and expression

of wild-type and mutant receptor proteins. *EMBO J.* **5**, 2513–2522.
Dejean, A., Bouguelberet, L., Grzeschik, K. H. and Tiollais, P. (1986) Hepatitis B virus DNA integration in a sequence homologous to v-erb-A and steroid receptor genes in a hepatocellular carcinoma. *Nature* **322**, 70–72.
de Thé, H., Marchio, A., Tiollais, P. and Dejean, A. (1987) A novel steroid thyroid hormone receptor-related gene inappropriately expressed in human hepatocellular carcinoma. *Nature* **330**, 667–670.
Druege, P. M., Klein-Hitpass, L., Green, S., Stack, G., Chambon, P. and Ryffel, U. G. (1986) Introduction of estrogen-responsiveness into mammalian cell lines. *Nucleic Acids Res.* **14**, 9329–9337.
Feldman, D., Do, Y., Burshell, A., Stathis, P. and Loose, D. S. (1982) An estrogen-binding protein and endogenous ligand in *Saccharomyces cerevisiae*: possible hormone receptor system. *Science* **218**, 297–298.
Feldman, D., Stathis, P., Hirst, M. A., Stover, E. P., Do, Y. and Kurz, W. (1984a) *Saccharomyces cerevisiae* produces a yeast substance that exhibits estrogenic activity in mammalian systems. *Science* **224**, 1109–1111.
Feldman, D., Toekes, L. G., Stathis, P., Miller, S., Kurz, W. and Harvey, D. (1984b) Identification of 17 β-estradiol as the estrogenic substance in *Saccharomyces cerevisiae. Proc. Natl. Acad. Sci. USA* **81**, 4722–4726.
Gehring, U., Segnitz, B., Foellmer, B. and Francke, U. (1985) Assignment of the human gene for the glucocorticoid receptor to chromosome 5. *Proc. Natl. Acad. Sci. USA* **82**, 3751–3755.
Giguère, V., Hollenberg, S. M., Rosenfeld, M. G. and Evans, R. M. (1986) Functional domains of the human glucocorticoid receptor. *Cell* **46**, 645–652.
Giguère, V., Ong, E. S., Segui, P. and Evans, R. M. (1987) Identification of a receptor for the morphogen retinoic acid. *Nature* **330**, 624–629.
Gininger, E. and Ptashne, M. (1988) Cooperative DNA binding of the yeast transcriptional activator, GAL4. *Proc. Natl. Acad. Sci. USA*, in press.
Glass, C. K., Franco, R., Weinberger, C., Albert, V. R., Evans, R. M. and Rosenfeld, M. G. (1987) A c-erb-A binding site in rat growth hormone gene mediates trans-activation by thyroid hormone. *Nature* **329**, 738–741.
Godowski, P. J., Rusconi, S., Miesfeld, R. and Yamamoto, K. R. (1987) Glucocorticoid receptor mutants that are constitutive activators of transcriptional enhancement. *Nature* **325**, 365–368.
Govindan, M. V., Devic, M., Green, S., Gronemeyer, H. and Chambon, P. (1985) Cloning of the human glucocorticoid receptor cDNA. *Nucleic Acids Res.* **13**, 8293–8304.
Green, S. and Chambon, P. (1986) A superfamily of potentially oncogenic hormone receptors. *Nature* **324**, 615–617.
Green, S. and Chambon, P. (1987) Oestradiol induction of a glucocorticoid-responsive gene by a chimeric receptor. *Nature* **325**, 75–78.
Green, S., Walter, P., Kumar, V., Krust, A., Bornert, J. M., Argos, P. and Chambon, P. (1986) Human oestrogen receptor cDNA: sequence, expression and homology to v-erbA. *Nature* **320**, 134–139.
Green, S., Gronemeyer, H. and Chambon, P. (1987) Structure and function of steroid hormone receptors. In: Sluyser, M. (ed.) *Oncogenes and Hormones in Breast Cancer*. Ellis Horwood, Chichester, pp. 7–28.

Greene, G. L., Gilna, P., Waterfield, M., Baker, A., Hort, Y. and Shine, J. (1986) Sequence and expression of human estrogen receptor complementary DNA. *Science* **231**, 1150–1154.

Gronemeyer, H., Harry, P. and Alberga, A. (1983) A reappraisal of ecdysteroid binding in *Drosophila. Mol. Cell. Endocrinology* **32**, 171–178.

Gronemeyer, H., Turcotte, B., Quirin-Stricker, C., Bocquel, M. T., Meyer, M. E., Krozowski, Z., Jeltsch, J. M., Lerouge, T., Garnier, J. M. and Chambon, P. (1987) The chicken progesterone receptor: sequence, expression and functional analysis. *EMBO J.* **6**, 3985–3994.

Hartshorne, T. A., Blumberg, H. and Young, E. T. (1986) Sequence homology of the yeast regulatory protein ADR1 with *Xenopus* transcription factor TFIIIA. *Nature* **320**, 283—287.

Hollenberg, S. M., Weinberger, C., Ong, E. S., Cerelli, G., Oro, A., Lebo, R., Thompson, E. B., Rosenfeld, M. G. and Evans, R. M. (1985) Primary structure and expression of a functional human glucocorticoid receptor cDNA. *Nature* **318**, 635–641.

Hollenberg, S. M., Giguère, V., Segui, P. and Evans, R. M. (1987) Colocalization of DNA-binding and transcriptional activation functions in the human glucocorticoid receptor. *Cell* **49**, 39–46.

Jantzen, H. M., Strähle, U., Gloss, B., Stewart, F., Schmid, W., Boshart, M., Miksicek, R. and Schütz, G. (1987) Cooperativity of glucocorticoid response elements located far upstream of the tyrosine aminotransferase gene. *Cell* **42**, 29–38.

Jeltsch, J. M., Krozowski, Z., Quirin-Stricker, C., Gronemeyer, H., Simpson, R. J., Garnier, J. M., Krust, A., Jacob, F. and Chambon, P. (1986) Cloning of the chicken progesterone receptor. *Proc. Natl. Acad. Sci. USA* **83**, 5424–5428.

Jeltsch, J. M., Roberts, M., Schatz, C., Garnier, J. M., Brown, A. M. C. and Chambon, P. (1987) Structure of the human oestrogen-responsive gene pS2. *Nucleic Acids Res.* **15**, 1401–1414.

Joab, I., Radanyi, C., Renoir, M., Buchou, T., Catelli, M. G., Binart, N., Mester, J. and Baulieu, E. E. (1984) Common non-hormone binding component in non-transformed chick oviduct receptors of four steroid hormones. *Nature* **308**, 850–853.

Johnston, M. (1987) Genetic evidence that zinc is an essential co-factor in the DNA binding domain of GAL4 protein. *Nature* **328**, 353–355.

Kammerer, B., Guyonvarch, A. and Hubert, J. C. (1984) Yeast regulatory gene PPR1. I. Nucleotide sequence, restriction map and codon usage. *J. Mol. Biol.* **180**, 239–250.

Klein-Hitpass, L., Schropp, M. Wagner, U. and Ryffel, G. U. (1986) An estrogen-responsive element derived from the 5′ flanking region of the *Xenopus* vitellogenin A2 gene functions in transfected human cells. *Cell* **46**, 1053–1061.

Kleinsek, D. A., Beattie, W. G., Tsai, M. J. and O'Malley, B. W. (1986) Molecular cloning of a steroid-regulated 108K heat shock protein gene from hen oviduct. *Nucleic Acids Res.* **14**, 10053–10069.

Klock, G., Strähle, U. and Schütz, G. (1987) Oestrogen and glucocorticoid responsive elements are closely related but distinct. *Nature* **329**, 734–736.

Klug, A. and Rhodes, D. (1987) "Zinc fingers": a novel protein motif for nucleic acid recognition. *TIBS* **12**, 464–469.

Koike, S., Sakai, M. and Muramatsu, M. (1987) Molecular cloning and characterization of rat estrogen receptor cDNA. *Nucleic Acids Res.* **15**, 2499–2513.

Kozak, M. (1984a) Compilation and analysis of sequences upstream from the translational start site in eukaryotic mRNAs. *Nucleic Acids Res.* **12**, 857–872.

Kozak, M. (1984b) Selection of initiation sites by eucaryotic ribosomes: effect of inserting AUG triplets upstream from the coding sequence for preproinsulin. *Nucleic Acids Res.* **12**, 3873–3893.

Kozak, M. (1986) Bifunctional messenger RNAs in eukaryotes. *Cell* **47**, 481–483.

Krust, A., Green, S., Argos, P., Kumar, V., Walter, P., Bornert, J. M. and Chambon, P. (1986) The chicken oestrogen receptor sequence: homology with v-erbA and the human oestrogen and glucocorticoid receptors. *EMBO J.* **5**, 891–897.

Kulomaa, M. S., Weigel, N. L., Kleinsek, D. A., Beattie, W. G., Conneely, O. M., March, C., Zarucki-Schulz, T., Schrader, W. Y. and O'Malley, B. W. (1986) Amino acid sequence of a chicken heat shock protein derived from the complementary DNA nucleotide sequence. *Biochemistry* **25**, 6244–6251.

Kumar, V., Green, S., Staub, A. and Chambon, P. (1986) Localisation of the oestradiol-binding and putative DNA-binding domains of the human oestrogen receptor. *EMBO J.* **5**, 2231–2236.

Kyte, J. and Doolittle, R. F. (1982) A simple method for displaying the hydropathic character of a protein. *J. Mol. Biol.* **157**, 105–1322.

Lathe, R. (1985) Synthetic oligonucleotide probes deduced from amino acid sequence data: theoretical and practical considerations. *J. Mol. Biol.* **183**, 1–12.

Laughon, A. and Gesteland, R. F. (1984) Primary structure of the *Saccharomyces cerevisiae* GAL4 gene. *Mol. Cell. Biol.* **4**, 260–267.

Law, M. L., Kao, F. T., Wei, Q., Hartz, J. A., Greene, G. L., Zarucki-Schultz, T., Conneely, O. M., Jones, C., Puck, T. T., O'Malley, B. W. and Horwitz, K. B. (1987) The progesterone receptor gene maps to human chromosome band 11q13, the site of the mammary oncogene int-2. *Proc. Natl. Acad. Sci. USA* **84**, 2877–2881.

Liu, C. C., Simonsen, C. C. and Levinson, A. D. (1984) Initiation of translation at internal AUG codons in mammalian cells. *Nature* **309**, 82–85.

Lohmar, P. H. and Toft, D. O. (1975) Inhibition of the binding of progesterone receptor to nuclei: effects of o-phenanthroline and rifamycin AF/013. *Biochem. Biophys. Res. Comm.* **67**, 8.

Loosfelt, H., Atger, M., Misrahi, M., Guiochon-Mantel, A., Meriel, C., Logeat, F., Benarous, R. and Milgrom, E. (1986) Cloning and sequence analysis of rabbit progesterone-receptor complementary DNA. *Proc. Natl. Acad. Sci. USA* **83**, 9045–9049.

Masiakowski, P., Breathnach, R., Bloch, J., Gannon, F., Krust, A. and Chambon, P. (1982) Cloning of cDNA sequences of hormone-regulated genes from the MCF-7 human breast cancer cell line. *Nucleic Acids Res.* **10**, 7895–7903.

Mattei, M. G., Krust, A., Stropp, U., Mattei, J. F. and Chambon, P. (1987) Assignment of the human progesterone receptor to the q22 band of chromosome 11. *Human Genetics*, submitted.

Maxwell, B. L., McDonnell, D. P., Conneely, O. M., Schulz, T. Z., Greene, G. L.

and O'Malley, B. W. (1987) Structural organization and regulation of the chicken estrogen receptor. *Mol. Endocrinology* **1**, 25–35.

McDonnell, D. P., Mangelsdorf, D. J., Pike, J. W., Haussler, M. R. and O'Malley, B. W. (1987) Molecular cloning of complementary DNA encoding the avian receptor for vitamin D. *Science* **235**, 1214–1217.

Messenguy, F., Dubois, E. and Descamps, F. (1986) Nucleotide sequence of the ARGRII regulatory gene and amino acid sequence homologies between ARGRII PPRI and GAL4 regulatory proteins. *Eur. J. Biochem.* **157**, 77–81.

Miesfeld, R., Okret, S., Wikström, A. C., Wrange, Ö., Gustafsson, J. A. and Yamamoto, K. R. (1984) Characterization of a steroid hormone receptor gene and mRNA in wild-type and mutant cells. *Nature* **312**, 779–781.

Miesfeld, R., Rusconi, S., Godowski, P. J., Maler, B. A., Okret, S., Wikström, A. C., Gustafsson, J. A. and Yamamoto, K. R. (1986) Genetic complementation of a glucocorticoid receptor deficiency by expression of cloned receptor cDNA. *Cell* **46**, 389–399.

Miesfeld, R., Godowski, P. J., Maler, B. A. and Yamamoto, K. R. (1987) Glucocorticoid receptor mutants that define a small region sufficient for enhancer activation. *Science* **236**, 423–427.

Miller, J., McLachlan, A. D. and Klug, A. (1985) Repetitive zinc-binding domains in the protein transcription factor IIIA from *Xenopus* oocytes. *EMBO J.* **4**, 1609–1614.

Misrahi, M., Atger, M., d'Auriol, L., Loosfelt, H., Meriel, C., Fridlansky, F., Guiochon-Mantel, A., Galibert, F. and Milgrom, E. (1987) Complete amino acid sequence of the human progesterone receptor deduced from cloned cDNA. *Biochem. Biophys. Res. Comm.* **143**, 740–748.

Mueller, P. P. and Hinnebusch, A. G. (1986) Multiple upstream AUG codons mediate translational control of GCN4. *Cell* **45**, 201–207.

Ohno, S., Kawasaki, H., Imajoh, S. and Suzuki, K. (1987) Tissue-specific expression of three distinct types of rabbit protein kinase C. *Nature* **325**, 161–166.

Okret, S., Poellinger, L., Dong, Y. and Gustafsson, J. A. (1986) Down-regulation of glucocorticoid receptor mRNA by glucocorticoid hormones and recognition by the receptor of a specific binding sequence within a receptor cDNA clone. *Proc. Natl. Acad. Sci. USA* **83**, 5899–5903.

Oppenheimer, J. H. and Samuels, H. H. (eds.) (1983) *Molecular Basis of Thyroid Hormone Action*. Academic Press, New York.

Pabo, C. O. and Sauer, R. T. (1984) Protein–DNA recognition. *Ann. Rev. Biochem.* **53**, 293–321.

Payvar, F. and Wrange, O. (1983) Relative selectivities and efficiencies of DNA binding by purified intact and protease-cleaved glucocorticoid receptor. In: Eriksson, H. and Gustafsson, J. A. (eds.) *Steroid Hormone Receptors: Structure and Function*, Nobel Symposium Vol. 57. North-Holland, Amsterdam, pp. 267–284.

Petkovich, M., Brand, N. J., Krust, A. and Chambon, P. (1987) A human retinoic acid receptor which belongs to the family of nuclear receptors. *Nature* **330**, 444–450.

Ptashne, M. (1986) Gene regulation by proteins acting nearby and at a distance. *Nature* **322**, 697–701.

Rosenberg, U. B., Schröder, C., Preiss, A., Kienlin, A., Côté, S., Riede, I. and Jäckle, H. (1986) Structural homology of the product of the Drosophila Krüppel gene with *Xenopus* transcription factor IIIA. *Nature* **319**, 336–339.

Rusconi, S. and Yamamoto, K. R. (1987) Functional dissection of the hormone and DNA binding activities of the glucocorticoid receptor. *EMBO J.* **6**, 1309–1315.

Rusconi, S., Miesfeld, R., Godowski, P. J., Vanderbilt, J. N., Maler, B. A. and Yamamoto, K. R. (1986) Functional analysis of cloned glucocorticoid receptor sequences. In: Reznikoff, W. S., Burgess, R. R., Dahlenberg, J. E., Gross, C. A., Record, T. N. and Wickens, N. P. (eds.) *RNA Polymerase and the Regulation of Transcription*, Sixteenth Steenbock Symposium. Elsevier, New York.

Sabbah, M., Redeuilh, G., Secco, C. and Baulieu, E. E. (1987) The binding activity of estrogen receptor to DNA and heat-shock protein (M_r 90,000) is dependent on receptor-bound metal. *J. Biol. Chem.* **262**, 8631–8635.

Sap, J., Munoz, A., Dam, K., Ghysdael, J., Leutz, A., Beug, H. and Vennström, B. (1986) The c-erbA protein is a high affinity receptor for thyroid hormone. *Nature* **324**, 635–640.

Sargan, D. R., Tsai, M. J. and O'Malley, B. W. (1986) Hsp108, a novel heat shock inducible protein of chicken. *Biochemistry* **25**, 6252–6258.

Schmidt, T. J., Sekula, B. C. and Litwack, G. (1981) The effects of 1,10-phenanthroline on the binding of activated rat hepatic glucocorticoid-receptor complexes to deoxyribonucleic acid–cellulose. *Endocrinology* **109**, 803.

Schuh, R., Aicher, W., Gaul, U., Côté, S., Preiss, A., Maier, D., Seifert, E., Nauber, U., Schröder, C., Kemler, R. and Jäckle, H. (1986) A conserved family of nuclear proteins containing structural elements of the finger protein encoded by Krüppel, a *Drosophila* segmentation gene. *Cell* **47**, 1025–1032.

Shyamala, G. and Yeh, Y. F. (1975) Is the estrogen receptor of mammary glands a metallo-protein? *Biochem. Biophys. Res. Comm.* **64**, 408.

Simpson, R. J., Grego, B., Govindan, M. V. and Gronemeyer, H. (1987) Peptide sequencing of the chick oviduct progesterone receptor form B. *Mol. Cell. Endocrinol.* **52**, 177–184.

Spurr, H. K., Solomon, E., Jansson, M., Sheer, D., Goodfellow, P. N., Bodner, W. F. and Vennström, B. (1984) Chromosomal localization of the human homologues to the oncogenes *erb* A and B. *EMBO J.* **3**, 159–163.

Strähle, U., Klock, G. and Schütz, G. (1987) A DNA sequence of 15 base pairs is sufficient to mediate both glucocorticoid and progesterone induction of gene expression. *Proc. Natl. Acad. Sci. USA* **84**, 7871–7875.

Tautz, D., Lehmann, R., Schnürch, H., Schuh, R., Seifert, E., Kienlin, A., Jones, K. and Jäckle, H. (1987) Finger protein of novel structure encoded by hunchback, a second member of the gap class of *Drosophila* segmentation genes. *Nature* **327**, 383–389.

Thompson, C. C., Weinberger, C., Lebo, R. and Evans, R. M. (1987) Identification of a novel thyroid hormone receptor expressed in the mammalian central nervous system. *Science* **237**, 1610–1614.

Tzamarias, D., Alexandraki, D. and Thireos, G. (1986) Multiple cis-acting elements modulate the translational efficiency of GCN4 mRNA in yeast. *Proc. Natl. Acad. Sci. USA* **83**, 4849–4853.

Vincent, A. (1986) TFIIIA and homologous genes. The 'finger' proteins. *Nucleic Acids Res.* **14**, 4385–4391.

Vincent, A., Colot, H. V. and Rosbash, M. (1985) Sequence and structure of the Serendipity locus of *Drosophila melanogaster. J. Mol. Biol.* **186**, 149–166.

von der Ahe, D., Renoir, J. M., Buchou, T., Baulieu, E. E. and Beato, M. (1986) Receptors for glucocorticosteroid and progesterone recognize distinct features of a DNA regulatory element. *Proc. Natl. Acad. Sci. USA* **83**, 2817–2821.

Walter, P., Green, S., Greene, G., Krust, A., Bornert, J. M., Jeltsch, J. M., Staub, A., Jensen, E., Scrace, G., Waterfield, M. and Chambon, P. (1985) Cloning of the human oestrogen receptor cDNA. *Proc. Natl. Acad. Sci. USA* **82**, 7889–7893.

Webster, N., Jin, J. R., Green, S., Hollis, M. and Chambon, P. (1988) The yeast UAS_G is a transcriptional enhancer in human HeLa cells in the presence of the GAL4 trans-activator. *Cell*, in press.

Weiler, I. J., Lew, D. and Shapiro, D. J. (1987) The *Xenopus laevis* estrogen receptor: sequence homology with human and avian receptors and identification of multiple estrogen receptor messenger ribonucleic acids. *Mol. Endocrinology*, in press.

Weinberger, C., Hollenberg, S. M., Ong, E. S., Harmon, J. M., Brower, S. T., Cidlowski, J., Thompson, E. B., Rosenfeld, M. G. and Evans, R. M. (1985a) Identification of human glucocorticoid receptor complementary DNA clones by epitope selection. *Science* **228**, 740–741.

Weinberger, C., Hollenberg, S. M., Rosenfeld, M. G. and Evans, R. M. (1985b) Domain structure of human glucocorticoid receptor and its relationship to the v-erb-A oncogene product. *Nature* **318**, 670–672.

Weinberger, C., Thompson, C. C., Ong, E. S., Lebo, R. and Evans, R. M. (1986) The c-erbA gene encodes a thyroid hormone receptor. *Nature* **324**, 641–646.

Westphal, H. M., Mugele, K., Beato, M. and Gehring, U. (1984) Immunochemical characterization of wild-type and variant glucocorticoid receptors by monoclonal antibodies. *EMBO J.* **3**, 1493–1498.

White, R., Lees, J. A., Needhan, M. and Parker, M. (1987) Structural organization and expression of the mouse estrogen receptor. *Mol. Endocrinol.* **1**, 735–744.

Willmann, T. and Beato, M. (1986) Steroid-free glucocorticoid receptor binds specifically to mouse mammary tumour virus DNA. *Nature* **324**, 688–691.

Yamamoto, K. R. (1985) Steroid receptor regulated transcription of specific genes and gene networks. In: Campbell, A., Herskowitz, I. and Sandler, L. M. (eds.) *Annual Reviews of Genetics*, Vol. 19. Annual Reviews Inc., pp. 209–252.

Zarucki-Schulz, T., Kulomaa, M. S., Headon, D. R., Weigel, N. L., Baez, M., Edwards, D. P., McGuires, W. L., Schrader, W. T. and O'Malley, B. W. (1984) Molecular cloning of a cDNA for the chicken progesterone receptor B antigen. *Proc. Natl. Acad. Sci. USA* **81**, 6358–6362.

16

Analysis of the human glucocorticoid receptor gene promoter

M. V. Govindan[a], M. Burelle[a], C. Cantin[a], M. Devic[c], Y. Lachance[a], G. Leblanc[a], C. Lefebvre[a], P. Patel[b] and U. Stropp[c]
[a]MRC Group in Molecular Endocrinology and [b]Department of Microbiology, Laval University, Québec, Canada, and [c]Institut de Chimie Biologique, Faculté de Médecine, Strasbourg, France

I. INTRODUCTION

Steroid hormone receptors are a family of regulatory proteins, which control gene expression by their dependence on the binding of a specific ligand. Activation or inhibition of transcription of target genes results from the interaction of the hormone–receptor complex with the promoter elements of these genes (Yamamoto,

1985). The ability of the purified glucocorticoid receptor to interact with defined sequences in different hormonally regulated genes has been well documented (Payvar *et al.*, 1981; Govindan *et al.*, 1982; Geise *et al.*, 1982; Pfahl, 1982; Scheidereit *et al.*, 1983; Scheidereit and Beato, 1984; Karin *et al.*, 1984; Renkawitz *et al.*, 1984). Studies of steroid hormone receptors are of paramount interest not only for elucidating how the receptor interacts specifically with its ligand but also for understanding the mechanism leading to the regulation of transcription through specific DNA–protein interactions. To investigate the regulation of the expression of the human glucocorticoid receptor gene at the molecular level, we have identified and characterized the 5′ flanking region of this gene.

II. DETERMINATION OF THE mRNA 5′ END

A. Southern analysis

The genomic organization of the human glucocorticoid receptor gene was analyzed by digestion of human MCF-7 DNA with several restriction enzymes followed by Southern analysis (Fig. 1). The same blot was hybridized with nick-translated human

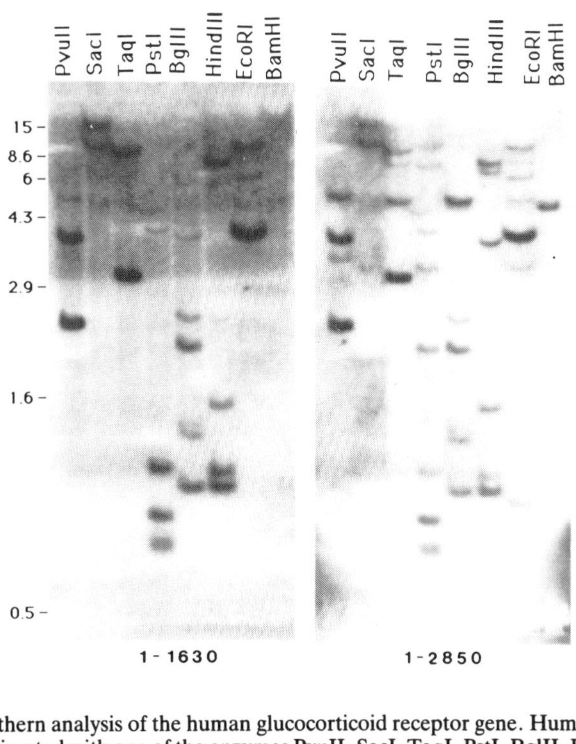

Fig. 1 — Southern analysis of the human glucocorticoid receptor gene. Human genomic DNA (30 μg) was digested with one of the enzymes PvuII, SacI, TaqI, PstI, BglII, HindIII, EcoRI or BamHI and electrophoresed on a 1.5% agarose gel. The DNA was transferred to diazobenzyl-oxymethyl (DBM) paper and hybridized with the nick-translated HGR 1-1630 and HGR 1-2850 cDNA probes. The position of the DNA size markers are indicated in kilobases.

glucocorticoid receptor cDNA sequences either from position 1-1630 (left) or from 1-2850 (right). The cDNA fragment from position 1-1630 encodes amino acid sequences from the amino terminus to the end of the v-erbA homology region (see Chapter 15), whereas the cDNA fragment from position 1-2850 contains the entire open reading frame of the human glucocorticoid receptor. From the data presented in Fig. 1 it appears that the length of the human glucocorticoid receptor is in the order of 20–30 kb.

B. Isolation of 5′-specific human glucocorticoid receptor genomic clones

Genomic clones containing the 5′ end of the human glucocorticoid receptor gene were isolated from a human λEMBL-3 genomic library (Frischauf *et al.*, 1983) using the oligonucleotide 5′-ATCTCCCCTCTCCTGAGCAAGCACACTGCT-3′ corresponding to amino acids 16–25 of the human glucocorticoid receptor. The first exon of the human glucocorticoid receptor gene was detected with a synthetic oligonucleotide (called 338 in Fig. 2) complementary to the glucocorticoid receptor 5′ non-translated region. One of the positive clones isolated contained three EcoRI fragments which were 4.5, 3.5 and 0.8 kb long. The 3.5 kb EcoRI–EcoRI genomic subclone contained the sequences corresponding to the amino terminus of the human glucocorticoid receptor (amino acids 1–131), including 14 nucleotides of the 5′ non-coding region. The 4.5 kb EcoRI–EcoRI genomic subclone contained a 1600 bp SmaI–SmaI fragment which hybridized to the oligo '338' and was found to contain a part of the 5′ untranslated region of the human glucocorticoid receptor cDNA. Sequence analysis of the SmaI–SmaI subclone revealed that one exon is located within this fragment. This exon contains the 5′ untranslated region (position +1 to +86 in Fig. 2) and ends with the nucleotides CG. The adjacent intron starts with the nucleotides GT (see Fig. 2). The sequence between the start of the open reading frame of the human glucocorticoid receptor and the first exon was not found in the SmaI–SmaI fragment. These nucleotides may be present in a separate exon.

C. Primer extension

To identify the 5′ boundary of the human glucocorticoid receptor RNA, the 5′ end-labeled oligonucleotide '338' (Fig. 2), corresponding to the anti-coding strand, was prepared and hybridized to LNCaP (human prostatic carcinoma cells) and MCF-7 (human breast tumor cells) polyA$^+$ RNA. The primer was extended with reverse transcriptase and the sizes of the resulting products were determined by gel electrophoresis under denaturing conditions, followed by autoradiography. The most abundant extension product has a length of 88 bases for both LNCaP and MCF-7 mRNAs (Fig. 3, lanes 6 and 8 respectively). Minor extension products 163, 136, 132, 120, 110 and 39 bases in length were observed in which the 136 bases long product appears to be more predominant in LNCaP than in MCF-7 cells mRNA. This suggests that the human glucocorticoid receptor gene transcription may be heterogeneous. At least two of these start sites, giving rise to the 136 and 39 bases long signals which are unique to LNCaP cells, may indicate that amplified and/or rearranged human glucocorticoid receptor genes are present in LNCaP cells. The predominant extended product in both MCF-7 and LNCaP is 88 bases long; thus the major human glucocorticoid receptor population is transcribed from the same gene and transcription is occurring at the same start site in both cell types.

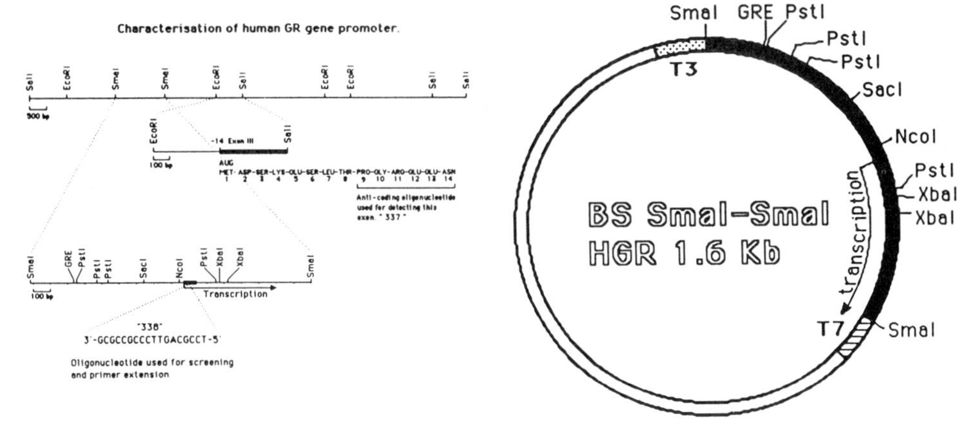


```
     -661       -651       -641       -631       -621       -611       -601       -591       -581
CCCAATGTGC TTTACACTTT ACTAGACATC CGGCTCGATC ACTGTTCTGT GCAACTTGCT GCAGACGGAT AACCAATTTC TCAGAGCACA
                                              GRE

     -571       -561       -551       -541       -531       -521       -511       -501       -491
TAGCTATCGA CGCGATTGAT GTGAGCGCAC ACTTCCGGCA GGACGTGGCT ATGGGGGGCG AAGTGCGCAT CTGACGAGGT CCGGCACCCT

     -481       -471       -461       -451       -441       -431       -421       -411       -401
GCAGCCTCGG CCGGTCGGGC CCCCGAGGCT GGGGGCAGAC CCGGCGCCCC TGCAGTTGCC AACGTCACCA ACAGTGCATC GTTCCCCGCC
                                                                                             -405

     -391       -381       -371       -361       -351       -341       -331       -321       -311
CGCGCGGCCC TCGGGCGGGG AGCGGCCGGG GGTGGAGTGG AGCGCGTGTG TGCGAGTGTG TGCCCGTGCG CCGCCTCCAC GCCCTCCCCT
             -388

     -301       -291       -281       -271       -261       -251       -241       -231       -221
CGGTCCCGCT CGCTCCGAGG CCGGGCTGCC CTTCGCTGTC CGCGCTCTCT TCCCTCCGCC GCCGCCCCTC CATTTTGCGA GCTCTTCTGT
                                                                     -249

     -211       -201       -191       -181       -171       -161       -151       -141       -131
GACGGACCCA GTCACGCCTC TCGGGATGGA TTCTTGGTGG AAGGAGACCC GCACCGGAGC GGCGCGAAGC AGCTGGGACC GGGACGGCAC

     -121       -111       -101        -91        -81        -71        -61        -51        -41
GCGCGCCCGG AAGCCCCGAC CCGCGGAGCC GGCCGGGCCG GAGGGCTGGC TTGTCAGCTG GGCAATGGGA GACTTTCTTA AATAGGCTCT
                                        -95

      -31        -21        -11         -1        +11        +21        +31        +41    *   +51
CCCCCCACCC ATGGAGAAAG GGGCGGCTGT TTACTTCCTT TTTTTAGAAA AAAAAAATAT ATTTCCCTCC TGCTCCTTCT GCGTTCACAA
      NcoI       -20                        Transcription                   *

  *   +61        +71      * +81  exon       +91       +101       +111  *    +121       +131       +141
GCTAAGTTGT TTATCTCGGT CGCGGCGGGA ACTGCGGTATAGCTAATCGAA GAGTATTGCT GACAACGGTT CAACTTTCCG TGATGGCACA
                   3' GCGCCGCCCT TGACGCCT 5'     *
                        primer 338      intron

     +151       +161       +171       +181       +191       +201       +211
GACTCTTTTA CTCGTTGGCC TCCACTGACC TATAAAAACA CTTCTGCAGG ATTCTGGCGT ACCGTTCCTG TCTAGA
                                                                             XbaI
```

Fig. 2 — Top, partial structure of the human glucocorticoid receptor gene. Top left, a restriction map of the promoter region showing the start site of transcription, located in the SmaI–SmaI fragment which has been introduced into the Bluescribe vector (described on the right), and the open reading frame starting in exon III. Bottom, nucleotide sequence of the 5′ portion of HGR gene where the arrow corresponds to the transcription initiation site.

D. Nuclease S1 mapping

The SmaI–SmaI (1.6 kb) genomic subfragment (Fig. 2) inserted into the BSM13 (BS(SmaI–SmaI 1.6)) at the 3′ to 5′ orientation to the T7-RNA polymerase and linearized with NcoI (31 bp 5′ to the transcription initiation; see Fig. 2) was used in synthesizing a uniformly [^{32}P]CTP labeled anti-mRNA probe (see Fig. 2). When this

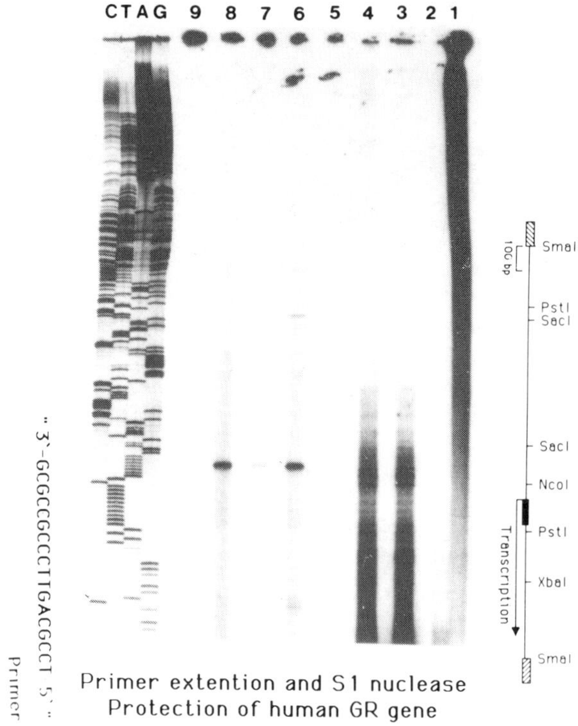

Fig. 3 — Mapping of the human glucocorticoid receptor mRNA cap site. The SmaI–SmaI fragment of the HGR gene (Fig. 2) was subcloned into M13mp18 and BSM13. The sequencing of the M13mp18 clone with oligonucleotide 5′-TCCGCAGTTCCCGCCGCG-3′ '338' as primer is shown on the left side. The BS plasmid containing the insert in the 5′ to 3′ orientation with regard to T3-RNA promoter was linearized at the NcoI site and a uniformly [^{32}P]CTP labeled anti-mRNA probe was synthesized using T7-RNA polymerase. 2 μg of each polyA$^+$ RNA from LNCaP (lane 3) and MCF-7 (lane 4) were hybridized with the anti-mRNA probe and digested with S1 nuclease, and the protected fragments were separated on 7 M urea–6% polyacrylamide sequencing gel. Lane 1 contained the labeled probe alone; lane 2 contained the probe hybridized to tRNA and S1 digested. Primer extension mapping was performed using [^{32}P]-end-labeled primer '338', reverse transcriptase and polyA$^+$ RNA from LNCaP (lanes 5 and 6) and MCF-7 (lanes 7 and 8) cells. The primer was extended in the presence and absence (lane 9) of RNA without treatment with S1 nuclease. The results obtained are presented schematically on the right.

probe was hybridized with 2 μg each of LNCaP and MCF-7 polyA$^+$ RNA and digested with nuclease S1, multiple protected fragments were detected. The major fragments had lengths of 88, 81 and 80 bases (Fig. 3). The largest of these protected fragments co-migrated with the major primer extension product demonstrating that the 5′ end of the primer is at the 3′ end of the exon.

The primer extension, nuclease S1 mapping and sequencing data (Figs. 2 and 3) together suggest that the human glucocorticoid receptor cDNAs isolated from an MCF-7 cDNA library (Govindan *et al.*, 1985, and unpublished results) extends to the extreme 5′ end of the human glucocorticoid receptor gene. Hollenberg *et al.* (1985) isolated human glucocorticoid receptor cDNA clones which contain identical 5′ untranslated cDNA sequences.

E. Sequence around the human glucocorticoid receptor promoter is G+C rich

The nucleotide sequence upstream of the human glucocorticoid receptor RNA start site contains neither a 'TATA box' nor a 'CAAT box' (Fig. 2). Most other characterized eukaryotic genes have these two sequence elements approximately 30 bp and 80 bp respectively upstream of the RNA initiation site (Breathnach and Chambon, 1981). Since *in vitro* transcription studies have shown that the 'TATA box' serves to fix the start site of transcription (Fromm and Berg, 1982), it seems surprising that human glucocorticoid receptor gene without a 'TATA box' has a predominant transcriptional start site (Fig. 3). It has been demonstrated that the EGF receptor gene, which possesses similar structural features to the human glucocorticoid receptor gene and has numerous transcriptional initiation sites (Ishii *et al.*, 1985), contains neither a 'TATA box' nor a 'CAAT box'. The sequence between −661 and −1 (Fig. 2), containing the putative promoter, has a G+C content of 72.5%. Further analysis of the human glucocorticoid receptor gene 5′ region shows that, similarly to the EGF receptor, a specific sequence CCGCCC and its inverted repeat GGGCGG are present many times (see Fig. 2) in the putative promoter region. The element CCGCCC is the same sequence that is repeated six times within the simian virus 40 (SV40) early promoter (Fromm and Berg, 1982). The 5′ region of the human glucocorticoid receptor gene therefore differs from the corresponding regions of most other eukaryotic genes analyzed to date but shares some common features with the SV40 early promoter, the EGF receptor promoter, the mouse metallothionein promoter and the hydroxy methylglutaryl-CoA reductase (HMG-CoA reductase) promoter (Ishii *et al.*, 1985; Serfling *et al.*, 1985).

The SV40 early promoter contains a GC-rich region, about 80 bp upstream of the RNA start site, which is essential for transcription (Fromm and Berg, 1982; Benoist and Chambon, 1981; Hansen and Sharp, 1983; Wasylyk and Chambon, 1983, Tjian, 1981) and contains six copies of the CCGCCC motif (Ishii *et al.*, 1985; Serfling *et al.*, 1985). The gene HMG-CoA reductase does not possess a typical 'TATA box', has five transcriptional start sites (Reynolds *et al.*, 1984) and contains three repeats of the sequence CCGCCC at the 5′ flanking region. The c-myc proto-oncogene promoter also contains the CCGCCC motif and is GC rich (see Fig. 6 of Reynolds *et al.* (1984)). The 5′ flanking region of the EGF receptor gene is GC rich and contains five CCGCC repeats. The mouse metallothionein-I gene promoters appear to include two CCGCCC elements present in an inverted configuration (Serfling *et al.*, 1985). Like the corresponding regions of these genes, the 5′ flanking region of the human glucocorticoid receptor gene is GC rich. Within this area, there are two repeats of the CCGCCC sequence and three of the inverted elements GGGCGG. Interestingly, the sequence of the human glucocorticoid receptor gene apparently contains in addition a putative glucocorticoid receptor binding site (TGTTCT; see Fig. 2) at position −621.

A recent report demonstrates that the human HeLa cell transcription factor Sp1 that stimulates *in vitro* transcription of the SV40 early promoter binds to specific SV40 early promoter region containing CCGCCC sequences (Gidoni *et al.*, 1984). Because Sp1 binding protein is found in uninfected cells, it seems likely that it plays a role in the expression of cellular as well as viral genes (Dynan and Tjian, 1983). An attractive hypothesis is that some 'growth control' genes such as human glucocorticoid receptor gene that contain the repeat CCGCCC in the promoter region are regulated by binding of the Sp1 protein.

III. FUNCTIONAL STUDIES

A. *In vitro* transcription of the human glucocorticoid receptor gene

To determine whether the cloned genomic DNA containing the 5′ flanking region of the human glucocorticoid receptor gene is active in initiating transcription (Fig. 4), we performed an *in vitro* transcription assay using extracts from HeLa cells (Gibco-BRL). The plasmid containing the 4.5 kb EcoRI–EcoRI genomic subfragment was linearized with XbaI, 211 bp downstream from the initiation start site, and then used as template (Fig. 2). Transcription products were purified on CsCl gradients, hybridized to uniformly [^{32}P]CTP labeled anti-mRNA probe synthesized from the NcoI linearized BS SmaI–SmaI 1.6) template and digested with nuclease S1. The protected fragments were fractionated under denaturing conditions on a 7 M urea–6% polyacrylamide gel. A major protected fragment of 211 bp was observed (Fig. 4). This demonstrates that *in vitro* as well as *in vivo* transcription initiation occurs at the same position, corroborating the results obtained above by primer extension and nuclease S1 mapping (Fig. 3).

B. Regulation of the human glucocorticoid receptor gene studied with human glucocorticoid receptor–chloramphenicol acetyl transferase chimeric plasmids

The glucocorticoid receptor is known to be down regulated by glucocorticoids and extensive studies on the level of mRNA are reported using rat glucocorticoid receptor cDNA probes. It has been also shown that a putative glucocorticoid receptor binding site is present in the 3′ untranslated region of rat glucocorticoid receptor cDNA (Okret *et al.*, 1986), which may be involved in the down regulation of glucocorticoid receptor by glucocorticoids. To study the regulation of human glucocorticoid receptor at the molecular level, a plasmid containing the human glucocorticoid receptor gene promoter inserted upstream of the bacterial chloramphenicol acetyl transferase (CAT) gene was constructed (pHGR4.5 CAT in Fig. 5). As a control an MMTV–LTR–CAT plasmid was used (MMTV–CAT in Fig. 5). The HGR–CAT as well as MMTV–CAT were transfected into COS-1 cells by calcium phosphate precipitation (Gorman *et al.*, 1982). After transfection, the cells were grown in hormone-depleted medium for 48 h and separated into two batches. The first batch was used as non-treated controls and the second batch of plates was treated for 12 h with 1 μM dexamethasone. The cells were harvested and cell extracts were prepared. The CAT activity was determined [^{14}C]chloramphenicol and the acetylated products were separated by thin-layer chromatography followed by autoradiography.

As it has been demonstrated (Miesfeld *et al.*, 1986), the cells transfected with MMTV–CAT showed a glucocorticoid-dependent stimulation of the CAT activity.

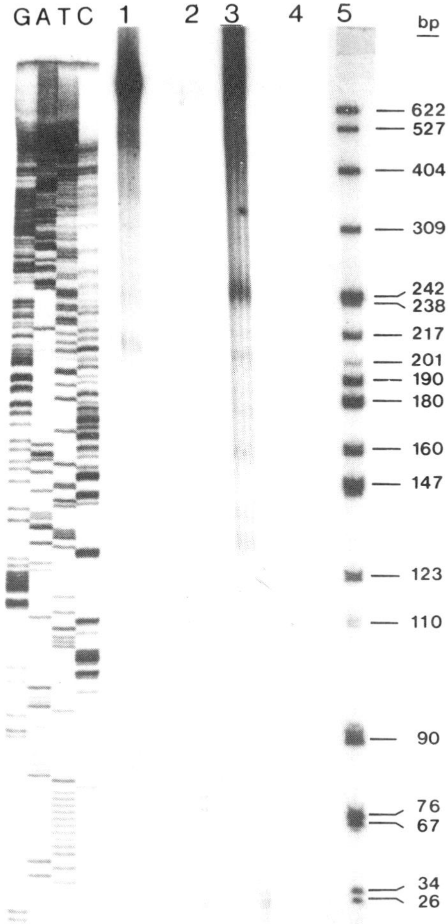

Fig. 4 — *In vitro* transcription of the human glucocorticoid receptor gene. The plasmid containing the 4.5 kb EcoRI–EcoRI genomic fragment, linearized with XbaI (see Fig. 2), was used as template for *in vitro* transcription. RNAs were synthesized using the HeLa cell *in vitro* transcription kit (BRL). RNAs were purified on CsCl gradient as described by Weingartener and Keller (1981). The transcripts were hybridized with the anti-mRNA probe (Fig. 3) and digested with S1 nuclease, and the protracted fragments were analyzed on sequencing gels (Fig. 3). Lane 1, probe alone. Lane 2, probe hybridized to tRNA and S1 nuclease treated. *In vitro* transcribed RNA in the absence (lane 3) and in the presence (lane 4) of α-amanitin, hybridized with the probe and treated with S1 nuclease. Lane 5 contains the labeled pBR322 digested with MspI.

There was a basal level of CAT activity observed in the non-treated MMTV–CAT transfected cells, whereas the CAT activity increased many-fold in the hormone-treated cells (Fig. 5A). However, the opposite results are observed in the HGR–CAT transfected cells (Figs. 5B and 5C). The CAT activity is at least 5 times higher in the non-hormone-treated HGR–CAT transfected cells than in the hor-

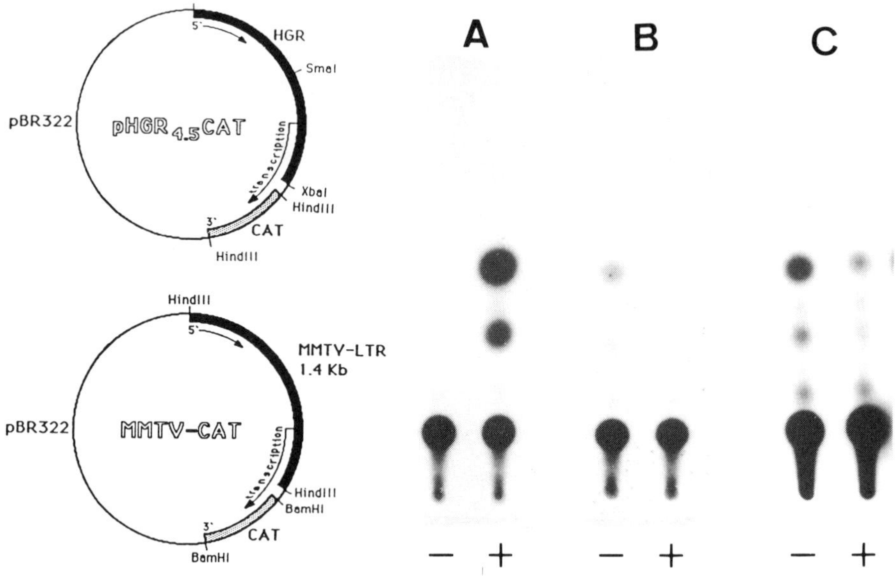

Fig. 5 — Transfection of COS-1 cells with HGR–CAT chimeric plasmids. Semiconfluent COS-1 cells were transfected with constructs containing MMTV–CAT (A) and HGR–CAT (B and C). The cells were grown in hormone-depleted medium for 48 h. 12 h before harvesting, dexamethasone (1 μM) was added to duplicate culture plates. Extracts were prepared and CAT activity was determined using the procedure described by Gorman *et al.* (1982). − indicates absence of hormone, + indicates presence of 1 μM dexamethasone. B and C show autoradiograms obtained after exposure for 1 day and 5 days respectively.

mone-treated HGR–CAT transfected cells. This is the first demonstration of the regulation of a regulatory protein gene, such as human glucocorticoid receptor, by a glucocorticoid ligand. These experiments define that the sequences required for the negative regulation of human glucocorticoid receptor are present within a 4 kb EcoRI–XbaI fragment, which contains the transcription start site, 5′ flanking region, two CCGCCC and three GGGCGG elements, and a putative glucocorticoid receptor binding site approximately 620 bp upstream of the transcription start site. These findings support the hypothesis that some ‘growth control’ genes such as human glucocorticoid receptor containing the repeat CCGCCC are regulated by binding of the Sp1 protein (Gidoni *et al.*, 1984) in the promoter region. The human glucocorticoid receptor gene may be induced by the Sp1 protein in the absence of its steroidal ligand. Subsequent to ligand binding, the hormone receptor complex may interact with the glucocorticoid receptor gene promoter at the putative glucocorticoid response element. We postulate that the binding of the glucocorticoid receptor to the TGTTCT motif in the presence of Sp1 protein bound to the CCGCCC and GGGCGG motifs results in steric hindrance. These steric hindrances in turn disturb the interaction with the transcription machinery resulting in the decrease of glucocorticoid receptor transcription. Thus the specific features of the 5′ flanking region of human glucocorticoid receptor gene may prove to be important for understanding the regulation of regulatory protein expression.

IV. SUMMARY

We have isolated the promoter region of the human glucocorticoid receptor gene from the λEMBL-3 human genomic library using synthetic oligonucleotides corresponding to the 5′ end of human glucocorticoid receptor cDNA. The clone was characterized by restriction digestion, hybridization with human glucocorticoid receptor cDNA probes and by DNA sequence analysis and found to contain an open reading frame corresponding to the human glucocorticoid receptor cDNA from amino acids 1–131 separated by a large intron from the 5′ non-coding sequences. The promoter region of the human glucocorticoid receptor has been defined by primer extension and nuclease S1 mapping using mRNA from human prostatic carcinoma cells (LNCaP) and human breast tumor cells (MCF-7) and by *in vitro* transcription using human glucocorticoid receptor–DNA fragments as template. The nucleotide sequence of the promoter region contains neither a 'TATA box' nor a 'CAAT box' and is very rich in GC. Gene transfer studies with human glucocorticoid receptor–CAT chimeric plasmids into COS cells showed that the hormonal regulatory sequences of human glucocorticoid receptor is contained within a 4 kb EcoRI–XbaI fragment.

ACKNOWLEDGMENTS

We thank Professor P. Chambon for his encouragement and interest in this project. This project is supported by an MRC Group grant for the MRC Group in Molecular Endocrinology, National Cancer Institute grant (M.V.G.).

REFERENCES

Benoist, C. and Chambon, P. (1981) *In vivo* sequence requirement of the SV40 early promoter region. *Nature* **290,** 304–310.

Breathnach, R. and Chambon, P. (1981) Organization and expression of eukaryotic split genes coding for proteins. *Ann. Rev. Biochem.* **50,** 349–383.

Dynan, W. S. and Tjian, R. (1983) Isolation of transcription factors that discriminate between different promoters recognized by RNA polymerase II. *Cell* **32,** 669–680.

Frischauf, A. M., Lehrach, H., Poustka, A. and Murray, N. E. (1983) Lambda replacement vectors carrying polylinker sequences. *J. Mol. Biol.* **170,** 827–842.

Fromm, M. and Berg, P. (1982) Deletion mapping of DNA regions required for SV40 early region promoter function *in vivo. J. Mol. Appl. Genet.* **1,** 457–481.

Geise, S., Scheidereit, C., Westphal, H., Hynes, N. E., Groner, B. and Beato, M. (1982) Glucocorticoid receptors recognize DNA sequences in and around murine mammary tumour virus DNA. *EMBO J.* **1,** 1613–1619.

Gidoni, D., Dynan, W. S. and Tjian, R. (1984) Multiple specific contacts between a mammalian transcription factor and its cognate promoters. *Nature* **312,** 409–413.

Gorman, M. C., Moffat, F. L. and Howard, B. H. (1982) Recombinant genomes which express chloramphenicol acetyltransferase in mammalian cells. *Mol. Cell. Biol.* **2,** 1044–1051.

Govindan, M. V., Spiess, E. and Majors, J. (1982) Purified glucocorticoid receptor-hormone complex from rat liver cytosol binds specifically to cloned mouse mammary tumor virus long-terminal repeats *in vitro*. *Proc. Natl. Acad. Sci. USA* **79,** 5157–5161.

Govindan, M. V., Devic, M., Green, S., Gronemeyer, H. and Chambon, P. (1985) Cloning of the human glucocorticoid receptor cDNA. *Nucleic Acids Res.* **13,** 8293–8304.

Hansen, U. and Sharp, P. A. (1983) Sequences controlling *in vitro* transcription of SV40 promoters. *EMBO J.* **2,** 2293–2303.

Hollenberg, S. M., Weinberger, C., Ong, E. S., Cerelli, G., Oro, A., Lebo, R., Thompson, E. B., Rosenfeld, M. G. and Evans, R. M. (1985) Primary structure and expression of a functional human glucocorticoid receptor cDNA. *Nature* **318,** 635–641.

Ishii, S., Xu, Y. H., Stratton, R. H., Roe, B. A., Merlino, G. T. and Pastan, I. (1985) Characterization and sequence of the promoter region of the human epidermal growth factor receptor gene. *Proc. Natl. Acad. Sci. USA* **82,** 4920–4924.

Karin, M., Haslinger, A., Holtgreve, H., Richards, R. I., Krauter, P., Westphal, H. and Beato, M. (1984) Characterization of DNA sequences through which cadmium and glucocorticoid hormones induce human metallothionein-IIA gene. *Nature* **308,** 513–519.

Miesfeld, R., Ruscini, S., Godowski, P. J., Maler, B. A., Okret, S., Wikström, A. C., Gustafsson, J. A. and Yamamoto, K. R. (1986) Genetic complementation of a glucocorticoid receptor deficiency by expression of cloned receptor cDNA. *Cell* **46,** 389–399.

Okret, S., Poellinger, L., Dong, Y. and Gustafsson, J. A. (1986) Down-regulation of glucocorticoid receptor mRNA by glucocorticoid hormones and recognition by the receptor of a specific binding sequence within a receptor cDNA clone. *Proc. Natl. Acad. Sci. USA* **83,** 5899–5903.

Payvar, F., Wrange, O., Carlstedt-Duk, J. Okret, S., Gustafsson, J. and Yamamoto, K. R. (1981) Purified glucocorticoid receptors bind selectively *in vitro* to a cloned DNA fragment whose transcription is regulated by glucocorticoids *in vivo*. *Proc. Natl. Acad. Sci. USA* **78,** 6628–6632.

Pfahl, M. (1982) Specific binding of the glucocorticoid-receptor complex to the mouse mammary tumor proviral promoter region. *Cell* **31,** 475–482.

Renkawitz, R., Schutz, G., Von der Ahe, D. and Beato, M. (1984) Sequences in the promoter region of the chicken lysozyme gene required for steroid regulation and receptor binding. *Cell* **37,** 503–510.

Reynolds, G. A., Basu, S. K., Osborne, T. H., Chin, D. J., Gil, G., Brown, M. S., Goldstein, J. L. and Luskey, K. L. (1984) HMG CoA reductase: a negatively regulated gene with unusual promoter and 5′-untranslated regions. *Cell* **38,** 275–285.

Scheidereit, C. and Beato, M. (1984) Contacts between hormone receptor and DNA double helix within a glucocorticoid regulatory element of mouse mammary tumor virus. *Proc. Natl. Acad. Sci. USA* **81,** 3029–3033.

Serfling, E., Jasin, M. and Schaffner, W. (1985) Enhancers and eukaryotic gene transcription. *Trends Genet.* **1,** 224–230.

Tjian, R. (1981) T-antigen and the control of SV40 gene expression. *Cell* **26,** 1–2.

Wasylyk, B. and Chambon, P. (1983) Potentiator effect of the SV40 72-bp repeat on initiation of transcription from heterologous promoter elements. *Cold Spring Harbor Symposium on Quantitative Biology,* Vol. 47. pp. 921–934.

Weingartener, B. and Keller, W. (1981) Transcription and processing of adenoviral RNA by extracts from HeLa cells. *Proc. Natl. Acad. Sci. USA* **78,** 4092–4096.

Yamamoto, K. R. (1985) Steroid receptor regulated transcription of specific genes and gene networks. In: Campbel, A., Herskowitz, I. and Sandler, L. M. (eds.) *Annual Reviews on Genetics*, Vol. 19. Annual Reviews Inc., pp. 209–252.

17

Characterization of a c-*erbA*–thyroid hormone receptor cDNA and its viral homologue

B. Vennström
Differentiation Programme, EMBL, Heidelberg, FRG

I. INTRODUCTION

In this article the rationale and the experimental procedures used for identifying the c-*erbA* gene product as a thyroid hormone receptor will be reviewed.

The viral *erbA* gene was first discovered in the avian erythroblastosis virus (AEV) strain ES4, a retrovirus that induces both erythroblastosis and sarcomas when injected into chickens, and which transforms fibroblasts and erythroblasts *in vitro* (for review, see Graf and Beug, 1978). AEV-ES4 also contains another oncogene, v-*erbB*, which represents a mutated version of the EGF receptor (Downward *et al.*, 1984; Ullrich *et al.*, 1984). This gene is sufficient for transformation of both types of cells, whereas v-*erbA*, being non oncogenic on its own, enhances the transformed phenotype of erythroblasts transformed by v-*erbB* or other oncogenes (Frykberg *et al.*, 1983; Sealy *et al.*, 1983; Kahn *et al.*, 1986a).

A virus containing only v-*erbA* is non-oncogenic both *in vivo* and *in vitro*, but two specific effects of v-*erbA* in erythroblasts have been observed: it completely blocks the spontaneous differentiation exhibited by transformed erythroblasts, and it allows cell proliferation at pH and ionic conditions otherwise toxic to the cells (Kahn *et al.*, 1986b; Damm *et al.*, 1987). The effects of v-*erbA* in fibroblasts are less pronounced, but changes in cell morphology and enhancement of v-*erbB* induced focus formation have been observed (Frykberg *et al.*, 1983; Jansson *et al.*, 1987).

II. EXPRESSION OF THE c-*erbA* GENE

To understand better the oncogenic effects of v-*erbA*, an analysis of the expression of the c-*erbA* gene by the Northern procedure was done. The results revealed that two RNAs, 4.5 and 3.0 kb in length, were found in many cell types (Fig. 1), and that

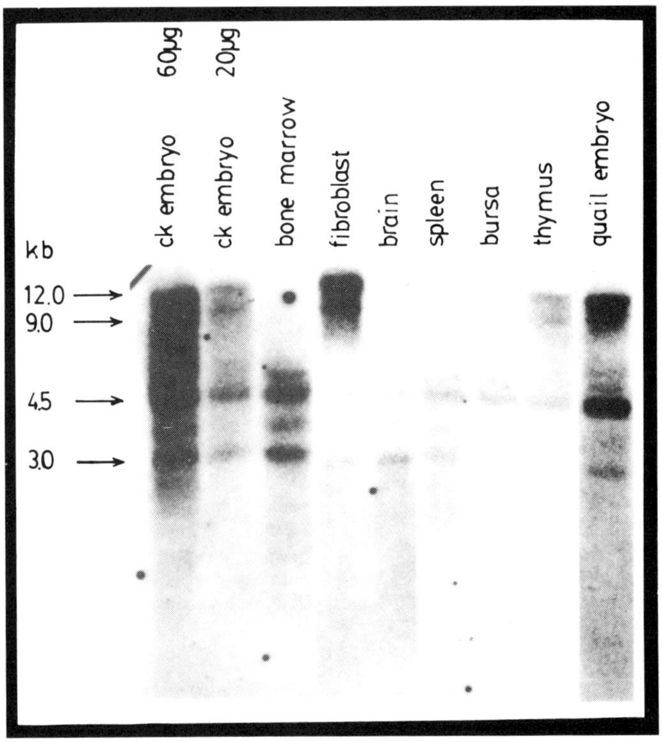

Fig. 1 — Expression of the 4.5 and 3.0 kb c-*erbA* mRNAs in chicken and quail tissues. Polyadenylated RNA, isolated from 10-day-old embryos or from tissues of 3-week-old chicks, was separated on denaturing gels, transferred to filter paper and hybridized with a probe recognizing both c-*erbA* and c-*erbB* (12 and 9 kb) RNAs. 25 μg RNA was analyzed unless indicated otherwise.

additional *erbA* related RNAs could be detected in bone marrow cells. Whether or not these RNAs derive from the same c-*erbA* locus as the 4.5 and 3.0 kb RNAs is unclear. Additional experiments have shown that the 4.5 and 3.0 kb RNAs are present in low abundance (5–10 copies per cell) and are associated with polyribosomes, suggesting that they are bona fide mRNAs (Vennström and Bishop, 1982; B. Vennström, unpublished data).

III. CLONING AND IDENTIFICATION OF THE c-*erbA* GENE

Analysis of the chicken c-*erbA* locus has proven difficult since repetitive elements in the proximity of its 3′ end have impaired its cloning in phage λ (Vennström and Bishop, 1982). We therefore chose to study the structure of the c-*erbA* gene product by isolation of cDNA clones. A cDNA library made with RNA from 11-day-old chicken embryos was consequently screened with a v-*erbA* probe, and the resulting positive clones were subjected to nucleotide sequence analysis. These results as well as *in vitro* synthesis of the encoded protein showed that the c-*erbA* protein has a molecular weight of 46 kd (Sap *et al.*, 1986). A comparison of the nucleotide sequences of the viral and cellular *erbA* genes revealed several amino acid differences. v-*erbA* has 13 amino acid substitutions of which two are located in the putative DNA binding domain and 11 in other regions (see Fig. 2). Furthermore, the v-*erbA*

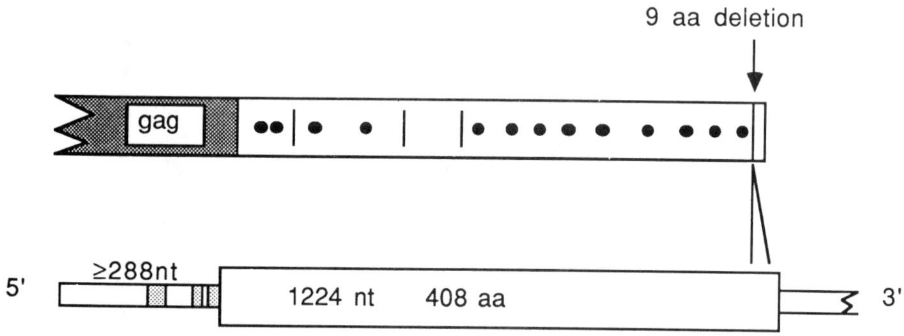

Fig. 2—Comparison of the cellular and viral *erbA* sequences. The 5′ non-coding sequence of c-*erbA* is at least 288 nucleotides long, and contains three short open reading frames closed by stop codons as indicated by stipled boxes. The dots in v-*erbA* indicate amino acid substitutions.

protein has lost nine amino acids located three amino acids before the C terminus. Finally, the N terminus of the c-*erbA* polypeptide is 12 amino acids longer than the *erbA* specific part of the v-*erbA* protein, since the latter is fused to the retroviral *gag* gene at this position.

Previously published data (Greene *et al.*, 1986; Hollenberg *et al.*, 1985; Weinberger *et al.*, 1985; Krust *et al.*, 1986; Green *et al.*, 1986) showed that receptors for steroids were in part homologous to the v-*erbA* protein. A further search in the literature revealed that none of the steroid receptors nor the vitamin D_3 receptor had a size compatible with that of the c-*erbA* protein. However, Samuels and collaborators had demonstrated using photoaffinity labeling with modified triiodothyronine (T_3) that one form of the nuclear receptor for thyroid hormone has a molecular weight of 47 kd (Pascual *et al.*, 1982). We therefore tested whether the *in vitro* synthesized c-*erbA* protein would bind $^{125}I\text{-}T_3$ in *in vitro* binding assays. The results showed that the c-*erbA* protein bound the ligand with an affinity $k_d \approx 0.2–0.3$ nM similar to that of the native thyroid hormone receptor. In addition, the c-*erbA* protein bound analogues of thyroid hormone with all the characteristics that typify the thyroid hormone receptor (Sap *et al.*, 1986), unambiguously demonstrating that the c-*erbA* protein is a high affinity nuclear receptor for thyroid hormone. Similar results have been obtained by Weinberger *et al.* (1986) studying human c-*erbA* genes.

IV. THE v-*erbA* PROTEIN IS DEFECTIVE

The v-*erbA* protein is defective in binding thyroid hormones (Sap *et al.*, 1986), presumably as a result of the many mutations it has accumulated. To identify which of the mutation(s) in v-*erbA* that conferred the inability to bind hormone, a series of recombinants between the viral and cellular genes was constructed utilizing restriction sites common to both genes. The sites were chosen so that the influence on hormone binding of the C-terminal deletion as well as the amino acid substitutions in v-*erbA* could be tested. The chimeric genes, cloned into plasmids containing the prokaryotic promoter, were transcribed and translated *in vitro*, and the resulting proteins tested for binding of $^{125}I\text{-}T3$. The results showed that the 9 amino acid deletion in the C terminus of v-*erbA* considerably diminished but did not completely abolish ligand binding. In addition, the c-*erbA* constructs containing the entire putative ligand binding domain of v-*erbA* bound no ligand. Taken together, the results so far suggest that both the amino acid substitutions and the C-terminal deletion in v-*erbA* diminish T3 binding; testing of additional chimeras as well as determination of dissociation constants for ligand binding for all the recombinant proteins are, however, required for a complete evaluation of the effects of the mutations. In addition, an analysis of their properties *in vivo* could reveal whether the mutations in the putative DNA binding region of the v-*erbA* protein have altered its specificity for the nucleotide sequences recognized as well as the effects on transcription.

In conclusion, the data suggest that the hormone binding site of the c-*erbA* protein is located in the homologous domain previously shown to be the ligand binding region of steroid receptors (see Chapter 15) and, furthermore, underscore the relatedness between the steroid and thyroid hormone receptors.

V. ANALYSIS OF MUTANT v-*erbA* GENES

We have also characterized the v-*erbA* oncogenes of *td*359, a transformation-defective mutant of AEV unable to transform erythroblasts, and a revertant r12, obtained after *in vivo* passage of the mutant. Biological testing of their v-*erbA* genes showed that v-$erbA^{td359}$ is inactive, whereas v-$erbA^{r12}$ has an effect on erythroblasts even stronger than that of the wt gene (Damm *et al.*, 1987).

Nucleotide sequence analysis showed that the v-*erbA* protein of *td*359 contained among other substitutions one amino acid change (proline to arginine) located in the short 'hinge' (Fig. 3) region at position 144 (Damm *et al.*, 1987). The v-*erbA* protein

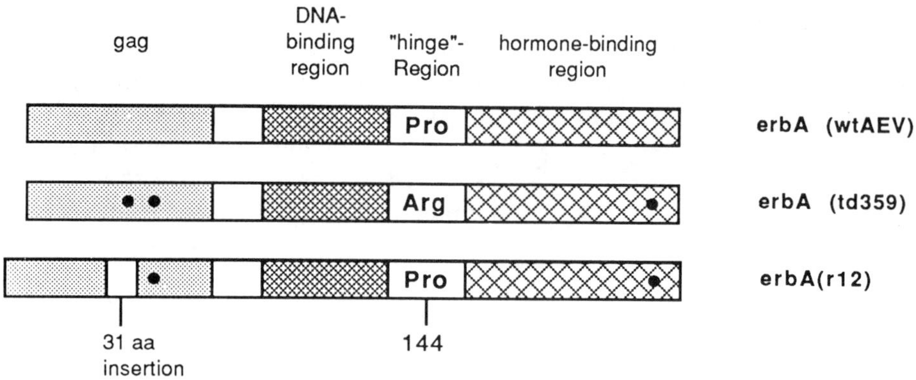

Fig. 3 — Comparison of the wt, *td*359 and r12 v-*erbA* sequences. The black dots indicate mutations.

of r12 had reverted to a proline in this position and had in addition acquired a 31 amino acid insertion of unknown origin in its *gag* domain.

The importance of the mutations in each v-*erbA* gene was determined by constructing v-*erbA* genes containing chimeric *td*359 and r12 v-*erbA* genes, followed by insertion into AEV . The results of subsequent transformation experiments and testing of the effects of the chimeric genes on erythroblasts showed that he corresponding changes at position 144 were responsible for the inactivation of the *td*359 gene and the reactivation in r12. In addition, the modification of the *gag* region in v-$erbA^{r12}$ appears to be responsible for the 'superactivity' of the r12 protein. The data also suggest that the 'hinge' region serves an important role in the structure and function of the v-*erbA* protein.

VI. HUMAN c-*erbA*–THYROID HORMONE RECEPTOR GENES

Results published as early as 1983 (Jansson *et al.*, 1983) suggested that the human genome contains at least two *erbA* related genes, and subsequent chromosome mapping experiments located the gene most homologous to v-*erbA* to chromosome 17p21-q24 (Dayton *et al.*, 1984; Spurr *et al.*, 1983; Zabel *et al.*, 1984). Both genes are transcribed and are therefore unlikely to represent pseudogenes. Weinberger *et al.* (1986), studying probably more distantly related c-*erbA*–thyroid hormone receptor genes, found three crosshybridizing sequences on chromosome 3, of which at least one codes for a high affinity thyroid hormone receptor, a second is highly homologous to c-*erbA* and is transcribed (R. Evans, personal communication), whereas the nature of the third is still unclear. Consequently, the human genome appears to encode at least three distinct thyroid hormone receptors. Preliminary evidence suggests that also the chicken genome contains more than one gene for thyroid hormone receptors (B. Vennström, unpublished).

VII. CONCLUSIONS

The cloning of cDNAs for thyroid hormone receptors has shown that several distinct receptors exist. Previous analysis, based on binding of radiolabeled hormone by nuclei from various tissues, gave no evidence of multiple receptors (Oppenheimer and Samuels, 1983). The recent results are surprising, particularly in view of the fact that there appears to be only one receptor for each of the steroid hormones. However, thyroid hormone has very diverse effects on gene expression, cell metabolism and differentiation, and it is possible that different receptors have distinct functions in these processes. The isolation of cDNAs for these receptors will undoubtedly facilitate an understanding and dissection of the broad biological effects of thyroid hormone.

Our data show that the v-*erbA* protein represents a ligand independent form of the thyroid hormone receptor. The fact that both the deletion in v-*erbA* and the point mutations are required for abolishing hormone binding suggests that they have accumulated gradually, and that the acquisition of hormone independence has increased the oncogenic activity of v-*erbA*.

Unpublished data (H. Beug ad U. Gehring) also demonstrate that the v-*erbA* protein binds DNA, suggesting that it acts like the thyroid hormone receptor by binding to regulatory elements of specific target genes. The nucleotide sequences recognized by v-*erbA* might, however, be distinct from those binding to the c-*erbA* protein, since the viral protein contains two amino acid substitutions in its putative DNA binding domain. In addition, the v-*erbA* protein might constitutively either up or down regulate the expression of its target genes. If their expression (or, alternatively, absence thereof) is important for differentiation of erythroblasts, the effects of v-*erbA* on erythroblasts are readily explained. A study of the synthesis of ion transporters in AEV transformed erythroblasts appears worthwhile, since v-*erbA* affects the ionic requirements of erythroblast growth, and thyroid hormone is known to increase synthesis of the Na–K ATPase.

ACKNOWLEDGMENTS

I thank my colleagues and coworkers Hartmut Beug, Klaus Damm, Thomas Graf, Alberto Munoz and Jan Sap for rewarding discussions and generous disclosure of unpublished data.

REFERENCES

Damm, K., Beug, H., Graf, T. and Vennström, B. (1987) A single point mutation in *erbA* restores the erythroid transforming potential of a mutant avian erythroblastosis virus (AEV) defective in both *erbA* and *erbB* oncogenes. *EMBO J.* **6,** 375–382.

Dayton, A. I., Selden, J. R., Laws, G., Dorney, D. J., Finan, J., Tripputi, P., Emanuel, B. S., Rovera, G., Nowell, P. C. and Groce, C. M. (1984) A human c-*erbA* oncogene homologue is closely proximal to the chromosome 17 breakpoint in acute promyelocytic leukemia. *Proc. Natl. Acad. Sci. USA* **81,** 4495–4499.

Frykberg, L., Palmieri, S., Beug, H., Graf, T., Hayman, M. J. and Vennström, B. (1983) Transforming capacities of avian erythroblastosis virus mutants deleted in the *erbA* or *erbB* oncogenes. *Cell* **32,** 227–238.

Graf, T. and Beug, H. (1978) Avian leukemia viruses: Interaction with their target cells *in vivo* and *in vitro*. *Biochim. Biophys. Acta* **516,** 269–299.

Green, S., Walter, P., Kumar, V., Krust, A., Bornert, J.-M., Argos, P. and Chambon, P. (1986) Human oestrogen receptor cDNA: sequence, expression and homology to v-*erbA*. *Nature* **320,** 134–139.

Greene, G. L., Gilna, P., Waterfield, M., Baker, A., Hort, Y. and Shine, J. (1986) Sequence and expression of human estrogen receptor complementary DNA. *Science* **231,** 1150–1154.

Hollenberg, S. M., Weinberger, C., Ong, E. S., Cerelli, G., Oro, A., Lebo, R., Thompson, E. B., Rosenfeld, M. G. and Evans, R. M. (1985) Primary structure and expression of a functional human glucocorticoid receptor cDNA. *Nature* **318,** 635–641.

Jansson, M., Philipson, L. and Vennström, B. (1983) Isolation and characterization of multiple human genes homologous to the oncogenes of avian erythroblastosis virus. *EMBO J.* **2,** 561–565.

Kahn, P., Frykberg, L., Brady, C., Stanley, I. J., Beug, H., Vennström, B. and Graf, T. (1986a) v-*erbA* cooperates with sarcoma oncogenes in leukemic cell transformation. *Cell* **45,** 349–356.

Kahn, P., Frykberg, L., Graf, T., Vennström, B. and Beug, H. (1986b) Cooperativity between v-*erbA* and v-*src*-related oncogenes in erythroid cell transformation. In Deinhardt, F. (ed.) *Proceedings* Springer, Berlin, pp. 131–134.

Krust, A., Green, S., Argos, P., Kumar, V., Walter, P., Bornert, J.-M. and Chambon, P. (1986) The chicken oestrogen receptor sequence: homology with v-*erbA* and the human oestrogen and glucocorticoid receptors. *EMBO J.* **5,** 891–897.

Oppenheimer, J. H. and Samuels, H. H. (eds.) (1983) *Molecular Basis of Thyroid Hormone Action*. Academic Press, New York, 1983.

Pascual, A., Casanova, J. and Samuels, H. H. (1982) Photoaffinity labeling of thyroid hormone nuclear receptors in intact cells. *J. Biol. Chem.* **257,** 9640–9647.

Sap, J., Munoz, A., Damm, K., Goldberg, Y., Ghysdael, J., Leutz, A., Beug, H. and Vennström, B. (1986) The c-*erbA* protein is a high affinity receptor for thyroid hormone. *Nature* **324,** 635–640.

Sealy, L., Privalsky, M. L., Moscovici, G., Moscovici, C. and Bishop, J. M. (1983) Site-specific mutagenesis of avian erythroblastosis virus: *erbB* is required for oncogenicity. *Virology* **130,** 155–178.

Spurr, N. K., Solomon, E., Jansson, M., Sheer, D., Goodfellow, P. N., Bodmer, W. F. and Vennström, B. (1984). Chromosomal localisation of the human homologoes to the oncogenes *erbA* and *B*. *EMBO J.* **3,** 159–163.

Ullrich, A., Coussens, L., Hayflick, J. S., Dull, T. J., Gray, A., Tam, A. W., Lee, J., Yarden, Y., Liberman, T. A., Schlessinger, J., Downward, J., Mayes, E. L. V., Whittle, N., Waterfield, M. D. and Seeburg, P. H. (1984) Human epidermal growth factor receptor cDNA sequence and aberrant expression of the amplified gene in A431 epidermoid carcinoma cells. *Nature* **309,** 418–425.

Vennström, B. and Bishop, J. M. (1982) Isolation and characterization of chicken DNA homologous to the two putative oncogenes of avian erythroblastosis virus. *Cell* **28,** 135–143.

Weinberger, C., Hollenberg, S. M., Rosenfeld, M. G. and Evans, R. M. (1985). Domain structure of human glucocorticoid receptor and its relationship to the v-

erbA oncogene product. *Nature* **318,** 670–672.
Weinberger, C., Thompson, C., Ong, E., Lebo, R., Gruol, D. and Evans, R. (1986) The c-*erbA* gene encodes a thyroid hormone receptor. *Nature* **324,** 641–646.
Zabel, B. U., Fournier, R. E., Lalley, P. A., Naylor, S. L. and Sakaguchi, A. Y. (1984) Cellular homologs of the avian erythroblastosis virus *erbA* and *erbB* genes are synthetic in mouse but asynthenic in man. *Proc. Natl. Acad. Sci. USA* **81,** 4874–4878.

Index